WIRELESS SENSOR NETWORKS

BICENTENNIAL
1807
WILEY
2007
BICENTENNIAL

THE WILEY BICENTENNIAL—KNOWLEDGE FOR GENERATIONS

*E*ach generation has its unique needs and aspirations. When Charles Wiley first opened his small printing shop in lower Manhattan in 1807, it was a generation of boundless potential searching for an identity. And we were there, helping to define a new American literary tradition. Over half a century later, in the midst of the Second Industrial Revolution, it was a generation focused on building the future. Once again, we were there, supplying the critical scientific, technical, and engineering knowledge that helped frame the world. Throughout the 20th Century, and into the new millennium, nations began to reach out beyond their own borders and a new international community was born. Wiley was there, expanding its operations around the world to enable a global exchange of ideas, opinions, and know-how.

For 200 years, Wiley has been an integral part of each generation's journey, enabling the flow of information and understanding necessary to meet their needs and fulfill their aspirations. Today, bold new technologies are changing the way we live and learn. Wiley will be there, providing you the must-have knowledge you need to imagine new worlds, new possibilities, and new opportunities.

Generations come and go, but you can always count on Wiley to provide you the knowledge you need, when and where you need it!

WILLIAM J. PESCE
PRESIDENT AND CHIEF EXECUTIVE OFFICER

PETER BOOTH WILEY
CHAIRMAN OF THE BOARD

WIRELESS SENSOR NETWORKS

Technology, Protocols, and Applications

KAZEM SOHRABY
DANIEL MINOLI
TAIEB ZNATI

WILEY-
INTERSCIENCE

A JOHN WILEY & SONS, INC., PUBLICATION

Published by John Wiley & Sons, Inc., Hoboken, New Jersey.
Published simultaneously in Canada.

For general information on our other products and services or for technical support, please contact our
Customer Care Department within the United States at (800) 762-2974, outside the United States at
(317) 572-3993 or fax (317) 572-4002.

Wiley also publishes its books in a variety of electronic formats. Some content that appears in print may
not be available in electronic formats. For more information about Wiley products, visit our web site at
www.wiley.com.

Library of Congress Cataloging-in-Publication Data:

Sohraby, Kazem.
   Wireless sensor networks: technology, protocols, and applications / by Kazem
Sohraby, Daniel Minoli, Taieb Znati.
      p. cm.
   ISBN 978-0-471-74300-2
   1. Sensor networks. 2. Wireless LANs. I. Minoli, Daniel, 1952– II. Znati,
Taieb F.
III. Title.
   TK7872. D48S64 2007
   681'. 2–dc22
                                                            2006042143
Printed in the United States of America

10 9 8 7 6 5 4 3 2 1

CONTENTS

8 Middleware for Wireless Sensor Networks   246

9 Network Management for Wireless Sensor Networks   262

PREFACE

The convergence of the Internet, communications, and information technologies, coupled with recent engineering advances, is paving the way for a new generation of inexpensive sensors and actuators, capable of achieving a high order of spatial and temporal resolution and accuracy. The technology for sensing and control includes sensor arrays, electric and magnetic field sensors, seismic sensors, radio-wave frequency sensors, electrooptic and infrared sensors, laser radars, and location and navigation sensors.

Advances in the areas of sensor design, materials, and concepts will further decrease the size, weight, and cost of sensors and sensor arrays by orders of magnitude and will increase their spatial and temporal resolution and accuracy. In the very near future, it will become possible to integrate millions of sensors into systems to improve performance and lifetime, and decrease life-cycle costs. According to current market projections, more than half a billion nodes will ship for wireless sensor applications in 2010.

The technology for sensing and control now has the potential for significant advances, not only in science and engineering, but equally important, on a broad range of applications relating to critical infrastructure protection and security, health care, the environment, energy, food safety, production processing, quality of life, and the economy. In addition to reducing costs and increasing efficiencies for industries and businesses, wireless sensor networking is expected to bring consumers a new generation of conveniences, including, but not limited to, remote-controlled heating and lighting, medical monitoring, automated grocery checkout, personal health diagnosis, automated automobile checkups, and child care.

This book is intended to be a high-quality textbook that provides a carefully designed exposition of the important aspects of wireless sensor networks. The text provides thorough coverage of wireless sensor networks, including applications, communication and networking protocols, middleware, security, and management. The book is targeted toward networking professionals, managers, and practitioners who want to understand the benefits of this new technology and plan for its use and deployment. It can also be used to support an introductory course in the field of wireless sensor networks at the advanced undergraduate or graduate levels.

At this time there is a limited number of textbooks on the subject of wireless sensor networks. Furthermore, most of these books are written with a specific focus on selected subjects related to the field. As such, the coverage of many important topics in these books is either inadequate or missing. With the ever-increasing popularity of wireless sensor networks and their tremendous potential to penetrate multiple aspects of our lives, we believe that this book is timely and addresses the needs of a growing community of engineers, network professionals and managers, and educators. The book is not so encyclopedic as to overwhelm nonexperts in the field. The text is kept to a reasonable length, and a concerted effort has been made to make the coverage comprehensive and self-contained, and the material easily understandable and exciting to read.

Acknowledgments

First author would like to acknowledge the contributions of his postdoctoral fellow, Dr. Chonggang Wang, while at the University of Arkansas, in the preparation of some of the material in this book.

ABOUT THE AUTHORS

Daniel Minoli has many years of telecom, networking, and IT experience with end users, carriers, academia, and venture capitalists, including work at ARPA think tanks, Bell Telephone Laboratories, ITT, Prudential Securities, Bell Communications Research (Bellcore/Telcordia), AT&T, Capital One Financial, SES Americom, New York University, Rutgers University, Stevens Institute, and Societé General de Financiament de Quebec (1975–2001). Recently, he played a founding role in the launching of two networking companies through the high-tech incubator *Leading Edge Networks Inc.*, which he ran in the early 2000s: Global Wireless Services, a provider of broadband hotspot mobile Internet and hotspot VoIP services to high-end marinas; and InfoPort Communications Group, an optical and gigabit Ethernet metropolitan carrier supporting Data Center/SAN/channel extension and Grid Computing network access services (2001–2003). Currently, he is working on IPTV, DVB-H, satellite technology and (wireless) emergency communications systems.

Mr. Minoli has worked extensively in the field of wireless and over the years has published approximately 20 papers on the topic. His work in wireless started in the mid-1970s with extensive efforts on ARPA-sponsored research on wireless packet networks. In the early 1980s he was involved in the design of high-resilience radio networks. In the mid-1980s he was involved in designing and deploying VSAT networks, including work on correlated traffic profiles. Recently, he has been involved with the novel design of Wi-Fi hotspot networks for interference-laden public places such as marinas, and has written the first book on the market on hotspot networking: *Hotspot Networks—Wi-Fi for Public Access Locations* (McGraw-Hill, 2003). He has also been involved in the planning and deployment of high-density enterprise IEEE 802.11b/g/e/i systems and VoWi-Fi. He recently acted as an expert witness in a (successful) $11 billion lawsuit regarding a wireless air-to-ground communication system for airplane-based telephony and information services. He has also done work on wireless networking applications of nanotechnology (quantum cascade lasers for free-space optics) and has just published a book on that topic with Wiley (2005).

Mr. Minoli is the author of a number of books on information technology, telecommunications, and data communications. He has also written columns for *ComputerWorld, NetworkWorld*, and *Network Computing* (1985–1995). He has spoken at 80 industry conferences and has taught at New York University (Information Technology Institute), Rutgers University, Stevens Institute of Technology, and Monmouth University (1984–2003). He was a technology analyst at-large for Gartner/DataPro (1985–2001). On their behalf, based on extensive hands-on work at financial firms and carriers, he tracked technologies and authored numerous CTO/CIO-level technical/architectural scans in the area of telephony and data communications systems, including topics on security, disaster recovery, IT outsourcing, network management, LANs, WANs (ATM and MPLS), wireless (LAN and public hotspot), VoIP, network design/economics, carrier networks (such as metro Ethernet and CWDM/DWDM), and e-commerce. Over the years he has advised venture capitalists for investments of $150 million in a dozen high-tech companies.

Dr. Kazem Sohraby is a professor of electrical engineering in the College of Engineering at the University of Arkansas, Fayetteville, where he also serves as professor and head at the Department of Computer Science and Computer Engineering.

Prior to the University of Arkansas engagement, Dr. Sohraby was with Bell Laboratories, Lucent Technologies, and AT&T Bell Labs. He has also served as director of the interdisciplinary academic program on telecommunications management at Stevens Institute of Technology, and before that as head of the Network Planning Department at Computer Sciences Corporation. At Bell Labs he played a key role in the research and development of high-tech communications, computing, network management, security, and other information technologies area. He spend most of his career at Bell Labs in the Advanced Communications Technologies Center, the Mathematical Sciences Research Center (Mathematics of Networks and Systems), and in forward-looking organizations working on future-generation switching and transmission technologies. In its golden age, Bell Labs was the world leader in research and development of new computing and communications technologies, and has created innumerous innovations in the advancement of communications and computer networking. Dr. Sohraby's contributions at Bell Labs, demonstrated by over 20 patents filed on his behalf and many of his publications, represent an outstanding benchmark in computer and communications technologies leadership.

Dr. Sohraby has generated numerous publications, including a book entitled *Control and Performance in Packet, Circuit, and ATM Networks* (Kluwer Publishers, 1995). He is a distinguished lecturer of the IEEE Communications Society and served as its president's representative on the Committee on Communications and Information Policy (CCIP). He served on the Education Committee of the IEEE

Communications Society, and is on the editorial boards of several publications. Dr. Sohraby received the B.S., M.S., and Ph.D. degrees in electrical engineering, has a graduate education in computer science, and received an M.B.A. degree from the Wharton School of the University of Pennsylvania.

Dr. Taieb Znati is professor in the Department of Computer Science, with a joint appointment in the telecommunication program (DIS) and in computer engineering (EE) at the University of Pittsburgh. Prof. Znati's interests include routing and congestion control in high-speed networks, multicasting, access protocols in local and metropolitan area networks, quality of service support in wired and wireless networks, performance analysis of network protocols, multimedia applications, distributed systems, and agent-based internet applications. Recent work has focused on the design and analysis of network protocols for wired and wireless communications, sensor networks, network security, agent-based technology with collaborative environments, and middleware. He is coeditor of the book *Wireless Sensor Networks* (Kluwer Publishers, 2004) and has published extensively on the topic.

Prof. Znati earned a Ph.D. degree in computer science, September 1988, at Michigan State University. He also has a Master of Science degree in computer science from Purdue University, December 1981. In addition, he earned other academic degrees in Europe. Currently, he is a professor in the Department of Computer Science, with a joint appointment in the telecommunication program (School of Library and Information Science), at the University of Pittsburgh. He recently took a leave from the university to serve as senior program director for networking research at the National Science Foundation. He is also the ITR coordinating committee chair. In the late 1990s he was an associate professor in the Department of Computer Science, with a joint appointment in the telecommunication program (School of Library and Information Science) at the University of Pittsburgh. In the early 1990s he was an assistant professor at the same institution. During the 1980s he held a number of industry positions, including the position of system manager for the management of VAX VMS-cluster daily operations at the Case Center for Computer-Aided Design at Michigan State University. He also held the position of network coordinator, with responsibility for the development of networking plans for the College of Engineering at Michigan State University.

Prof. Znati has chaired several conferences and workshops, including conferences and workshops on wireless sensor networks. He is on the editorial board of several scientific journals in networking and distributed systems. He is frequently invited to present lectures and tutorials and to participate in panels related to networking and distributed multimedia topics in the United States and abroad.

1

INTRODUCTION AND OVERVIEW OF WIRELESS SENSOR NETWORKS

1.1 INTRODUCTION

A *sensor network*[1] is an infrastructure comprised of sensing (measuring), computing, and communication elements that gives an administrator the ability to instrument, observe, and react to events and phenomena in a specified environment. The
administrator typically is a civil, governmental, commercial, or industrial entity.
The environment can be the physical world, a biological system, or an information
technology (IT) framework. Network(ed) sensor systems are seen by observers as
an important technology that will experience major deployment in the next few
years for a plethora of applications, not the least being national security
[1.1–1.3]. Typical applications include, but are not limited to, data collection,
monitoring, surveillance, and medical telemetry. In addition to sensing, one is
often also interested in control and activation.

There are four basic components in a sensor network: (1) an assembly of distributed or localized sensors; (2) an interconnecting network (usually, but not always,
wireless-based); (3) a central point of information clustering; and (4) a set of computing resources at the central point (or beyond) to handle data correlation, event
trending, status querying, and data mining. In this context, the sensing and computation nodes are considered part of the sensor network; in fact, some of the computing

---

[1]Although the terms *networked sensors* and *network of sensors* are perhaps grammatically more correct
than the term *sensor network*, generally in this book we employ the de facto nomenclature *sensor network*.

Wireless Sensor Networks: Technology, Protocols, and Applications, by Kazem Sohraby, Daniel Minoli,
and Taieb Znati

may be done in the network itself. Because of the potentially large quantity of data collected, algorithmic methods for data management play an important role in sensor networks. The computation and communication infrastructure associated with sensor networks is often specific to this environment and rooted in the device- and application-based nature of these networks. For example, unlike most other settings, in-network processing is desirable in sensor networks; furthermore, node power (and/or battery life) is a key design consideration. The information collected is typically parametric in nature, but with the emergence of low-bit-rate video [e.g., Moving Pictures Expert Group 4 (MPEG-4)] and imaging algorithms, some systems also support these types of media.

In this book we provide an exposition of the fundamental aspects of *wireless sensor networks* (WSNs). We cover wireless sensor network technology, applications, communication techniques, networking protocols, middleware, security, and system management. There already is an extensive bibliography of research on this topic; the reader may wish, for example, to consult [1.4] for an up-to-date list. We seek to systematize the extensive paper and conference literature that has evolved in the past decade or so into a cohesive treatment of the topic. The book is targeted to communications developers, managers, and practitioners who seek to understand the benefits of this new technology and plan for its use and deployment.

1.1.1   Background of Sensor Network Technology

Researchers see WSNs as an "exciting emerging domain of deeply networked systems of low-power wireless motes[2] with a tiny amount of CPU and memory, and large federated networks for high-resolution sensing of the environment" [1.93]. Sensors in a WSN have a variety of purposes, functions, and capabilities. The field is now advancing under the *push* of recent technological advances and the *pull* of a myriad of potential applications. The radar networks used in air traffic control, the national electrical power grid, and nationwide weather stations deployed over a regular topographic mesh are all examples of early-deployment sensor networks; all of these systems, however, use specialized computers and communication protocols and consequently, are very expensive. Much less expensive WSNs are now being planned for novel applications in physical security, health care, and commerce. Sensor networking is a multidisciplinary area that involves, among others, radio and networking, signal processing, artificial intelligence, database management, systems architectures for operator-friendly infrastructure administration, resource optimization, power management algorithms, and platform technology (hardware and software, such as operating systems) [1.5]. The applications, networking principles, and protocols for these systems are just beginning to be developed [1.48]. The near-ubiquity of the Internet, the advancements in wireless and wireline communications technologies, the network build-out (particularly

---

[2]The terms *sensor node*, *wireless node*, *smart dust*, *mote*, and *COTS* (commercial off the shelf) *mote* are used somewhat interchangeably; the most general terms, however, are *sensor node* and *wireless node*.

in the wireless case), the developments in IT (such as high-power processors, large random-access memory chips, digital signal processing, and grid computing), coupled with recent engineering advances, are in the aggregate opening the door to a new generation of low-cost sensors and actuators that are capable of achieving high-grade spatial and temporal resolution.

The technology for sensing and control includes electric and magnetic field sensors; radio-wave frequency sensors; optical-, electrooptic-, and infrared sensors; radars; lasers; location/navigation sensors; seismic and pressure-wave sensors; environmental parameter sensors (e.g., wind, humidity, heat); and biochemical national security–oriented sensors. Today's sensors can be described as "smart" inexpensive devices equipped with multiple onboard sensing elements; they are low-cost low-power untethered multifunctional nodes that are logically homed to a central sink node. Sensor devices, or wireless nodes (WNs), are also (sometimes) called *motes* [1.91]. A stated commercial goal is to develop complete microelectromechanical systems (MEMSs)–based sensor systems at a volume of 1 mm^3 [1.93]. Sensors are internetworked via a series of multihop short-distance low-power wireless links (particularly within a defined *sensor field*); they typically utilize the Internet or some other network for long-haul delivery of information to a point (or points) of final data aggregation and analysis. In general, within the sensor field, WSNs employ contention-oriented random-access channel sharing and transmission techniques that are now incorporated in the IEEE 802 family of standards; indeed, these techniques were originally developed in the late 1960s and 1970s expressly for wireless (not cabled) environments and for large sets of dispersed nodes with limited channel-management intelligence [1.6]. However, other channel-management techniques are also available.

Sensors are typically deployed in a high-density manner and in large quantities: A WSN consists of densely distributed nodes that support sensing, signal processing [1.7], embedded computing, and connectivity; sensors are logically linked by self-organizing means [1.8–1.11] (sensors that are deployed in short-hop point-to-point master–slave pair arrangements are also of interest). WNs typically transmit information to collecting (monitoring) stations that aggregate some or all of the information. WSNs have unique characteristics, such as, but not limited to, power constraints and limited battery life for the WNs, redundant data acquisition, low duty cycle, and, many-to-one flows. Consequently, new design methodologies are needed across a set of disciplines including, but not limited to, information transport, network and operational management, confidentiality, integrity, availability, and, in-network/local processing [1.12]. In some cases it is challenging to collect (extract) data from WNs because connectivity to and from the WNs may be intermittent due to a low-battery status (e.g., if these are dependent on sunlight to recharge) or other WN malfunction.[3] Furthermore, a lightweight protocol stack is desired. Often, a very large number of client units (say 64k or more) need to be supported by the system and by the addressing apparatus.

---

[3]Special statistical algorithms may be employed to correct from biases caused by erratic or poorly placed WNs [1.91].

Sensors span several orders of magnitude in physical size; they (or, at least some of their components) range from nanoscopic-scale devices to mesoscopic-scale devices at one end, and from microscopic-scale devices to macroscopic-scale devices at the other end. *Nanoscopic* (also known as *nanoscale*) refers to objects or devices on the order of 1 to 100 nm in diameter; mesoscopic scale refers to objects between 100 and 10,000 nm in diameter; the microscopic scale ranges from 10 to 1000 μm, and the macroscopic scale is at the millimeter-to-meter range. At the low end of the scale, one finds, among others, biological sensors, small passive microsensors (such as Smart Dust[4]), and "lab-on-a-chip" assemblies. At the other end of the scale one finds platforms such as, but not limited to, identity tags, toll collection devices, controllable weather data collection sensors, bioterrorism sensors, radars, and undersea submarine traffic sensors based on sonars.[5] Some refer to the latest generation of sensors, especially the miniaturized sensors that are directly embedded in some physical infrastructure, as *microsensors*. A sensor network supports any type of generic sensor; more narrowly, networked microsensors are a subset of the general family of sensor networks [1.13]. Microsensors with onboard processing and wireless interfaces can be utilized to study and monitor a variety of phenomena and environments at close proximity.

Sensors can be simple point elements or can be multipoint detection arrays. Typically, nodes are equipped with one or more application-specific sensors and with on-node signal processing capabilities for extraction and manipulation (preprocessing) of physical environment information. *Embedded network sensing* refers to the synergistic incorporation of microsensors in structures or environments; embedded sensing enables spatially and temporally dense monitoring of the system under consideration (e.g., an environment, a building, a battlefield). Sensors may be passive and/or be self-powered; farther down the power-consumption chain, some sensors may require relatively low power from a battery or line feed [1.14–1.19]. At the high end of the power-consumption chain, some sensors may require very high power feeds (e.g., for radars).

Sensors facilitate the instrumenting and controlling of factories, offices, homes, vehicles, cities, and the ambiance, especially as commercial off-the-shelf technology becomes available. With sensor network technology (specifically, with embedded networked sensing), ships, aircraft, and buildings can "self-detect" structural faults (e.g., fatigue-induced cracks). Places of public assembly can be instrumented to detect airborne agents such as toxins and to trace the source of the contamination should any be present (this can also be done for ground and underground situations). Earthquake-oriented sensors in buildings can locate potential survivors and can help assess structural damage; tsunami-alerting sensors are useful for nations with extensive coastlines. Sensors also find extensive applicability on the battlefield for reconnaissance and surveillance [1.20].

---

[4]The Smart Dust mote is an autonomous sensing, computing, and communication system that uses the optical visible spectrum for transmission [1.89]. They are tiny inexpensive sensors developed by UC–Berkeley engineers (see also Chapter 2).

[5]Although satellites can be used to support sensing, we do not include them explicitly in the technical discussion.

In this book we emphasize the emergence of open standards in support of WSNs; standardization drives commercialization of the technology. "New things" generally start out as advanced research projects pursued at government and/or academic labs. Typically, pure and/or applied research goes on for a number of years. At this early stage, specialized, one-of-a-kind, complex, and noninterworking prototypes, pilots, or deployments are common. Eventually, however, if a new thing is to become a ubiquitous technology, commercial-level open standards, chipsets, and products are needed, which must meet commercial service- and operational-level agreements in terms of reliability, cost, usability, durability, and simplicity. Following is a sample classification of research topics by frequency of publication based on a fair-sized sample of recent scientific WSN articles.

Deployment	9.70%
Target tracking	7.27%
Localization	6.06%
Data gathering	6.06%
Routing and aggregation	5.76%
Security	5.76%
MAC protocols	4.85%
Querying and databases	4.24%
Time synchronization	3.64%
Applications	3.33%
Robust routing	3.33%
Lifetime optimization	3.33%
Hardware	2.73%
Transport layer	2.73%
Distributed algorithms	2.73%
Resource-aware routing	2.42%
Storage	2.42%
Middleware and task allocation	2.42%
Calibration	2.12%
Wireless radio and link characteristics	2.12%
Network monitoring	2.12%
Geographic routing	1.82%
Compression	1.82%
Taxonomy	1.52%
Capacity	1.52%
Link-layer techniques	1.21%
Topology control	1.21%
Mobile nodes	1.21%
Detection and estimation	1.21%
Diffuse phenomena	0.91%
Programming	0.91%
Power control	0.61%
Software	0.61%
Autonomic routing	0.30%

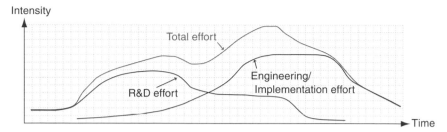

Figure 1.1   Shift and progression in emphasis over time in support of commercialization.

To appreciate the importance and criticality of simplicity-fostering standards in making a technology a pervasive reality, one need only study the progression of late-1960s wireless random-access systems (e.g., [1.21–1.23]) to the present-day LANs and WLAN/2.5G/3G systems (e.g., [1.6]); or the early-1970s ARPAnet (e.g., among many, [1.24]) to the present-day Internet (e.g., [1.25]); or the mid-1970s Voice Over Packet (e.g., [1.26–1.30]) to the current Voice Over IP technology (e.g., [1.31,1.32]); or the late-1980s video compression (e.g., [1.33]) to the current MPEG-2 and MPEG-4 digital video transmission revolution (e.g., [1.34]). See Figure 1.1 for a pictorial representation of the shift in technical emphasis over time.

Indeed, at this juncture, sensor networking is becoming a burgeoning field; there is currently extensive interest in this discipline not only from academia and government, but also from developers, manufacturers, startup companies, investors, and original equipment manufacturers (OEMs). According to industry observers, the wireless sensor market is now poised to take off commercially. Current market reports indicate that more than half a billion nodes are expected to ship for wireless sensor applications by 2010, for a market worth more than $7 billion [1.35]. As an example, advanced radio-frequency integrated circuits (RFICs) are now available for $3 or less, and smart sensor integrated circuits have become commonplace [1.35]. In the next few years, advances in the areas of sensor design and materials that have taken place in the recent past will lead, almost assuredly, to significant reductions in the size, weight, power consumption, and cost of sensors and sensor arrays; these advances will also affect an increase in their spatial and temporal resolution, along with improved measuring accuracy.

Implementations of WSNs have to address a set of technical challenges; however, the move toward standardization will, in due course, minimize a number of these challenges by addressing the issues once and then result in off-the-shelf chipsets and components. A current research and development (R&D) challenge is to develop low-power communication with low-cost on-node processing and self-organizing connectivity/protocols; another critical challenge is the need for extended temporal operation of the sensing node despite a (typically) limited power supply (and/or battery life). In particular, the architecture of the radio, including the use of low-power circuitry, must be properly selected. In practical terms this implies low power consumption for transmission over low-bandwidth channels and low-power-consumption logic to preprocess and/or compress data. Energy-efficient wireless communications systems are being sought and are typical of WSNs. Low power consumption is a key factor in ensuring long operating horizons for non-power-fed systems (some systems can indeed be power-fed and/or rely on other power sources). Power efficiency in WSNs is generally accomplished in three ways:

1. Low-duty-cycle operation.
2. Local/in-network processing to reduce data volume (and hence transmission time).
3. Multihop networking reduces the requirement for long-range transmission since signal path loss is an inverse exponent with range or distance. Each node in the sensor network can act as a repeater, thereby reducing the link range coverage required and, in turn, the transmission power.

Conventional wireless networks are generally designed with link ranges on the order of tens, hundreds, or thousands of miles. The reduced link range and the compressed data payload in WSNs result in characteristic link budgets that differ from those of conventional systems. However, the power restrictions, along with the desire for low node cost, give rise to what developers call "profound design challenges" [1.36]. Cooperative signal processing between nodes in proximity may enhance sensitivity and specificity to environmental event detection [1.36,1.37]. New CMOS (complementary metal-oxide semiconductor) chipsets optimized for WSNs are the key to commercialization success and are, in fact, being developed.

In this book we taxonomize (commercial) sensor networks and systems into two categories:

- *Category 1 WSNs* (C1WSNs): almost invariably mesh-based systems with multihop radio connectivity among or between WNs, utilizing dynamic routing in both the wireless and wireline portions of the network. Military-theater systems typically belong to this category.
- *Category 2 WSNs* (C2WSNs): point-to-point or multipoint-to-point (star-based) systems generally with single-hop radio connectivity to WNs, utilizing static routing over the wireless network; typically, there will be only one route from the WNs to the companion terrestrial or wireline forwarding node (WNs are pendent nodes). Residential control systems typically belong to this category.

C1WSNs support highly distributed high-node-count applications (e.g., environmental monitoring, national security systems); C2WSNs typically support confined short-range spaces such as a home, a factory, a building, or the human body. C1WSNs are different in scope and/or reach from evolving wireless C2WSN technology for short-range low-data-rate wireless applications such as

RFID (radio-frequency identification) systems, light switches, fire and smoke detectors, thermostats, and, home appliances. C1WSNs tend to deal with large-scale multipoint-to-point systems with massive data flows, whereas C2WSNs tend to focus on short-range point-to-point, source-to-sink applications with uniquely defined transaction-based data flows.

For a number of years, vendors have made use of proprietary technology for collecting performance data from devices. In the early 2000s, sensor device suppliers were researching ways of introducing standardization. WNs typically transmit small volumes of simple data (e.g., "Is the temperature at the set level or lower?"). For *within-building* applications, designers ruled out Wi-Fi (wireless fidelity, IEEE 802.11b) standards for sensors as being too complex and supporting more bandwidth than is actually needed for typical sensors. Infrared systems require line of sight, which is not always achievable; Bluetooth (IEEE 802.15.1) technology was at first considered a possibility, but it was soon deemed too complex and expensive. This opened the door for a new standard IEEE 802.15.4 along with ZigBee (more specifically, ZigBee comprises the software layers above the newly adopted IEEE 802.15.4 standard and supports a plethora of applications). C2WSNs have lower layers of the communication protocol stack (Physical and Media Access Control), which are comparable to that of a personal area network (PAN), defined in the recently developed IEEE 802.15 standard: hence, the utilization of these IEEE standards for C2WSNs. IEEE 802.15.4 operates in the 2.4-GHz industrial, scientific, and medical (ISM) radio band and supports data transmission at rates up to 250 kbps at ranges from 30 to 200 ft. ZigBee/IEEE 802.15.4 is designed to complement wireless technologies such as Bluetooth, Wi-Fi, and ultrawideband (UWB), and is targeted at commercial point-to-point sensing applications where cabled connections are not possible and where ultralow power and low cost are requirements [1.35].

With the emergence of the ZigBee/IEEE 802.15.4 standard, systems are expected to transition to standards-based approaches, allowing sensors to transfer information in a standardized manner. C2WSNs (and C1WSN, for that matter) that operate *outside a building and over a broad geographic area* may make use of any number of other standardized radio technologies. The (low-data-rate) C2WSN market is expected to grow significantly in the near future: The volume of low-data-rate wireless devices is forecast to be three times the size of Wi-Fi by the turn of the decade, due to the expected deployment of the systems based on the ZigBee/IEEE 802.15.4 standard (industry observers expect the number of ZigBee-compliant nodes to increase from less than 1 million in 2005 to 100 million in 2008). A discussion of both categories of technology, C1WSNs and C2WSNs, is provided in this book, but the reader should keep in mind that the technical issues affecting these two areas are, to a large degree, different.

There is also considerable research in the area of mobile ad hoc networks (MANETs). WSNs are similar to MANETs in some ways; for example, both involve multihop communications. However, the applications and technical requirements for the two systems are significantly different in several respects [1.38–1.41,1.48]:

1. The typical mode of communication in WSN is from multiple data sources to a data recipient or sink (somewhat like a reverse multicast) rather than communication between a pair of nodes. In other words, sensor nodes use primarily multicast or broadcast communication, whereas most MANETs are based on point-to-point communications.

2. In most scenarios (applications) the sensors themselves are not mobile (although the sensed phenomena may be); this implies that the dynamics in the two types of networks are different.

3. Because the data being collected by multiple sensors are based on common phenomena, there is potentially a degree of redundancy in the data being communicated by the various sources in WSNs; this is not generally the case in MANETs.

4. Because the data being collected by multiple sensors are based on common phenomena, there is potentially some dependency on traffic event generation in WSNs, such that some typical random-access protocol models may be inadequate at the queueing-analysis level; this is generally not the case in MANETs.

5. A critical resource constraint in WSNs is energy; this is not always the case in MANETs, where the communicating devices handled by human users can be replaced or recharged relatively often. The scale of WSNs (cspccially, C1WSNs) and the necessity for unattended operation for periods reaching weeks or months implies that energy resources have to be managed very judiciously. This, in turn, precludes high-data-rate transmission.

6. The number of sensor nodes in a sensor network can be several orders of magnitude higher than the nodes in a MANET.

For these reasons the plethora of routing protocols that have been proposed for MANETs are not suitable for WSNs, and alternative approaches are required [1.48]. Note that MANETs per se are not discussed further in this book.

Others also study wireless mesh networks (WMNs) (see, e.g., [1.94] for an extensive tutorial). Wi-Fi-based WMNs are being applied as hot zones, which cover a broad area such as a downtown city district. Although WMNs have many of the same networking characteristics as WSNs, their application can, in principle, be more general. Also, a fairly large fraction of the commercial WSNs of the near future are expected to be of the C1WSN category, which does not (obligatorily) require or entail meshing. Like WSNs, WMNs can use off-the-shelf radio technology such as Wi-Fi, WiMax (worldwide interoperability for microwave access), and cellular 3G. As an observation, the topic of network mobility (NEMO) is unrelated to WSNs in general terms. NEMO is concerned with managing the mobility of an entire network, which changes, as a unit, its point of attachment to the Internet and thus its reachability in the topology. The mobile network includes one or more mobile routers which connect it to the global Internet. A mobile network is assumed to be a leaf network, i.e., it will not carry transit traffic [1.96]. As should be clear by now, the focus of this book is on WSNs; hence, we do not spend any time covering WMNs.

1.1.2   Applications of Sensor Networks

Traditionally, sensor networks have been used in the context of high-end applications such as radiation and nuclear-threat detection systems, "over-the-horizon" weapon sensors for ships, biomedical applications, habitat sensing, and seismic monitoring. More recently, interest has focusing on networked biological and chemical sensors for national security applications; furthermore, evolving interest extends to direct consumer applications. Existing and potential applications of sensor networks include, among others, military sensing, physical security, air traffic control, traffic surveillance, video surveillance, industrial and manufacturing automation, process control, inventory management, distributed robotics, weather sensing, environment monitoring, national border monitoring, and building and structures monitoring [1.13]. A short list of applications follows.

- Military applications

  - Monitoring inimical forces
  - Monitoring friendly forces and equipment
  - Military-theater or battlefield surveillance
  - Targeting
  - Battle damage assessment
  - Nuclear, biological, and chemical attack detection
  and more . . .

- Environmental applications

  - Microclimates
  - Forest fire detection
  - Flood detection
  - Precision agriculture
  and more . . .

- Health applications

  - Remote monitoring of physiological data
  - Tracking and monitoring doctors and patients inside a hospital
  - Drug administration
  - Elderly assistance
  and more . . .

- Home applications

  - Home automation
  - Instrumented environment
  - Automated meter reading
  and more . . .

- Commercial applications

  - Environmental control in industrial and office buildings
  - Inventory control
  - Vehicle tracking and detection
  - Traffic flow surveillance

  and more . . .

Chemical-, physical-, acoustic-, and image-based sensors can be utilized to study ecosystems (e.g., in support of global parameters such as temperature and micro-organism populations). Defense applications have fostered research and development in sensor networks during the past half-century. On the battlefield, sensors can be used to identify and/or track friendly or inimical objects, vehicles, aircraft, and personnel; here, a system of networked sensors can detect and track threats and can be utilized for weapon targeting and area denial [1.13,1.20]. "Smart" *disposable* microsensors can be deployed on the ground, in the air, under water, in (or on) human bodies, in vehicles, and inside buildings. Homes, buildings, and locales equipped with this technology are being called *smart spaces*.

Wireless sensors can be used where wireline systems cannot be deployed (e.g., a dangerous location or an area that might be contaminated with toxins or be subject to high temperatures). The rapid deployment, self-organization, and fault-tolerance characteristics of WSNs make them versatile for military *command*, *control*, *communications*, *intelligence*, *surveillance*, *reconnaissance*, and *targeting systems* [1.38]. Many of these features also make them ideal for national security. Sensor networking is also seen in the context of pervasive computing [1.42].

The deployment scope for sensing and control networks is poised for significant expansion in the next three to five years as we have already mentioned; this expansion relates not only to science and engineering applications but also to a plethora of "new" consumer applications. Industry players expect that in the near future it will become possible to integrate sensors into commercial products and systems to improve the performance and lifetime of a variety of products; industry planners also expect that with sensors one can decrease product life-cycle costs. Consumer applications include, but are not limited to, critical infrastructure protection and security, health care, the environment, energy, food safety, production processing, and quality of life [1.35]. WSNs are also expected to afford consumers a new set of conveniences, including remote-controlled home heating and lighting, personal health diagnosis, automated automobile maintenance telemetry, and automated in-marina boat-engine telemetry, to list just a few. The ultimate expectation is that eventually wireless sensor network technologies will enable consumers to keep track of their belongings, pets, and young children [1.35]. Ubiquitous high-reliability public-safety applications covering a multithreat management are also on the horizon.

Near-term commercial applications include, but are not limited to, industrial and building wireless sensor networks, appliance control [lighting, and heating, ventilation, and air conditioning (HVAC)], automotive sensors and actuators, home automation and networking, automatic meter reading/load management, consumer electronics/entertainment, and asset management. Commercial market segments include the following:

- Industrial monitoring and control
- Commercial building and control
- Process control
- Home automation
- Wireless automated meter reading (AMR) and load management (LM)
- Metropolitan operations (traffic, automatic tolls, fire, etc.)
- National security applications: chemical, biological, radiological, and nuclear wireless sensors
- Military sensors
- Environmental (land, air, sea) and agricultural wireless sensors

Suppliers and products tend to cluster according to these categories.

1.1.3   Focus of This Book

This book focuses on wireless sensor networks.[6,7] We look at basic WSN technology and supporting protocols, with emphasis placed on standardization. The treatise provides an exposition of the fundamental aspects of wireless sensor networks from a practical engineering perspective. The text provides an introductory up-to-date survey of WSNs, including applications, communication, technology, networking protocols, middleware, security, and management. Both C1WSNs and C2WSNs are addressed.

The present chapter aims at assessing, from an introductory perspective, sensor technology as a whole, including some of the recent history of the field. We also address some of the challenges to be faced and addressed by the evolving practice. In Chapter 2 we discuss near-term and longer-range applications of WSNs and look at network sensor applications for both business- and government-oriented applications. In Chapter 3 we look at basic sensor systems and provide a survey of sensor technology, including classification in terms of microsensors (tiny sensors), radar sensors, nanosensors, and other sensors. We address sensor functionality, sensing and actuation units, processing units, communication units, power units, and other application-dependent units. We also look at design issues, the operating environment and hardware constraints, transmission media, radio-frequency integrated circuits, power constraints, communications network interfaces, network architecture and protocols, network topology, performance issues, fault tolerance, scalability, and self-organization and mobility capabilities. Sensor arrays and networks are also discussed.

Chapter 4 begins a discussion of sensor network protocols. We address physical layer issues such as channel-related concerns, radio-frequency bands, bandwidth,

---

[6]Some sensor networks are not wireless; although many of the issues are similar, others are not. Our discussion focuses on the wireless situation.
[7]Control and actuation are covered here only in passing.

propagation modes (ground wave, sky wave, line of sight), and channel impairments (e.g., refraction, atmospheric absorption, fading, multipath, free space, Gaussian noise, Rayleigh fading, Rician fading). Reference is made to the gamut of off-the-shelf radio technologies that can be used for WSNs. Chapter 5 extends the topics introduced in Chapter 4 by covering medium access control protocols in some detail; we provide a survey of media access control (MAC) protocols for sensor networks, including the IEEE 802.11 family, the IEEE 802.15 family (e.g., Bluetooth and ZigBee), and other protocols. In Chapter 6 we discuss routing protocols in sensor networks, providing a survey of key routing protocols for sensor networks and discussing the main design issues (e.g., scalability, mobility, power awareness, self-organization, naming). In Chapter 7 we look at transport protocols, provide a survey of transport layer protocols for sensor networks, and discuss design requirements (e.g., error control, reliability, power awareness, delay guarantees).

Chapter 8 begins a discussion of sensor network middleware, operating systems (OSs), and application programming interfaces (APIs). Chapter 8 covers middleware for sensor networks, including data dissemination models (data aggregation and follow-on data dissemination protocols), compression techniques, and data storage. In Chapter 9 we examine sensor management, including naming and localization and maintenance and fault tolerance. In Chapter 10 we address operating systems for sensor networks. The discussion includes design factors (size constraints, power awareness, distribution and reconfiguration; and APIs and programming language paradigms). A survey of commercially available operating systems for sensor networks is provided. Chapter 11 covers performance and traffic management.

1.2 BASIC OVERVIEW OF THE TECHNOLOGY

In Section 1.1 we provided a high-level description of the approach, issues, and technologies associated with WSNs. Some additional details are provided in this section from a generic perspective; many of these issues and concepts are then discussed in greater detail in the chapters that follow. As we proceed, the reader should keep in mind that sensor networks deal with space and time: location, coverage, and data synchronization. Data are the intrinsic "currency" of a sensor network. Typically, there will be a large amount of time-stamped time-dependent data. Therefore, sensor networks often support in-network computation. Some sensor networks use source-node processing; others use a hierarchical processing architecture. Instead of sending the raw data to the nodes responsible for the data fusion, nodes often use their processing abilities locally to carry out basic computations, and then transmit only a subset of the data and/or partially processed data. In a hierarchical processing architecture, processing occurs at consecutive tiers until the information about events of interest reaches the appropriate decision-making and/or administrative point. Sensor nodes are almost invariably constrained in energy supply and radio channel transmission bandwidth; these constraints, in conjunction with a typical deployment of large number of sensor nodes, have posed a plethora of challenges to the design and management of WSNs. These challenges necessitate energy awareness at all layers of a communications protocol stack [1.92]. Some of the key technology and standards elements that are relevant to sensor networks are as follows:

- Sensors
  - Intrinsic functionality
  - Signal processing
  - Compression, forward error correction, encryption
  - Control/actuation
  - Clustering and in-network computation
  - Self-assembly
- Wireless radio technologies
  - Software-defined radios
  - Transmission range
  - Transmission impairments
  - Modulation techniques
  - Network topologies
- Standards (de jure)
  - IEEE 802.11a/b/g together with ancillary security protocols
  - IEEE 802.15.1 PAN/Bluetooth
  - IEEE 802.15.3 ultrawideband (UWB)
  - IEEE 802.15.4/ZigBee (IEEE 802.15.4 is the physical radio, and ZigBee is the logical network and application software)
  - IEEE 802.16 WiMax
  - IEEE 1451.5 (Wireless Sensor Working Group)
  - Mobile IP
- Standards (de facto)
  - Tiny OS (TinyOS is being developed by the University of California–Berkeley as an open-source software platform; the work is funded by DARPA and is undertaken in the context of the Network Embedded Systems Technology Research Project at UC–Berkeley in collaboration with the University of Virginia, Palo Alto Research Center, Ohio State University, and approximately 100 other organizations)
  - Tiny DB (a query-processing system for extracting information from a network of TinyOS sensors)
- Software applications
  - Operating systems
  - Network software

- Direct database connectivity software
- Middleware software
- Data management software

1.2.1 Basic Sensor Network Architectural Elements

In this section we briefly highlight the basic elements and design focus of sensor networks. These elements and design principles need to be placed in the context of the C1WSN sensor network environment, which is characterized by many (sometimes all) of the following factors: large sensor population (e.g., 64,000 or more client units need to be supported by the system and by the addressing apparatus), large streams of data, incomplete/uncertain data, high potential node failure; high potential link failure (interference), electrical power limitations, processing power limitations, multihop topology, lack of global knowledge about the network, and (often) limited administrative support for the network [1.43] (C2WSNs have many of these same limitations, but not all). Sensor network developments rely on advances in sensing, communication, and computing (data-handling algorithms, hardware, and software). As noted, to manage scarce WSN resources adequately, routing protocols for WSNs need to be energy-aware. Data-centric routing and in-network processing are important concepts that are associated intrinsically with sensor networks [1.44–1.48]. The end-to-end routing schemes that have been proposed in the literature for mobile ad hoc networks are not appropriate WSNs; data-centric technologies are needed that perform in-network aggregation of data to yield energy-efficient dissemination [1.48].

Sensor Types and Technology   A sensor network is composed of a large number of sensor nodes that are densely deployed [1.38,1.39]. To list just a few venues, sensor nodes may be deployed in an open space; on a battlefield in front of, or beyond, enemy lines; in the interior of industrial machinery; at the bottom of a body of water; in a biologically and/or chemically contaminated field; in a commercial building; in a home; or in or on a human body. A sensor node typically has embedded processing capabilities and onboard storage; the node can have one or more sensors operating in the acoustic, seismic, radio (radar), infrared, optical, magnetic, and chemical or biological domains. The node has communication interfaces, typically wireless links, to neighboring domains. The sensor node also often has location and positioning knowledge that is acquired through a global positioning system (GPS) or local positioning algorithm [1.13,1.49–1.52]. (Note, however, that GPS-based mechanisms may sometimes be too costly and/or the equipment may be too bulky.) Sensor nodes are scattered in a special domain called a *sensor field*. Each of the distributed sensor nodes typically has the capability to collect data, analyze them, and route them to a (designated) *sink* point. Figure 1.2 depicts a typical WSN arrangement. Although in many environments all WNs are assumed to have similar functionality, there are cases where one finds a heterogeneous environment in regard to the sensor functionality.

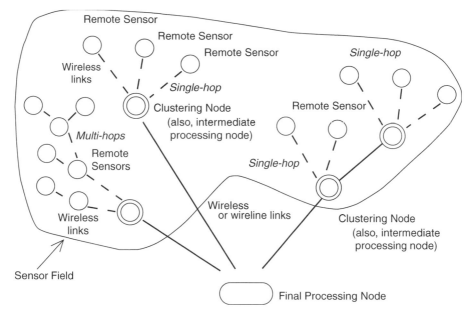

Figure 1.2   Typical sensor network arrangement.

The following are important issues pertaining to WSNs (see also Table 1.1): sensor type; sensor placement; sensor power consumption, operating environment, computational/sensing capabilities and signal processing, connectivity, and teleme-try or control of remote devices. It is critical to note in this context that node loca-tion and fine-grained time (stamping) are essential for proper operation of a sensor network; this is almost the opposite of the prevalent Internet architecture, where server location is immaterial to a large degree and where latency is often not a key consideration or explicit design objective. In sensor networks, fine-grained time synchronization and localization are needed to detect events of interest in the environment under observation. Location needs to be tracked both in local three-dimensional space (e.g., On what floor and in which quadrant is the smoke detected? What is the temperature of the atmosphere at height h?) and over a broader topography, to assess detection levels across a related set (array) of sensors (e.g., What is the wind direction for wind containing contaminated particles at mile-post i, $i+1$, $i+2$, etc., along a busy highway?). Localization is used for function-ality such as beamforming for localization of target and events, geographical forwarding, and geographical addressing [1.5].

Embedded sensor networks are predicated on three supporting components: embed-ding, networking, and sensing. *Embedding* implies the incorporation of numerous distributed devices to monitor the physical world and interact with it; the devices are untethered nodes of small form factors that are equipped with a control and communication subsystem. Spatially- and temporally-dense arrangements are com-mon. *Networking* implies the concept of physical and logical connectivity.

TABLE 1.1 Categorization of Issues Related to Sensors and Their Communication/Computing Architecture

Sensors	*Size:* Small [e.g., nanoscale electromechanical systems (MEMS)], medium [e.g., microscale electromechanical systems (MEMS)], and large (e.g., radars, satellites): cubic centimeters to cubic decimeters *Mobility:* stationary (e.g., seismic sensors), mobile (e.g., on robot vehicles) *Type:* passive (e.g., acoustic, seismic, video, infrared, magnetic) or active (e.g., radar, ladar)
Operating environment	*Monitoring requirement:* distributed (e.g., environmental monitoring) or localized (e.g., target tracking) *Number of sites:* sometimes small, but usually large (especially for C1WSNs) *Spatial coverage:* dense, spars: C1WSN: low-range multihop or C2WSN: low-range single-hop (point-to-point) *Deployment:* fixed and planned (e.g., factory networks) or ad hoc (e.g., air-dropped) *Environment:* benign (factory floor) or adverse (battlefield) *Nature:* cooperative (e.g., air traffic control) or noncooperative (e.g., military targets) *Composition:* homogeneous (same types of sensors) or heterogeneous (different types of sensors) *Energy availability:* constrained (e.g., in small sensors) or unconstrained (e.g., in large sensors)
Communication	*Networking:* wired (on occasion) or wireless (more common) *Bandwidth:* high (on occasion) or low (more typical)
Processing architecture	Centralized (all data sent to central site), distributed or in-network (located at sensor or other sides), or hybrid

Source: Modified from [1.13], with permission.

Logical connectivity has the goal of supporting coordination and other high-level tasks; physical connectivity is typically supported over a wireless radio link [1.53]. *Sensing* implies the presence of these capabilities in a tightly coupled environment, typically for the measurement of physical-world parameters. Some of the characteristic features of sensor networks include the following [1.38,1.39]:

- Sensor nodes are densely deployed.
- Sensor nodes are prone to failures.
- The topology of a sensor network changes very frequently.
- Sensor nodes are limited in power, computational capacities, and memory.
- Sensor nodes may not have global *identification* because of the large amount of overhead and the large number of sensors.

Sensor networks require sensing systems that are long-lived and environmentally resilient. Unattended, untethrered, self-powered low-duty-cycle systems are typical.

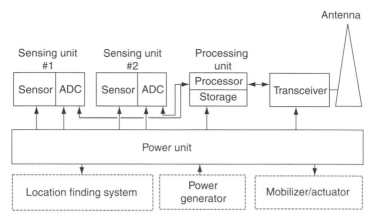

ADC = Analog-to-Digital Converter

Figure 1.3   Typical sensing node.

Power consumption is often an issue that needs to be taken into account as a design constraint. In most instances, communication circuitry and antennas are the primary elements that draw most of the energy [1.54–1.58]. Sensors are either passive or active devices. *Passive sensors* in element form include seismic-, acoustic-, strain-, humidity-, and temperature-measuring devices. Passive sensors in array form include optical- [visible, infrared 1 micron (μm), infrared 10 μm], and biochemical-measuring devices. Passive sensors tend to be low-energy devices. *Active sensors* include radar and sonar; these tend to be high-energy systems. The trend is toward VLSI (very large scale integration), integrated optoelectronics, and nanotechnology; work is under way in earnest in the biochemical arena. The components of a (remote) sensing node include (see Figure 1.3) the following:

- A sensing and actuation unit (single element or array)
- A processing unit
- A communication unit
- A power unit
- Other application-dependent units

Figure 1.4 depicts an example on an (ultra)miniature sensor.

In addition to (embedded) sensing there is a desire to build, deploy, and manage unattended or untethered embedded *control and actuation systems*, sometimes called *control networks*. Such a control system acts on the environment either in a self-autonomous manner or under the telemetry of a remote or centralized node. Key applications require more than just sensing: They need control and actuation. To the extent that we cover the topic in this book, *control* refers to some "minor" activity internal to the sensor (e.g., zoom, add an optical filter, rotate

Figure 1.4 Miniature sensor: the MacroMote, developed at UC–Berkeley. (Courtesy of UC–Berkeley.)

an antenna); *actuation* refers to a "major" activity external to the sensor itself (e.g., open a valve, emit some fluid into the environment, engage a motor to relocate somewhere else). Applications requiring control and/or actuation include transportation, high-tech agriculture, medical monitoring, drug delivery, battlefield interventions, and so on. In addition to standard concerns (e.g., reliability, security), actuation systems also have to take into account factors such as safety. The topic of WSN applications is revisited in Chapter 2.

Software (Operating Systems and Middleware) To support the node operation, it is important to have *open-source operating systems* designed specifically for WSNs. Such operating systems typically utilize a *component-based architecture* that enables rapid implementation and innovation while minimizing code size as required by the memory constraints endemic in sensor networks. TinyOS is one such example of a de facto standard, but not the only one. TinyOS's *component library* includes network protocols, distributed services, sensor drivers, and data acquisition tools; these can be used as-is or be further refined for a specific application. TinyOS's event-driven execution model enables fine-grained power management, yet allows the scheduling flexibility made necessary by the unpredictable nature of wireless communication and physical world interfaces. TinyOS has already been ported to over a dozen platforms and numerous sensor boards. A wide community uses TinyOS in simulation to develop and test various algorithms and protocols, and numerous groups are actively contributing code to establish standard interoperable network services [1.90]. This topic is revisited in Chapter 8.

Standards for Transport Protocols The goal of WSN engineers is to develop a cost-effective standards-based wireless networking solution that supports low-to-medium data rates, has low power consumption, and guarantees security and reliability [1.66–1.73]. The position of sensor nodes does not have be predetermined, allowing random deployment in inaccessible terrains or dynamic situations; however, this also means that sensor network protocols and algorithms must possess self-organizing capabilities [1.38,1.39]. For military and/or national security

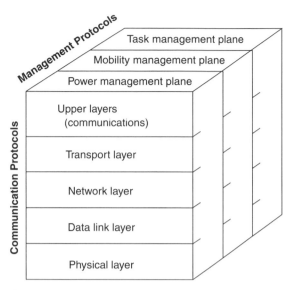

Figure 1.5   Generic protocol stack for sensor networks.

applications, sensor devices must be amenable to rapid deployment, the deployment must be supportable in an ad hoc fashion, and the environment is expected to be highly dynamic.

Researchers have developed many new protocols specifically designed for WSNs, where energy awareness is an essential consideration; focus has been given to the routing protocols, since they might differ from traditional networks (depending on the application and network architecture) [1.92]. Networking per se is an important architectural component of sensor networks, and standards play a major role in this context. Figure 1.5 depicts a generic protocol stack model that can be utilized to describe the communications apparatus (also see Table 1.2). Table 1.3 shows some typical lower-layer protocols that are in principle applicable to

TABLE 1.2   Possible WSN Protocol Stack[a]

Upper layers	In-network applications, including application processing, data aggregation, external querying query processing, and external database
Layer 4	Transport, including data dissemination and accumulation, caching, and storage
Layer 3	Networking, including adaptive topology management and topological routing
Layer 2	Link layer (contention): channel sharing (MAC), timing, and locality
Layer 1	Physical medium: communication channel, sensing, actuation, and signal processing

[a]Table modeled after [1.05].

TABLE 1.3 Possible Lower-Layer WSN Protocols

	GPRS/GSM 1xRTT/CDMA	IEEE 802.11b/g	IEEE 802.15.1	IEEE 802.15.4
Market name for standard	2.5G/3G	Wi-Fi	Bluetooth	ZigBee
Network target	WAN/MAN	WLAN and hotspot	PAN and DAN (desk area network)	WSN
Application focus	Wide area voice and data	Enterprise applications (data and VoIP)	Cable replacement	Monitoring and control
Bandwidth (Mbps)	0.064–0.128+	11–54	0.7	0.020–0.25
Transmission range (ft)	3000+	1–300+	1–30+	1–300+
Design factors	Reach and transmission quality	Enterprise support, scalability, and cost	Cost, ease of use	Reliability, power, and cost

WSNs; overall, a lightweight protocol stack is sought for WSNs. Issues here relate to the following:

1. Physical connectivity and coverage: How can one interconnect dispersed sensors in a cost-effective and reliable manner, and what medium should be used (e.g., wireless channels)?
2. Link characteristics and capacity, along with data compression (see, e.g., [1.59])
3. Networking security and communications reliability (including naturally occurring phenomena such as noise impairments, and malicious issues such as attacks, interference, and penetration)
4. Physical-, link-, network-, and transport-layer protocols, with an eye to reliable transport, congestion detection and avoidance, and scalable and robust communication (e.g., [1.60–1.64])
5. Communication mechanisms in what could be an environment with highly correlated and time-dependent arrivals (where many of the queueing assumptions used for system modeling could break down [1.6,1.65])

Although sensor electronics are becoming inexpensive, observers see the lack of networking standards as a potentially retardant factor in the commercial deployment of sensor networks. Because today there are still numerous proprietary network protocols, manufacturers have created vendor-specific and consequently, expensive products that will not work with products from other manufacturers.

The lack of open standards has not only prevented the possibility for interoperability but has also limited innovation. Evolving standards may provide, on a going-forward basis, a common framework on which developers can create applications that will leverage the hardware advances with radios and sensors. The goal of standards is to enable developers to design solutions that will lower installation and maintenance costs for a variety of sensors used in industrial, commercial, and residential settings [1.35]. As one example of an applicable standard, particularly for C2WSNs, the IEEE 802.15.4 specification for the physical, media access, and data link layers was formally ratified in 2003; at press time, ZigBee Alliance[8] members were defining a global specification for reliable, cost-effective, low-power wireless applications based on the IEEE 802.15.4 standard. Another standard of potential interest is the IEEE 802.16, also known as WiMax. This topic is revisited in Chapters 4, 5, 6, and 7.

Routing and Data Dissemination   Routing and data dissemination issues deal with data dissemination mechanisms for large-scale wireless networks, directed diffusion (see, e.g., [1.74]), data-centric routing [also known as data aggregation (see, e.g., [1.44])], adaptive routing, and other specialized routing mechanism. Routing protocols for WSNs generally fall into three groups: data-centric, hierarchical, and location-based. The concept of data aggregation is to combine the data arriving from different sources along the way (enroute). This allows one to eliminate redundancy, minimize the number of transmissions, and in turn, be parsimonious with energy consumption. This routing approach shifts the emphasis from the traditional *address-centric* approaches (finding short routes between pairs of addressable end nodes) to a *data-centric* approach (finding routes from multiple sources to a single destination that allows in-network consolidation of redundant data) [1.48]; see Table 1.4.

As already noted, there is interest in handling in-network processing, even while the data are being routed. Communications links may be expensive (not only from an electromagnetic spectrum perspective, but also in terms of the operational support of the requisite infrastructure); the bandwidth may be limited; and the power availability at the sensor may be limited and/or expensive in reference to supporting a high-capacity/high-range link (i.e., to feed a high-power antenna). It follows that one wants to perform data processing in the network, in proximity of the source of the data, and then only forward summarized, aggregated, fused, and/or synthesized results.

To support data-centric routing and directed diffusion, one needs to name the data (rather than the nodes) with relevant attributes such as (but not limited to)

---

[8]The ZigBee Alliance is a nonprofit industry consortium of leading semiconductor manufacturers, technology providers, OEMs, and end users worldwide. Membership is open to all. ZigBee Alliance members are defining a global specification for reliable, cost-effective, low-power wireless applications based on the IEEE 802.15.4 standard. Over 68 member companies are working actively to define the ZigBee specification, including six promoters (Honeywell, Invensys, Mitsubishi, Motorola, Philips, and Samsung) and participants that include semiconductor manufacturers, wireless IP providers, and OEMs.

TABLE 1.4   Summary of Routing Protocols Utilized in WSNs

Routing Protocol Category	Description	Examples
Data centric	The sink sends queries to certain WSN regions and waits for data from WNs located in the regions selected. Because data are being requested through queries, attribute-based naming is necessary to specify the properties of data. Due to the large number of nodes deployed, in many WSNs it is not practical to assign global identifiers to each node. This, along with potential random deployment of WNs, makes it challenging to select a specific (or a specific set of) WNs to be queried. Hence, data are typically transmitted from every WN with in the deployment region; this gives rise, however, to significant redundancy along with inefficiencies in terms of energy consumption. It follows that it is desirable to have routing protocols that will be able to select a set of sensor nodes and utilize data aggregation during the relaying of data. This has led to the development of data-centric routing (in traditional address-based routing, routes are created between addressable nodes managed in the network layer mechanism).	Sensor protocols for information via negotiation (SPIN) Directed diffusion Rumor routing Gradient-based routing (GBR) Constrained anisotropic diffusion routing (CADR) COUGAR ACQUIRE
Hierarchical	A single-tier (gateway or cluster-point) network can cause the gateway node to become overloaded, particularly as the density of sensors increases. This, in turn, can cause latency in event status delivery. To permit WSNs to deal with a large population of WNs and to cover a large area of interest, multipoint clustering has been proposed. The goal of hierarchical routing is to manage the energy consumption of WNs efficiently by establishing multihop communication within a particular cluster, and by performing data aggregation and fusion to decrease the number of transmitted packets to the sink.	Energy-adaptive clustering hierarchy (LEACH) Threshold-sensitive energy-efficient sensor network protocol (TEEN) and adaptive threshold-sensitive energy-efficient sensor network protocol (APTEEN) Power-efficient gathering in sensor information systems (PEGASIS)

(Continued)

TABLE 1.4   (*Continued*)

Routing Protocol Category	Description	Examples
Location based	Location information about the WNs can be utilized in routing data in an energy-efficient manner. Location information is used to calculate the distance between two given nodes so that energy consumption can be determined (or at least, estimated). For example, if the region to be sensed is known, the query can be diffused only to that specific region, limiting and/or eliminating the number of transmissions in the out-of-region space. Location-based routing is ideal for mobile ad hoc networks, but it can also be used for generic WSNs. (Note that non-energy-aware location-based protocols designed for wireless ad hoc networks, such as Cartesian and trajectory-based routing, are not desirable or ideal in WSNs.)	Minimum energy communication network (MECN) and small minimum energy communication network (SMECN) Geographic adaptive fidelity (GAF) Geographic and energy aware routing (GEAR)
QoS-oriented	Quality of service (QoS)–aware protocols consider end-to-end delay requirements in setting up the paths in the sensor network.	Sequential assignment routing (SAR) Stateless protocol for end-to-end delay (SPEED)

Source: Based partially on [1.92].

data type, time, and location. One needs to diffuse requests and responses over the network with application-cognizant routing; and one must support in-network data aggregation and processing [1.75,1.76]. Some view sensor networks as being peer to peer at the logical level, even though the physical communication topology is generally hierarchical; here one peer is the data source that "publishes" the data (could be a basic sensor node or an aggregation node) and the other peer is the data client that subscribes to a data content list. This topic is revisited in Chapter 6.

Sensor Network Organization and Tracking   Areas of interest involving network organization and tracking include distributed group management (maintaining organization in large-scale sensor networks); self-organization, including authentication, registration, and session establishment; and entity tracking: target detection, classification, and tracking. Dynamic sensor allocation (i.e., how to deal with impaired or unreliable sensors and/or how to "clean" and query noisy sensors) is also of interest. Some of the factors that come into play include the following: area of coverage (portion of topography of interest that is covered by sensors); detectability (probability that the sensor will detect an event such as a value variation or a moving object); and node coverage (portion of sensor population that is covered, in an overlapping sense, by other sensors that could be used in case of malfunction of the primary sensor). In case of control or actuation, factors include assessments as to where one needs to add new nodes (or to reorient or rotate a measuring probe) for optimal coverage and/or how to move a sensor (autonomously) to a new location for maximal coverage. This topic is revisited in Chapter 9.

Computation   Computation deals with data aggregation, data fusion, data analysis, computation hierarchy, grid computing (utility-based decision making in wireless sensor networks), and signal processing. We have already mentioned the desire for data-centric protocols that support in-network processing; however, it must be noted that per-node processing by itself is not sufficient: One needs interpretation of spatially distributed events and data related to those events. The network may be required to handle in-network processing based on the locality of the data, and queries must be directed automatically to the node or nodes that have the best view of the system (environment) in the context of the data queried. An area of recent research is networked information processing: how to extract useful, reliable, and timely information from the sensor network deployed; this implies leveraging the distributed computing environment created by these sensors for signal and information processing in the network and for dynamic and interactive querying and tasking the sensor network [1.13]. This topic is revisited in Chapter 10.

Data Management   Data management deals with data architectures; database management, including querying mechanisms; and data storage and warehousing. In a traditional environment (even in a traditional sensor network environment), data are collected to a centralized server for storage, against which queries are issued. In a more elaborate environment, particularly in support of true-real-time data querying, a mechanism can be deployed to support distributed data storage (possibly extending to clustering nodes) and to support distributed data querying [1.77–1.81]. In particular, one is interested in multiresolution/multitiered data storage and retrieval. The data need to be indexed for efficient temporal and spatial searching; at the same time, one wants to be able easily to generate global values associated with variables or requirements of interest. This topic is revisited in Chapter 8.

Security   Security deals with confidentiality (encryption), integrity (e.g., identity management, digital signatures), and availability (protection from denial of service).

Network Design Issues   We have already noted that in sensor networks, issues relate to reliable transport (possibly including encryption), bandwidth-and power-limited transmission, data-centric routing, in-network processing, and self-configuration. Design factors include operating environment and hardware constraints such as transmission media, radio-frequency integrated circuits, power constraints, communications network interfaces; and network architecture and protocols, including network topology and fault tolerance, scalability, self-organization, and mobility [1.82,1.83].

Sensor networks are generally self-configuring systems. The goal is to be able to adapt to unpredictable situations and states. Static or semidynamic topologies lend themselves easily to preconfiguration, but highly dynamic environments require self-configuration. In designing a sensor network, one is naturally looking for acceptable accuracy of information (even in the presence of failed nodes and/or links, and possibly conflicting or partial data); low network and computing latency; and optimal resource use (specifically, power and bandwidth). Work is under way to develop techniques that can be employed to deal with these and other pertinent issues, such as how to represent sensor data, how to structure sensor queries, how to adapt to changing node or network conditions, and how to manage a large network environment where nodes have limited network management functionality.

Sensor networks often employ data processing directly in the network itself. Part of the motivation is the potential for large pools of data being generated by the sensors. By utilizing computation close to the source of the data for trending, averaging, maxima and minima, or out-of-range activities, one is able to reduce the communication throughput that would otherwise be needed. Intrinsic to this is the development of localized algorithms that support global goals; it follows that forms of collaborative signal processing are desired.

Researchers are looking at new system architectures to manage interactions. Currently, many sensor systems suffer from being one-of-a-kind with piecemeal design approaches. This predicament leads to suboptimal economics, longevity, interoperability, scalability, and robustness. Standards will go a long way to address a number of these concerns. A number of researchers [1.5] are taking the position that the traditional approach and/or protocol suite is not adequate for embedded, energy-constrained, untethered, small-form-factor, unattended systems, because these systems cannot tolerate the communication overhead associated with the routing and naming intrinsic in the Internet suite of protocols. Proponents are making a pitch for special-purpose system functions in place of the general-purpose Internet functionality designed for elastic applications. In effect, resource constraints require a more streamlined and more tightly integrated communications layer than that possible with a TCP–IP or ISO (International Organization for Standardization) stack. This topic is revisited in Chapter 9 and 11.

1.2.2 Brief Historical Survey of Sensor Networks

The history of sensor networks spans four phases, described briefly below [1.13].

Phase 1: Cold-War Era Military Sensor Networks   During the cold war, extensive acoustic networks were developed in the United States for submarine surveillance; some of these sensors are still being used by the National Oceanographic and Atmospheric Administration (NOAA) to monitor seismic activity in the ocean. Also, networks of air defense radars were deployed to cover North America; to handle this, a battery of Airborne Warning and Control System (AWACS) planes operated as sensors.

Phase 2: Defense Advanced Research Projects Agency Initiatives The major impetus to research on sensor networks took place in the early 1980s with programs sponsored by the Defense Advanced Research Projects Agency (DARPA). The distributed sensor networks (DSN) work aimed at determining if newly developed TCP–IP protocols and ARPAnet's (the predecessor of the Internet) approach to communication could be used in the context of sensor networks. DSN postulated the existence of many low-cost spatially distributed sensing nodes that were designed to operate in a collaborative manner, yet be autonomous; the goal was for the net work to route information to the node that can best utilize the information [1.84,1.85]. The DSN program focused on distributed computing, signal processing, and tracking. Technology elements included acoustic sensors, high-level communication protocols, processing and algorithm calculations (e.g., self-location algorithms for sensors), and distributed software (dynamically modifiable distributed systems and language design) [1.13]. Researchers at Carnegie Mellon University focused on providing a network operating system for flexible transparent access to distributed resources, and researchers at the Massachusetts Institute of Technology focused on knowledge-based signal-processing techniques. Testbeds were developed for tracking multiple targets in a distributed environment; all components in the testbed network were custom built. Ongoing work in the 1980s resulted in the development of a multiple-hypothesis tracking algorithm to address difficult problems involving high target density, missing detections, and false alarms [1.86]; multiple-hypothesis tracking is now a standard approach to challenging tracking problems.

Phase 3: Military Applications Developed or Deployed in the 1980s and 1990s (These can properly be called first-generation commercial products.) Based on the results generated by the DARPA–DSN research and the testbeds developed, military planners set out in the 1980s and 1990s to adopt sensor network technology, making it a key component of network-centric warfare. An effort was made at the time to start employing commercial off the shelf (COTS) technology and common network interfaces, thereby reducing cost and development time. In traditional warfare environments each platforms "owns" its weapons in a fairly autonomous manner (distinct platforms operate independently). In network-centric warfare, weapon systems are not (necessarily) tightly affiliated with a specific platform; instead, through the use of distributed sensors, the weapon systems and platforms collaborate with each other over a sensor network, and information is sent to the appropriate node. Sensor networks can improve detection and tracking performance through multiple observations, geometric and phenomenological diversity, extended detection range, and faster response time [1.13]. An example of network-centric warfare include the cooperative engagement capability, a system that consists of multiple radars collecting data on air targets. Other sensor networks in the military arena include acoustic sensor arrays for antisubmarine warfare, such as the fixed distributed system and the advanced deployable system, and autonomous ground sensor systems such as the remote battlefield sensor system and the tactical remote sensor system.

Phase 4: Present-Day Sensor Network Research   (These can properly be called second-generation commercial products.) Advances in computing and communication that have taken place in the late 1990s and early 2000s have resulted in a new generation of sensor network technology. Evolving sensor networks represent a significant improvement over traditional sensors [1.38,1.39]. Inexpensive compact sensors based on a number of high-density technologies, including MEMS and (in the next few years) nanoscale electromechanical systems (NEMS), are appearing. Standardization is a key to wide-scale deployment of any technology, including WSN (e.g., Internet–Web, MPEG-4 digital video, wireless cellular, VoIP). Advances in IEEE 802.11a/b/g-based wireless networking and other wireless systems such as Bluetooth, ZigBee,[9] and WiMax are now facilitating reliable and ubiquitous connectivity. Inexpensive processors that have low power-consumption requirements make possible the deployment of sensors for a plethora of applications. Commercially-focused efforts are now directed at defining mesh, peer-to-peer, and cluster-tree network topologies with data security features and interoperable application profiles. Table 1.5 summarizes these generations of commercial products and alludes to a next-generation (third-generation) set of products.

TABLE 1.5   Commercial Generations of Sensor Networks

	First Generation (1980s–1990s)	Second Generation (Early 2000s)	Third Generation (Late 2000s)
Size	Attaché or larger	Paperback book or smaller	Small, even a dust particle
Weight	Pounds	Ounces	Grams or less
Deployment mode	Physically installed or air-dropped	Hand-placed	Embedded or "sprinkled," possibly nanotechnology-based
Node architecture	Separate sensing, processing, and communication	Integrated sensing, processing, and communication	Fully integrated sensing, processing, and communication
Protocols	Proprietary	Proprietary	Standard: Wi-Fi, ZigBee, WiMax, etc.
Topology	Point-to-point, star, and multihop	Client–server and peer-to-peer	Fully peer to peer
Power supply	Large batteries or line feed	AA batteries	Solar or possibly nanotechnology-based
Life span	Hours, days, and longer	Days to weeks	Months to years

[9]Although ZigBee proper comprises the software layers above the newly adopted IEEE 802.15.4 standard, at times we use *ZigBee* to mean "IEEE 802.15.4 with ZigBee middleware software running on top of the 802.15.4 MAC/PHY."

1.2.3 Challenges and Hurdles

For WSNs to become truly ubiquitous, a number of challenges and hurdles must be overcome. Challenges and limitations of wireless sensor networks include, but are not limited to, the following:

- Limited functional capabilities, including problems of size
- Power factors
- Node costs
- Environmental factors
- Transmission channel factors
- Topology management complexity and node distribution
- Standards versus proprietary solutions
- Scalability concerns [1.95]

Hardware Constraints A sensor may need to fit into a tight module on the order of $2 \times 5 \times 1$ cm or even as small as a $1 \times 1 \times 1$ cm. As shown in Figure 1.3, a sensor node is typically comprised of four key components and four optional components. The key components include a *power unit* (batteries and/or solar cells), a *sensing unit* (sensors and analog-to-digital converters), a *processing unit* (along with storage), and a *transceiver unit* (connects the node to the network). The optional components include a *location-finding system*, a *power generator*, a *control actuator,* and *other application-dependent elements.* The environmentally-intrinsic analog signals measured by the sensors are converted to digital signals by analog-to-digital converters and then are supplied to the processing unit. Sensor nodes may also have to be disposable, autonomous, and adaptive to the environment. R&D must be directed to solving the issue of reliable packaging of sensors despite the hardware constraints and challenges.

Power Consumption The sensor node lifetime typically exhibits a strong dependency on battery life. In many cases, the wireless sensor node has a limited power source (<500 mAh, 1.2 V), and replenishment of power may be limited or impossible altogether. Battery operation for sensors used in commercial applications is typically based on two AA alkaline cells or one Li-AA cell. It follows, as already noted, that power management and power conservation are critical functions for sensor networks, and one needs to design power-aware protocols and algorithms. The function of a sensor node in a sensor field is to detect events, perform local data processing, and transmit raw and/or processed data. Power consumption can therefore be allocated to three functional domains: *sensing, communication,* and *data processing,* each of which requires optimization. In the context of communications, in a multihop sensor network a node may play the dual role of data collection and processing and of being a data relay point. As can easily be understood, (excessive) rerouting and/or retransmission will require additional power.

Node Unit Costs Almost by definition, a sensor network consists of a large set of sensor nodes. It follows that the cost of an individual node is critical to the overall financial metric of the sensor network. Clearly, the cost of each sensor node has to be kept low for the global metrics to be acceptable. Current sensor systems based on Bluetooth technology cost about $10; however, Bluetooth is limited as a transmission technology in terms of both bandwidth and distance. However, the cost of a sensor node is generally targeted to be less than $1, which is lower than the current state-of-the-art technology.

Environment Sensor networks often are expected to operate in an unattended fashion in dispersed and/or remote geographic locations: Nodes may be deployed in harsh, hostile, or widely scattered environments. Such environments give rise to challenging management mechanisms. At the other end of the spectrum, sensor nodes are occasionally deployed densely either in close proximity with or directly inside the environment to be observed.

Transmission Channels Sensor networks often operate in a bandwidth- and performance-constrained multihop wireless communications medium. These wireless communications links operate in the radio, infrared, or optical range. Some low-power radio-based sensor devices use a single-channel RF transceiver operating at 916 MHz [1.87]; some sensor systems use a Bluetooth-compatible 2.4-GHz transceiver with an integrated frequency synthesizer [1.88]; yet other systems use 2.4 GHz (IEEE 802.11b technology), 5.0 GHz (IEEE 802.11a technology), or possibly other bands (for IEEE 802.15.4/IEEE 802.16 and/or for international use). To facilitate global operation of these networks, the transmission channel selected must be available on a worldwide basis.

Connectivity and Topology Deploying and managing a high number of nodes in a relatively bounded environment requires special techniques. Hundreds to thousands of sensors in close proximity (feet) may be deployed in a sensor field. The density of sensors may be as high as 27 nodes/m^3 [1.88]. Sensor network applications require ad hoc networking techniques; although many protocols and algorithms have been proposed for traditional wireless ad hoc networks, they are not well suited to the unique features and application requirements of sensor networks [1.38,1.39]. Nodes could be deployed *en mass* or be injected in the sensor field individually (e.g., they could be deployed by dropping them from an helicopter, scattered by an artillery shell or rocket, or deployed individually by a human or a robot). Any time after deployment, topology changes may ensue, due to changes in sensor node position; power availability, dropouts, or brownouts; malfunctioning; reachability impairments; jamming; and so on. At some future time, additional sensor nodes may need to be deployed to replace malfunctioning nodes, for example; hence, although some sensor nodes may fail or be blocked due to lack of power or have physical damage or environmental interference, this failure should not affect the overall mission of the sensor network.

Standards As implied by the protocol stack of Figure 1.4, a suite of protocols and open standards are needed at the physical, link, network, and transport layers; in addition, other management protocols and standards are required (physical layer standards are also known as air interface standards). Historically, sensor networks have used network- and application-specific protocols. This has had the effect of slowing cost-effective commercial deployment on a wide scale. Standards are now beginning to be incorporated into sensor networks. The highest degree of standardization has occurred at the lower layers. Within-building WSNs now tend to look to use ZigBee/IEEE802.15.4; WSNs that are in the open (outside buildings and over a broad geography) may find other technologies useful. In particular, IEEE-based wireless LAN standards have been given consideration. IEEE 802.11 supports 1- or 2-Mbps transmission in the 2.4-GHz band using either frequency-hopping spread spectrum or direct-sequence spread spectrum. IEEE 802.11a is an extension of 802.11 that provides up to 54 Mbps in the 5-GHz band and uses orthogonal frequency-division multiplexing encoding. IEEE 802.11b is an extension to 802.11 that provides 11-Mbps transmission in the 2.4-GHz band using DSSS. IEEE 802.11g provides up to 54 Mbps in the 2.4-GHz band. Extensions of these standards were also under way at the time of this writing (e.g., IEEE 802.11n). Another transmission method is free-space optics operating in the 1-μm wavelength (infrared). Infrared is license-free line-of-sight technology that operates at short range (300 to 3000 m). The new WiMax standard (IEEE 802.16) may also be useful for metropolitan environments, as is the application of cellular third-generation technologies. Earlier we also mentioned the Smart Dust mote, which uses the visible optical spectrum to communicate.

1.3 CONCLUSION

In this chapter we introduced the basic concept of WSNs and supportive technologies. The chapters that follow address in much greater detail and technical depth the issues that have been highlighted here.

REFERENCES

[1.1] C. S. Raghavendra, K. M. Sivalingam, T. Znati Eds., *Wireless Sensor Networks*, Kluwer Academic, New York, 2004.

[1.2] E. Cayirci, R. Govindan, T. Znati, M. Srivastava, Editorial: "Wireless Sensor Networks," *Computer Networks: International Journal of Computer and Telecommunications Networking*, Vol. 43, No. 4, Nov. 2003.

[1.3] T. Znati, C. Raghavendra, K. Sivalingam, Guest editorial, Special Issue on Wireless Sensor Networks, *Mobile Networks and Applications*, Vol. 8, No. 4, Aug. 2003.

[1.4] B. Krishnamachari, "A Wireless Sensor Networks Bibliography," Autonomous Networks Research Group, University of Southern California–Los Angeles, http://ceng.usc.edu/~anrg/SensorNetBib.html#0103.

[1.5] J. Kurose, V. Lesser, E. de Sousa e Silva, A. Jayasumana, B. Liu, *Sensor Networks Seminar*, CMPSCI 791L, University of Massachusetts, Amherst, MA, Fall 2003.

[1.6] D. Minoli, *Hotspot Networks: Wi-Fi for Public Access Locations*, McGraw-Hill, New York, 2003.

[1.7] R. Nowak, U. Mitra, "Boundary Estimation in Sensor Networks: Theory and Methods," *Proceedings of the 2nd Workshop on Information Processing in Sensor Networks* (IPSN'03), Palo Alto, CA, Apr. 2003.

[1.8] K. Sohraby et al., "Protocols for Self-Organization of a Wireless Sensor Network," *IEEE Personal Communications*, Vol. 7, No. 5, Oct. 2000, pp. 16ff.

[1.9] A. Cerpa, D. Estrin, "ASCENT: Adaptive Self-Configuring Sensor Networks Topologies," *Proceedings of the 21st Annual Joint Conference of the IEEE Computer and Communications Societies* (InfoCom'02), New York, Vol. 3, June 2002.

[1.10] H. Gupta et al., "Connected Sensor Cover: Self-Organization of Sensor Networks for Efficient Query Execution," *Proceedings of the 4th ACM International Symposium on Mobile Ad Hoc Networking and Computing* (MobiHoc'03), Annapolis, MD, June 2003.

[1.11] Q. Huang *et al.*, "Fast Authenticated Key Establishment Protocols for Self-Organizing Sensor Networks," *Proceedings of the 2nd Workshop on Sensor Networks and Applications* (WSNA'03), San Diego, CA, Sept. 2003.

[1.12] R. Kumar et al., "Computation Hierarchy for In-Network Processing," *Proceedings of the 2nd Workshop on Sensor Networks and Applications* (WSNA'03), San Diego, CA, Sept. 2003.

[1.13] C.-Y. Chong, S. P. Kumar, "Sensor Networks: Evolution, Opportunities, and Challenges," *Proceedings of the IEEE*, Vol. 91, No. 8, Aug. 2003, pp. 1247ff.

[1.14] A. Salhieh et al., "Power Efficient Topologies for Wireless Sensor Networks," *Proceedings of the 2001 International Conference on Parallel Processing* (ICPP'01), Valencia, Spain, Sept. 2001, pp. 156ff.

[1.15] A. Sinha, A. Chandrakasan, "Dynamic Power Management in Wireless Sensor Networks," *IEEE Design and Test of Computers*, Vol. 18, No. 2, Mar. 2001.

[1.16] D. Rakhmatov, S. Vrudhula, "Energy Management for Battery-Powered Embedded Systems," *ACM Transactions on Embedded Computing Systems*, Vol. 2, No. 3, Aug. 2003.

[1.17] J. M. Rabaey *et al.*, "PicoRadios for Wireless Sensor Networks: The Next Challenge in Ultra-Low Power Design," *Proceedings of the 7th IEEE International Symposium on Computers and Communications* (ISCC'02), July 2002.

[1.18] M. Kubisch *et al.*, "Distributed Algorithms for Transmission Power Control in Wireless Sensor Networks," *Proceedings of the IEEE Wireless Communications and Networking Conference* (WCNC'03), Vol. 1, Mar. 2003.

[1.19] S. Coleri *et al.*, "Power Efficient System for Sensor Networks," *Proceedings of the 8th IEEE International Symposium on Computers and Communication* (ISCC'03), July 2003.

[1.20] T. Clouqueur et al., "Sensor Deployment Strategy for Target Detection," *Proceedings of the 1st Workshop on Sensor Networks and Applications* (WSNA'02), Atlanta, GA, Sept. 2002.

[1.21] D. Minoli, I. Gitman, "On Connectivity in Mobile Packet Radio Networks," *28th IEEE Vehicular Technology Conference Record*, Mar. 1978, pp. 105–109.

[1.22] D. Minoli, W. Nakamine, "A Taxanomy and Comparison of Random Access Protocols for Computer Networks," *Networks'80 Conference Record*, 1980. Included in S. Ramani, Ed., *Data Communication and Computer Networks*, pp. 187–206.

[1.23] D. Minoli, I. Gitman, "Combinatorial Issues in Mobile Packet Radio," *IEEE Transactions on Communication*, Vol. 26, Dec. 1978, pp. 1821–1826.

[1.24] D. Minoli, "Intelligent Terminal Standards Could Enhance Net Efficiency," *Data Communications*, Vol. 8, No. 11, Nov. 1979, pp. 59–68.

[1.25] D. Minoli, A. Schmidt, *Internet Architectures*, Wiley, New York, 1999.

[1.26] D. Minoli, "Packetized Speech Networks, Part 1: Overview," *Australian Electronics Engineer*, Apr. 1979, pp. 38–52.

[1.27] D. Minoli, "Packetized Speech Networks, Part 2: Queueing Model," *Australian Electronics Engineer*, July 1979, pp. 68–76.

[1.28] D. Minoli, "Packetized Speech Network, Part 3: Delay Behavior and Performance Characteristics," *Australian Electronics Engineer*, Aug. 1979, pp. 59–68.

[1.29] D. Minoli, "Issues in Packet Voice Communication," *Proceedings of the IEE*, Vol. 126, No. 8, Aug. 1979, pp. 729–740.

[1.30] D. Minoli, "Digital Voice Communication over Digital Radio Links," *SIGCOMM Computer Communications Review*, Vol. 9, No. 4, Oct. 1979, pp. 6–22.

[1.31] D. Minoli, E. Minoli, *Delivering Voice over IP and the Internet*, 2nd ed., Wiley, New York, 2002.

[1.32] D. Minoli, *Voice over MPLS*, McGraw-Hill, New York, 2002.

[1.33] D. Minoli, "Putting Video on Desktops," *Computerworld*, Oct. 1986, pp. 35ff.

[1.34] D. Minoli, E. Minoli, et al., "Digital Video," in *The Telecommunications Handbook*, K. Terplan and P. Morreale, Eds, IEEE Press, Piscataway, NJ, 2000.

[1.35] M. Hatler, *Wireless Sensor Networks: Mass Market Opportunities*, ON World, Inc., San Diego, CA, Feb. 22, 2004.

[1.36] T.-H. Lin, W. J. Kaiser, G. J. Pottie, "Integrated Low-Power Communication System Design for Wireless Sensor Networks," *IEEE Communications*, Dec. 2004, pp. 142ff.

[1.37] C. M. Okino, M. G. Corr, "Statistically-Accurate Sensor Networking," *Proceedings of the IEEE Wireless Communications and Networking Conference* (WCNC'02), Vol. 1, Mar. 2002.

[1.38] I. F. Akyildiz et al., "A Survey of Sensor Networks," *IEEE Communications*, Aug. 2002, pp. 102ff.

[1.39] W. Su et al., "Communication Protocols for Sensor Networks," in *Wireless Sensor Networks*, C. S. Raghavendra, K. Sivalingam, and T. Znati, Eds., Kluwer Academic, New York, 2004.

[1.40] A. B. McDonald, T. Znati, "Session A: Routing—Predicting Node Proximity in Ad-Hoc Networks: A Least Overhead Adaptive Model for Selecting Stable Routes," *Proceedings of the 1st ACM International Symposium on Mobile Ad Hoc Networking and Computing*, (MobiHoc'00), Nov. 2000.

[1.41] G. H. Lynn, T. Znati, "ROMR: Robust Multicast Routing in Mobile Ad-Hoc Networks," doctoral dissertation, University of Pittsburgh, Jan. 2003.

[1.42] D. Estrin, D. Culler, K. Pister, "Connecting the Physical World with Pervasive Networks," *IEEE Pervasive Computing*, Jan.–Mar. 2002.

[1.43] J. Pan et al., "Topology Control for Wireless Sensor Networks," *Proceedings of the 9th ACM Conference on Mobile Computing and Networking* (MobiCom'03), San Diego, CA, Sept. 2003.

[1.44] A. Boukerche et al., "Energy-Aware Data-Centric Routing in Microsensor Networks," *International Symposium on Modeling, Analysis and Simulation of Wireless and Mobile Systems,* San Diego, CA, Sept. 2003.

[1.45] D. Braginsky, D. Estrin, "Rumor Routing Algorithm for Sensor Networks," *Proceedings of the 1st Workshop on Sensor Networks and Applications* (WSNA'02), Atlanta, GA, Oct. 2002.

[1.46] D. Ganesan et al., "Highly-Resilient, Energy-Efficient Multipath Routing in Wireless Sensor Networks," *Mobile Computing and Communications Review,* Vol. 1, No. 2, 2002.

[1.47] R. Kannan, et al., "Sensor-Centric Quality of Routing in Sensor Networks," *Proceedings of the 22nd Annual Joint Conference of the IEEE Computer and Communications Societies* (InfoCom'03), Vol. 1, Apr. 2003.

[1.48] B. Krishnamachari, D. Estrin, S. Wicker, "Modelling Data-Centric Routing in Wireless Sensor Networks," *Proceedings of the 21st Annual Joint Conference of the IEEE Computer and Communications Societies* (InfoCom'02), New York, June 2002.

[1.49] N. Bulusu et al., "Adaptive Beacon Placement," *Proceedings of the 21st International Conference on Distributed Computing Systems* (ICDCS'21), Phoenix, AZ, Apr. 2001, pp. 489ff.

[1.50] P. Bergamo, G. Mazzini, "Localization in Sensor Networks with Fading and Mobility," *Proceedings of the 13th IEEE International Symposium on Personal, Indoor and Mobile Radio Communications* (PIMRC'02), Vol. 2, Sept. 2002.

[1.51] Q. Li et al., "Distributed Algorithms for Guiding Navigation Across a Sensor Network," *Proceedings of the 9th ACM International Conference on Mobile Computing and Networking* (MobiCom'03), San Diego, CA, Sept. 2003.

[1.52] R. Iyengar, B. Sikdar, "Scalable and Distributed GPS-Free Positioning for Sensor Networks," *Proceedings of the IEEE International Conference on Communications,* (ICC'03), Vol. 1, 2003.

[1.53] P. Morreale, K. Sohraby, B. Li, Y. Lin, Guest editorial: "Active, Programmable, and Mobile Code Networking," *IEEE Communications,* Vol. 38, No. 3, Mar. 2000, pp. 122–123.

[1.54] M. Gerla et al., "The Mars Sensor Network: Efficient, Energy Aware Communications," *Proceedings of the IEEE Military Communications Conference* (MilCom'01): *Communications for Network-Centric Operations—Creating the Information Force,* McLean, VA, Vol. 1, Oct. 2001.

[1.55] M. Haenggi, "Energy-Balancing Strategies for Wireless Sensor Networks," *Proceedings of the 2003 International Symposium on Circuits and Systems* (ISCAS'03), Vol. 4, May 2003, pp. 25ff.

[1.56] J. Zhu, S. Papavassiliou, "On the Connectivity Modeling and the Tradeoffs Between Reliability and Energy Efficiency in Large Scale Wireless Sensor Networks," *Proceedings of the IEEE Wireless Communications and Networking Conference* (WCNC'03), Vol. 2, Mar. 2003.

[1.57] R. Min, A. Chandrakasan, "Top Five Myths About the Energy Consumption of Wireless Communication," *ACM Mobile Computing and Communications Review,* Vol. 7, No. 1, Jan. 2003.

[1.58] S. Lindsey et al., "Data Gathering in Sensor Networks Using the Energy Delay Metric," presented at the International Workshop on Parallel and Distributed Computing Issues in Wireless Networks and Mobile Computing, San Francisco, CA, Apr. 2001.

[1.59] J. Kusuma et al., "Distributed Compression for Sensor Networks," *Proceedings of the International Conference on Image Processing* (ICIP'01), Vol. 1, Oct. 2001.

[1.60] C. Wang, B. Li, K. Sohraby, "A Simple Mechanism on MAC Layer to Improve the Performance of IEEE 802.11 DCF," *Proceedings of the 1st Annual International Conference on Broadband*, (Broadnets'04), San Jose, CA, Oct. 2004.

[1.61] L. C. Zhong et al., "Data Link Layer Design for Wireless Sensor Networks," *Proceedings of the IEEE Military Communications Conference* (MilCom'01): Communications for Network-Centric Operations—Creating the Information Force, McLean, VA, Vol. 1, Oct. 2001.

[1.62] C. Y. Wan et al., "A Reliable Transport Protocol for Wireless Sensor Networks," *Proceedings of the 1st Workshop on Sensor Networks and Applications* (WSNA'02), Atlanta, GA, Sept. 2002.

[1.63] F. Stann, J. Heidemann, "A Reliable Data Transport in Sensor Networks," *Proceedings of the 1st IEEE International Workshop on Sensor Network Protocols and Applications* (SNPA'03), Anchorage, AK, May 2003.

[1.64] R. R. Kompella, A. C. Snoeren, "Practical Lazy Scheduling in Sensor Networks," *Proceedings of the 1st ACM Conference on Embedded Networked Sensor Systems* (SenSys'03), Los Angeles, Nov. 2003.

[1.65] D. Minoli, "Aloha Channels Throughput Degradation," *1986 Computer Networking Symposium Conference Record*, pp. 151–159.

[1.66] Promotional materials, ZigBee Alliance, Bishop Ranch, CA.

[1.67] E. Shih et al., "Physical Layer Driven Protocol and Algorithm Design for Energy-Efficient Wireless Sensor Networks," *Proceedings of the 7th ACM International Conference on Mobile Computing and Networking* (MobiCom'01), Rome, Italy, July 2001, pp. 272–287.

[1.68] J. Kulik et al., "Negotiation-Based Protocols for Disseminating Information in Wireless Sensor Networks," *Proceedings of the 5th ACM/IEEE International Conference on Mobile Computing and Networking* (MobiCom'99), Seattle, WA, Aug. 1999.

[1.69] R. Kannan et al., "Energy and Rate Based MAC Protocol for Wireless Sensor Networks," *SIGMOD Record*, Vol. 32, No. 4, Dec. 2003.

[1.70] W. R. Heinzelman et al., "Adaptive Protocols for Information Dissemination in Wireless Sensor Networks," *Proceedings of the 5th ACM/IEEE International Conference on Mobile Computing and Networking* (MobiCom'99), Seattle, WA, Aug. 1999, pp. 174ff.

[1.71] W. R. Heinzelman et al., "Energy-Efficient Communication Protocol for Wireless Microsensor Networks," *Proceedings of the 33rd Hawaii International Conference on System Sciences* (HICSS'00), Maui, HI, Jan. 2000.

[1.72] L. Zhou, Z. J. Haas, "Securing Ad Hoc Networks," *IEEE Networks*, Special Issue on Network Security, Dec. 1999.

[1.73] S. Slijepcevic et al., "On Communication Security in Wireless Ad-Hoc Sensor Networks," *Proceedings of the IEEE 11th International Workshops on Enabling*

Technologies: Infrastructure for Collaborative Enterprises (WETICE'02), Pittsburgh, PA, June 2002.

[1.74] C. Intanagonwiwat et al., "Directed Diffusion: A Scalable and Robust Communication Paradigm for Sensor Networks," *Proceedings of the 6th ACM International Conference on Mobile Computing and Networks* (MobiCom'00), Boston, MA, Aug. 2000.

[1.75] B. Krishnamachari et al., "The Impact of Data Aggregation in Wireless Sensor Networks," *International Workshop on Distributed Event-Based Systems* (DEBS'02), Vienna, Austria, July 2002.

[1.76] B. Przydatek et al., "Secure Information Aggregation in Sensor Networks," *Proceedings of the 1st ACM Conference on Embedded Networked Sensor Systems* (SenSys'03), Los Angeles, Nov. 2003.

[1.77] G. Gupta, M. Younis, "Performance Evaluation of Load-Balanced Clustering of Wireless Sensor Networks," *Proceedings of the 10th International Conference on Telecommunications* (ICT'03), Vol. 2, Mar. 2003.

[1.78] S. Ratnasamy et al., "A Geographic Hash Table for Data-Centric Storage," *Proceedings of the 1st Workshop on Sensors Networks and Applications* (WSNA'02), Atlanta, GA, Sept. 2002.

[1.79] S. Ratnasamy et al., "Data-Centric Storage in Sensornets with GHT: A Geographic Hash Table," Special Issue on Wireless Sensor Networks, *Mobile Networks and Applications*, Aug. 2003, pp. 427ff.

[1.80] Y. Yao, J. Gehrke, "Query Processing for Sensor Networks," *Proceedings of the 1st Biennial Conference on Innovative Data Systems Research (CIDR'03), Asilomar, CA, Jan. 2003.*

[1.81] Y. Yao, J. Gehrke, "The Cougar Approach to In-Network Query Processing in Sensor Networks," *SIGMOD Record*, Vol. 31, No. 1, Mar. 2002.

[1.82] S. Adlakha, M. Srivastava, "Critical Density Thresholds for Coverage in Wireless Sensor Networks," *Proceedings of the IEEE Wireless Communications and Networking Conference* (WCNC'03), Vol. 3, Mar. 2003.

[1.83] S. Dhillon, K. Chakrabarty, "Sensor Placement for Effective Coverage and Surveillance in Distributed Sensor Networks," *Proceedings of the IEEE Wireless Communications and Networking Conference* (WCNC'03), Vol. 3, Mar. 2003.

[1.84] *Proceedings of the Distributed Sensor Nets Workshop*, Department of Computer Science, Carnegie Mellon University, Pittsburgh, PA, 1978.

[1.85] R. F. Sproull, D. Cohen, "High-Level Protocols," *Proceedings of the IEEE*, Vol. 66, Nov. 1978, pp. 1371–1386.

[1.86] C. Y. Chong et al., "Distributed Multitarget Multisensor Tracking," in *Multitarget Multisensor Tracking: Advanced Applications*, Artech House, Norwood, MA, 1990.

[1.87] A. Woo, D. Culler, "A Transmission Control Scheme for Media Access in Sensor Networks," *Proceedings of the 7th ACM International Conference on Mobile Computing and Networking* (MobiCom'01), Rome, Italy, July 2001, pp. 221–235.

[1.88] E. Shih et al., "Physical Layer Driven Protocol and Algorithm Design for Energy-Efficient Wireless Sensor Networks," *Proceedings of the 7th ACM International Conference on Mobile Computing and Networking* (MobiCom'01), Rome, Italy, July 2001, pp. 272–286.

[1.89] J. M. Kahn et al., "Next Century Challenges: Mobile Networking for Smart Dust," *Proceedings of the 5th ACM International Conference on Mobile Computing and Networking* (MobiCom'99), Seattle, WA, Aug. 1999, pp. 270–278.

[1.90] Mission statement, TinyOS Community Forum, http://www.tinyos.net/.

[1.91] A. Deshpande et al., "Model-Driven Data Acquisition in Sensor Networks," *Proceedings of the 30th International Conference on Very Large Data Bases*, 2004, pp. 588–599.

[1.92] K. Akkaya, M. Younis, "A Survey on Routing Protocols for Wireless Sensor Networks," Department of Computer Science and Electrical Engineering, University of Maryland, Baltimore, MD, Aug. 18, 2003.

[1.93] M. Welsh, D. Malan, B. Duncan, T. Fulford-Jones, S. Moulton, "Wireless Sensor Networks for Emergency Medical Care," presented at GE Global Research Conference, Harvard University and Boston University School of Medicine, Boston, MA, Mar. 8, 2004.

[1.94] I. Akyildiz, X. Wang, "A Survey on/of Wireless Mesh Networks," *IEEE Radio Communications*, Sept. 2005, pp. S23–S30.

[1.95] E. Egea-Lopez, J. Vales-Alonso et al., "Simulation Scalability Issues in Wireless Sensor Networks," *IEEE Communications*, July 2006, pp. 64ff.

[1.96] V. Devarapalli, R. Wakikawa, A. Petrescu, P. Thubert, "Network Mobility (NEMO) Basic Support Protocol," *IETF RFC 3963*, January 2005.

2

APPLICATIONS OF WIRELESS SENSOR NETWORKS

2.1 INTRODUCTION

WSNs are collections of compact-size, relatively inexpensive computational nodes that measure local environmental conditions or other parameters and forward such information to a central point for appropriate processing. WSNs nodes (WNs) can sense the environment, can communicate with neighboring nodes, and can, in many cases, perform basic computations on the data being collected. WSNs support a wide range of useful applications. In this chapter we identify some of these applications; the chapter is not intended to be exhaustive, simply illustrative.

2.2 BACKGROUND

In Chapter 1 we taxonomized (commercial) sensor networks and systems into two basic categories:

- *Category 1 WSNs* (C1WSNs): almost invariably mesh-based systems with multihop radio connectivity among or between WNs, utilizing dynamic routing in both the wireless and wireline portions of the network. Military-theater systems typically belong to this category.
- *Category 2 WSNs* (C2WSNs): point-to-point or multipoint-to-point (star-based) systems generally with single-hop radio connectivity to WNs, utilizing

Wireless Sensor Networks: Technology, Protocols, and Applications, by Kazem Sohraby, Daniel Minoli, and Taieb Znati

static routing over the wireless network; typically, there will be only one route from the WNs to the companion terrestrial/wireline forwarding node (WNs are pendent nodes). Residential control systems typically belong to this category.

C2WSNs are networks in which end devices (sensors) are one radio hop away from a terrestrially homed forwarding node (see Figure 2.1). The forwarding node (call it a *wireless router*) is connected to the terrestrial network via either a landline or a point-to-point wireless link. The important characterizations are that (1) sensor nodes (i.e., the WNs) do not support communications on behalf of any other sensor nodes; (2) the forwarding node supports only static routing to the terrestrial network, and/or only one physical link to the terrestrial network is present; (3) the radio link is measured in hundreds of meters; and (4) the forwarding node does not support data processing or reduction on behalf of the sensor nodes. In effect, these are relatively simple wireless systems.

C1WSNs are networks in which end devices (sensors) are permitted to be more than one radio hop away from a routing or forwarding node (see Figure 2.2). The forwarding node is a wireless router that supports dynamic routing (i.e., it has a mechanism that is used to find the best route to the destination out of a possible set of more than one route); wireless routers are often connected over wireless links. The important characterizations are that (1) sensor nodes can support communications

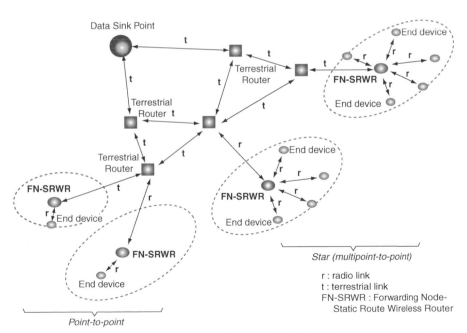

Figure 2.1 Category 2 WSNs: point-to-point, generally-singlehop systems utilizing static routing.

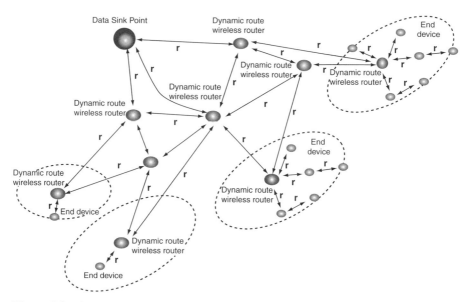

Figure 2.2 Category 1 WSNs: multipoint-to-point, multihop systems utilizing dynamic routing.

on behalf of other sensor nodes by acting as repeaters; (2) the forwarding node supports dynamic routing and more than one physical link to the rest of the network is physically and logically present; (3) the radio links are measured in thousands of meters; and (4) the forwarding node can support data processing or reduction on behalf of the sensor nodes. These are relatively complex and "meshy" wireless systems.

Some refer to the two types of behavior as *cooperative* (when a node forwards information on behalf of another node) or *noncooperative* (when a node handles only its own communication) [2.54] (see Figure 2.3). The two categories of WSNs are intended to be mutually exclusive by definition.[1] WSNs (particularly C1WSNs) typically consist of hundreds (even thousands) of inexpensive WNs.

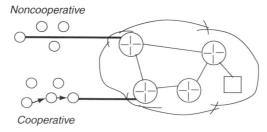

Figure 2.3 Cooperative and noncooperative nodes.

[1]The mutual exclusivity is reasonably well but not perfectly described by the definitions provided for the two classes.

The WNs have computational power and sensing capabilities and typically operate in an unattended mode; they are battery-, piezoelectrically-, or solar-powered. The technical implications of the network environment (multihop/dynamic routing versus singlehop/static routing) are discussed in subsequent chapters.

Although other classifications are possible, particularly for the technology itself (see Section 2.6), applications are discussed here according to the categorization just described. Namely, as a practical matter we look at applications supported by C1WSNs as being distinct from applications supported by C2WSNs.[2] Basically, we see two groups of applications: those that typically entail point-to-point systems and those that entail complex or dynamic mesh multihop systems. Category 1 applications are most-often supported by and delivered over C1WSNs. Category 2 applications are most-often supported by and delivered over C2WSNs.

Distributed WSNs using sensor and microsensor technology are expected to enable a plethora of applications for sensing and controlling the physical world for commercial as well as military purposes. Applications range from environmental control (e.g., tracking soil contamination, habitat monitoring), warehouse inventory, and health care at one end of the spectrum, to scientific and military uses at the other [2.1–2.4]. In recent years, particularly since the beginning of this decade, WSN research has undergone a revolution; the advances originating from this research promise to have a significant impact on a broad range of applications relating to national security, health care, the environment, energy, food safety, and manufacturing, to list just a few [2.5].

The range of potential applications is really limited only by the imagination; examples include tracking wild fires; microclimate assessment; monitoring animal populations; defense systems; enabling businesses to monitor and control workspaces; and allowing authorities to monitor for toxic chemicals, explosives, and biological agents, to list only a few [2.7]. Law-enforcement WSNs offer functional capabilities and enhancements in operational efficiency in civilian applications; this technology can also assist in the effort to increase alertness to potential terrorist threats [2.6]. National defense relies on accurate intelligence, surveillance, and reconnaissance (ISR). The utilization of a dense set of small affordable sensors that are deployed appropriately in the environment of interest has the potential to increase the dependability of ISR systems because of the fact that a large set of redundant sensors decreases the vulnerability of the system to failure. However, in applications such as these, the ability to combine information becomes a critical factor in managing network bandwidth and facilitating ultimate decision making [2.8].

There has been extensive academic research on WSNs over the recent past, as we noted in Chapter 1, but for all intents and purposes, until open technical standards take hold pervasively, applications and deployment will remain specialized. Although a topic is strictly at the research level, there is a lot of academic interest; however, as standards begin to take hold, the topic becomes more practical and

---

[2]To be exact, C2WSN-like applications could be supported (but perhaps not cost-effectively) by C1WSNs; however, C1WSN-like applications cannot generally be supported by C2WSNs.

the technical development becomes more pragmatic. We believe that, in fact, we have reached this inflection point with regard to WSNs. The convergence of the Internet, wireless communications, and information technologies with techniques for miniaturization has placed sensor technology at the threshold of an era of significant potential growth [2.5]. WSN hardware, particularly low-cost processors, miniature sensors, and low-power radio modules, are now becoming available under the thrust of emerging standards; further improvements in cost and capabilities are expected in the next few years, fostering additional deployment and applications. Sensor networks typically operate at 900 MHz (868- and 915-MHz bands); commercially evolving systems will operate (via IEEE 802.11b or IEEE 802.5.4) in the 2.4-GHz range. The market scope for WSNs is expected to see major expansion in the next three to five years; this expansion relates not only to science and engineering applications but also to a plethora of new consumer applications of the technology. In the remainder of the chapter we survey some of the applications of WNS technology.

2.3 RANGE OF APPLICATIONS

As noted, WSNs support a broad spectrum of applications, ranging from environmental sensing to vehicle tracking, from perimeter security to inventory management, and from habitat monitoring to battlefield management (see Table 2.1). For example, WSNs may be deployed outdoors in large *sensor fields* to detect and control the spread of wild fires, to detect and track enemy vehicles, or to support environmental monitoring, including precision agriculture [2.9–2.12,2.51]. Stakeholders are now focusing on developing applications that deliver measurable business value; the goal is to take the extensive body of research in this space and apply it to the real world [2.13–2.15]. With WSNs one can monitor and control factories, offices, homes, vehicles, cities, the ambiance, and the environment. For example, one can detect structural faults (e.g., fatigue-induced cracks) in ships, aircraft, and buildings; public-assembly locations can be equipped to detect toxins and to trace the source of the contamination. Volcanic eruption, earthquake detection, and tsunami alerting—applications that generally require WNs deployed in remote, even difficult-to-reach locations—can be useful environmental-monitoring systems.

The following is a recent view expressed by the National Science Foundation [2.5] on WSNs:

> *Emerging technologies will likely lead to a decrease in the size, weight and cost of sensors and sensor arrays by orders of magnitude, and they will lead to an increase the sensors' spatial and temporal resolution and accuracy. Large numbers of sensors may be integrated into systems to improve performance and lifetime, and decrease life-cycle costs. Communications networks provide rapid access to information and computing, eliminating the barriers of distance and time for telemedicine, transportation, tracking endangered species, detecting toxic agents, and monitoring the security of civil and engineering infrastructures. The coming years will likely see a growing reliance on and need for more powerful sensor systems, with increased performance and functionality.*

TABLE 2.1 Applications Mentioned in This Chapter

Air traffic control
Appliance control (lighting and HVAC)
Area and theater monitoring (military)
Assembly line and workflow
Asset management (e.g., container tracking)
Automated automobile maintenance telemetry
Automatic control of multiple home systems to improve conservation, convenience, and safety
Automatic meter reading
Automating control of multiple systems to improve conservation, flexibility, and security
Automotive sensors and actuators
Auto-to-auto applications (FCC recently approved specific frequencies for highway sensor
    and auto-to-auto applications; range is about 100 m [2.55])
Battlefield management
Battlefield reconnaissance and surveillance
Biological monitoring for agents
Biomedical applications
Blinds, drapery, and shade controls
Body-worn medical sensors
Borders monitoring (Mexican and Canadian borders)
Bridge and highway monitoring (safety)
Building and structures monitoring
Building automation (security, HVAC, automated meter reading, lighting control, access control)
Building energy monitoring and control
Capturing highly detailed electric, water, and gas utility usage data
Centibots (DARPA): embedded mobile sensor nodes; 100 robots mapping, tracking, and
    guarding an environment in a coherent manner
Chemical, biological, radiological, and nuclear wireless sensors (sensors for toxic chemicals,
    explosives, and biological agents)
Civil engineering applications
Collection of long-term databases of clinical data (enables correlation of biosensor readings
    with other patient information)
Combat field surveillance
Commercial applications
Commercial building control
Configuring and running multiple home control systems from a single remote control
Consumer applications
Consumer electronics and entertainment (TV, VCR, DVD/CD)
Consumers' ability to keep track of their belongings, pets, and young children
Control of temperature
Controlling the spread of wild fires
Critical infrastructure protection and security
Defense systems
Detecting an impulsive event (e.g., a footstep or gunshot) or vehicle (e.g., wheeled or tracked,
    light or heavy)
Detecting structural faults in aircraft
Detecting structural faults in buildings (e.g., fatigue-induced cracks)
Detecting structural faults in ships
Detecting toxic agents

(Continued)

TABLE 2.1 (*Continued*)

Detection and tracking of enemy vehicles
Disaster management
Distributed robotics
Distributed sensing (military)
Earthquake detection
Electricity load management
Embedding intelligence to optimize consumption of natural resources
E-money/point-of-sale applications (including kiosks)
Enabling businesses to monitor and control workspaces
Enabling deployment of wireless monitoring networks to enhance perimeter protection
Enabling extension and upgrading of building infrastructure with minimal effort
Enabling installation, upgrading, and networking of home control system without wires
Enabling networking and integration of data from multiple access control points
Enabling rapid reconfiguring of lighting systems to create adaptable workspaces
Energy management
Environmental (land, air, sea) and agricultural wireless sensors
Environmental control (e.g., tracking soil contamination, habitat monitoring)
Environmental monitoring, including precision agriculture
Environmental sensing applications
Equipment management services and preventive maintenance
Extending existing manufacturing and process control systems reliably
Facilitating the reception of automatic notification upon detection of unusual events
Farm sensor and actuator networks (monitoring soil moisture, feeding pigs, unmanned
    tractor control)
Flexible management of lighting, heating, and cooling systems from anywhere in the home
Food safety
Gas, water, and electric meters
Gateway or field service links to sensors and equipment (monitored to support preventive
    maintenance, status changes, diagnostics, energy use, etc.)
Habitat monitoring
Habitat sensing
Health care
Heartbeat sensors
Heating control
Helping automate data acquisition from remote sensors to reduce user intervention
Helping deploy monitoring networks to enhance employee and public safety
Helping identify inefficient operation or poorly performing equipment
Helping streamlining data collection for improved compliance reporting
Herd control from central location using sensor-based fences and remote-controlled gates
Home automation, including alarms (e.g., an alarm sensor that triggers a call to a security firm)
Home control applications to provide control, conservation, convenience, and safety
Home monitoring for chronic and elderly patients (collection of periodic or continuous data
    and upload to physicians)
Home security
Homeland Security Advanced Research Projects Agency, which has the goal of developing a
    national sensor net to detect biological, chemical, and nuclear agents
Hotel energy management

TABLE 2.1   (*Continued*)

HVAC control

iBadge (UCLA): used to track the behavior of children or patients (e.g., speech
  recording/replaying, position detection, direction detection, local climate:
  temperature, humidity, pressure)

iButton: a small computer chip enclosed in a stainless steel container that looks like a
  button containing up-to-date information that can travel with a person or object
  (e.g., be used wirelessly with an ATM or vending machine)

IEEE 802.15.4 mote (Telos is first 802.15.4-based mote; 2/2004; www.moteiv.com)

Improving asset management by continuous monitoring of critical equipment

Industrial and building automation

Industrial and building monitoring

Industrial and manufacturing automation

Industrial automation applications that provide control, conservation, and efficiency

Industrial control (asset management, process control, environmental, energy management)

Industrial monitoring and control

Integrating and centralizing management of lighting, heating, cooling, and security

Intrusion detection

Inventory control

Inventory management

Law enforcement

Lighting control

Localization

Manufacturing control

Mass-casualties management

Materials processing systems (heat, gas flow, cooling, chemical)

Medical disaster response

Medical sensing and monitoring

Metropolitan operations (traffic, automatic tolls, fire, etc.)

Microclimate assessment and monitoring

Military applications

Military command, control, communications, intelligence, and targeting systems

Military sensing

Military sensor networks to detect and gain information about enemy movements

Military tactical surveillance

Military vigilance for unknown troop and vehicle activity

Mobile robotics

Monitoring and controlling cities

Monitoring and controlling factories

Monitoring and controlling homes

Monitoring and controlling offices

Monitoring and controlling the ambiance

Monitoring and controlling the environment

Monitoring and controlling vehicles

Monitoring animal populations

Monitoring complex machinery and processes/condition-based maintenance (CBM)

Monitoring for explosives

Monitoring for toxic chemicals

                                                                              (*Continued*)

TABLE 2.1 (*Continued*)

Monitoring intersections
Monitoring on-truck and on-ship tamper of assets
Monitoring rooftops (military)
Monitoring the limb movements and muscle activity of stroke patients during rehabilitation
    exercise
Monitoring the security of civil and engineering infrastructures
Monitoring wild fires
Nanoscopic sensor applications (e.g., biomedics)
National defense
National security
Near field communication (NFC) as a "virtual connector" (NFC acts like RFID but requires
    close proximity to read, providing easy identification and security; wireless connectivity
    needed to transport data [2.55,2.58])
Nose-on-a-chip (Oak Ridge National Laboratory): a MEMS-based sensor that can detect 400
    types of gases and transmit information to a central control station, indicating the level
Perimeter security
Personal health diagnosis
Personal health care (patient monitoring, fitness monitoring)
Pervasive computing ("invisible computing," "ubiquitous computing")
Physical security
Pre-hospital and in-hospital emergency care
Preventive maintenance for equipment used by a semiconductor fabricator
Process control
Production processing
Providing detailed data to improve preventive maintenance programs
Public assembly locations monitoring
Public-safety applications
Quality-of-life applications
Radar used to profile soil composition in vineyards (UC–Berkeley)
Radiation and nuclear-threat detection systems
Real-time collection of data (e.g., to check temperature or monitor pollution levels)
Real-time continuous patient monitoring (e.g., pre-hospital, in-hospital, and ambulatory
    monitoring)
Reducing energy costs through optimized manufacturing processes
Reducing energy expenses through optimized HVAC management
Refrigeration cage or appliance monitoring
Remote underwater sampling station (RUSS) robots used to monitor municipal water
    supplies; the WNs are solar-powered robots that float on the surface and deploy
    descendable sensors underwater to sample temperature, oxygen, turbidity, light, and salt
    content; data are transmitted by cell phone to central lab and posted on the Web [2.55]
Remotely-controlled home heating and lighting
Remotely monitored assets, billing, and energy management
Residential control and monitoring applications
Residential/light commercial control (security, HVAC, lighting control, access control,
    lawn and garden irrigation)
RF-based localization
RFID tags

TABLE 2.1 (*Continued*)

Ring sensor (MIT): monitors the physiological status of the wearer and transmits the information to a medical professional over the Internet

Routing, naming, discovery, and security for wireless medical sensors, personal digital assistants, PCs, and other devices

Scientific applications

Security services (including peel-n'-stick security sensors)

Seismic accelerometers (devices able to measure movement)

Sensor networks for theme parks

Sensor networks to detect and characterize chemical, biological, radiological, nuclear, and explosive attacks and material

Sensor networks to detect and monitor environmental changes in plains, forests, and oceans

Sensors embedded in a glacier in Norway (pelletlike WNs are embedded 60 m inside a glacier and use collaborative methods to collect and transmit data) [2.55]

Sensors in chimneys to monitor creosote buildup

Smart bullet fired from a paintball gun (wireless transmitter and battery capable of a range of 70 m) [2.57]

Smart bricks: accelerometer/thermistor/etc. embedded in bricks (UIUC)

Smart kindergarten project (Mani Srivastava/UCLA): I-badges embedded in children's hats to track position, bearing, and record sound; classroom toys have sensors embedded to detect use

Smart structures that are able to self-diagnose potential problems and self-prioritize requisite repairs

Smoke, CO, and H_2O detectors

Stroke patient rehabilitation

Supermarket management

Supporting the straightforward installation of wireless sensors to monitor a wide variety of conditions

Telemedicine

Toxin detection

Tracing source of contamination

Tracking criminals

Tracking endangered species

Tracking wild fires

Traffic light sensors and control (using distributed greedy algorithms) [2.56]

Traffic flow and surveillance

Tsunami alerting

Turf cam microcameras (about 0.5 cm^3) placed throughout a football field [2.55]

Underfloor air distribution systems

Universal remote control to a set-top box

Vehicle tracking

Video surveillance

Virtual fence using a sensor or actuator as a collar (Dartmouth College is using Wi-Fi PDA collars)

Vital sign data, such as pulse oximetry and two-lead EKG (medical)

Volcanic eruptions

Warehouse inventory

Warehouses, fleet management, factory, supermarkets, and office complexes

(*Continued*)

TABLE 2.1 *(Continued)*

Water supply protection (detecting poisons such as ricin and other pathogens)
  via microfluidics and WSN-based sensors
Weapon sensors for ships
Weather monitoring
Weather sensing
Wi-Fi tags to track children [2.55]
Wildfire tracking and monitoring
Wireless automated meter reading and load management
Wireless lighting control (e.g., dimmable ballasts, controllable light switches,
  customizable lighting schemes, energy savings on bright days)
Wireless parking lot sensor networks to determine which parking spots are available
Wireless smoke and CO detectors
Wireless surveillance sensor networks for providing security in shopping malls and
  parking garages
Wireless traffic sensor networks to monitor vehicle traffic on highways or in
  congested locations
WolfPack (DARPA): distributed sensing and radio jamming device (a soda-can-sized pod
  deployed about 1 per 1 km^2 is designed to replace or supplement similar technologies
  that currently reside in aircraft; because of proximity to enemy radios, less power is
  required to jam signals; adhoc networking and multihop routing are used to control and
  retrieve data from the network, which can also monitor enemy communications in addition
  to jamming them; pods are designed to last for about two months) [2.55,2.59]
Workplace applications
WSN-based data logger system for redwood monitoring; 50 nodes installed by
  UC–Berkeley at UC Botanical Gardens
WSNs for winemaking: UC–Berkeley motes for real-time mesoclimate
  monitoring and historical analysis [2.55]

    Traditionally, WSNs have been used in the context of high-end applications such as radiation and nuclear-threat detection systems; weapon sensors for ships; battlefield reconnaissance and surveillance; military command, control, communications, intelligence, and targeting systems; biomedical applications; habitat sensing; and seismic monitoring [2.16,2.17]. Recently, interest has extended to networked biological and chemical sensors for national security applications; furthermore, evolving interest extends to direct consumer applications. Applications with potential growth in the near future include military sensing, physical security, process control, air traffic control, traffic surveillance, video surveillance, industrial and manufacturing automation, distributed robotics, weather sensing, environment monitoring, and building and structure monitoring [2.18]. Ubiquitous high-reliability public-safety applications covering multithreat management are also on the horizon. Habitat monitoring (e.g., Zebranet [2.19], SensorWebs [2.20]), defense systems (e.g., Self-Healing Land Mines [2.21]), and workplace applications of sensor networks [2.13] represent just a few other examples.

    The technology has progressed to a point where one can begin exploring WNS applications with an eye to the financial return on investment that a company could expect with the deployment of such a sensor network. Some pragmatists believe that WSN applications represent the next step in the evolution of sensor networking science, which so far has focused on research-level problems rather than on meeting business needs directly on a large scale [2.13–2.15]. Business establishments have already shown interest in sensor technology. For example, insurance companies have reportedly expressed interest in using sensors in chimneys to monitor the creosote buildup, with the goal of minimizing fire hazards; there also has been interest in monitoring the temperature of water pipes with the goal of preventing ice damage. The motivation for these applications is to reduce losses and related disbursements [2.22].

Consumer applications include, but are not limited to, critical infrastructure protection and security, health care, the environment, energy, food safety, production processing, and quality-of-life support [2.23]. WSNs are expected to afford consumers a new set of conveniences, including remote-controlled home heating and lighting, personal health diagnosis, and automated automobile maintenance telemetry, to list just a few. *Near-term* commercial applications include, but are not limited to, industrial and building monitoring, appliance control (lighting and HVAC), automotive sensors and actuators, home automation, automatic meter reading, electricity load management, consumer electronics and entertainment, and asset management. Specifically, these applications fall into the following categories:

- Commercial building control
- Environmental (land, air, sea) and agricultural wireless sensors
- Home automation, including alarms (e.g., an alarm sensor that triggers a call to a security firm)
- National security applications: chemical, biological, radiological, and nuclear wireless sensors (sensors for toxic chemicals, explosives, and biological agents)
- Industrial monitoring and control
- Metropolitan operations (traffic, automatic tolls, fire, etc.)
- Military sensors
- Process control
- Wireless automated meter reading and load management

Observers expect that in the *medium term*, one will be able to integrate sensors into commercial products and systems to improve the performance and lifetime of a variety of devices while decreasing product life-cycle costs. The ultimate expectation is that eventually, WSNs will enable consumers to keep track of their belongings, pets, and young children (called quality-of-life support) [2.23]. Anywhere there is a need to connect large numbers of sensors, the approach of using WSNs with some well-established local and metropolitan area technology (e.g., IEEE 802.11/.15/.16) makes economic sense [2.15].

Sensor networking is also seen in the context of pervasive computing. The terms *invisible computing*, *pervasive computing*, and *ubiquitous computing* are used by various researchers to describe this field. Invisible computing promises a world filled with networked[3] devices, not only desktop or laptop computers, but also cars, cell phones, RFID tags, and even kitchen utensils that communicate with each other [2.24]. The widespread distribution and availability of small-scale sensors, actuators, and embedded processors offers an opportunity for transforming the physical world into a computing platform [2.1]. Invisible computing is driven by advances in wireless technologies, WSNs, IP services, Internet, and VoIP technologies. Some claim that over the next decade, traffic from the edges of the network will be as heavy as the traffic flowing from servers to clients [2.24]. WSNs are one of the first real-world examples of pervasive computing, the notion that small, smart, inexpensive sensing and computing devices will soon permeate the environment [2.7]. The sections that follow provide additional details on these and other applications.

2.4 EXAMPLES OF CATEGORY 2 WSN APPLICATIONS

In this section we discuss a number of WSN applications that either fall in the C2WSN category and/or have a strong commercial focus. These applications tend to use point-to-point (sometimes star-based) topologies, generally with single-hop radio connectivity utilizing static routing. C2WSN technology is being targeted for a gamut of building automation, industrial, medical, residential control, and monitoring applications. Many of these applications are being contemplated in the context of the IEEE 802.15.4 (ZigBee) standard solution. ZigBee middleware provides interoperability and desirable RF performance characteristics, with chipsets implementing the standard-specified protocol stack being developed at press time.[4] Examples of applications include lighting controls; automatic meter reading; wireless smoke and CO detectors; HVAC control; heating control; home security; environmental controls; blind, drapery, and shade controls; medical sensing and monitoring; universal remote control to a set-top box that includes home control, industrial automation, and building automation.

There is increasing interest in connecting and controlling in real time all sort of devices, such as personal health care (patient monitoring, fitness monitoring); building automation (security, HVAC, AMR, lighting control, access control); residential/light commercial control (security, HVAC, lighting control, access control, lawn and garden irrigation); consumer electronics (TV, VCR, DVD/CD); PCs and peripherals (mouse, keyboard, joystick); industrial control (asset management, process control, environmental, energy management); and supermarket management. These applications are different from other wireless applications, such as

---

[3]According to IDC, the market for invisible computing in 2008 will be $674 billion, with $6 billion of that for RFID-like devices, $224 billion for mobile devices; and $196 billion for connection services [2.24].

[4]In-Stat predicts IEEE 802.15.4/ZigBee node and chipset annual shipments in 2008 to exceed 160 million [2.25].

enterprise wireless LANs (for which IEEE 802.11a/b/g/h/etc. standards are ideally suited), cable replacement (for which IEEE 802.15.1/Bluetooth standards are ideally suited), or metropolitan transport (for which IEEE 802.15.3/WiMax standards are ideally suited). The sensor environment under discussion has unique requirements: The primary drivers for this environment are simplicity, long battery life, networking capabilities, reliability, and low cost. The IEEE 802.15.4 standard has been developed precisely for these applications [2.26–2.28]. Prior to the emergence of this IEEE standard, no approach existed that addressed the unique needs of most remote monitoring and control and sensory network applications.

ZigBee[5] enables the broad-based deployment of wireless networks with low-cost, low-power solutions. Also, the standard offers the ability to run for years on inexpensive primary batteries for a typical monitoring application. ZigBee is capable of inexpensively supporting robust networking environments and is able to support, by design, a very large set of nodes. Issues of interest in the context of this standard cover appropriate worldwide frequencies and data rates, topologies, and security. It should be noted that ZigBee and Bluetooth protocols are substantially different and are designed for different purposes: ZigBee is designed for low- to very-low-duty-cycle static and dynamic environments with many active nodes; Bluetooth is designed for high QoS, a variety of duty cycles, and moderate data rates in networks with limited active nodes [2.26–2.28]. Many sensors utilized in commercial applications have a battery arrangement based on two AA alkaline cells or one Li-AA cell. Typically, the sensors have an oscillator waking up the main processor at a specified interval to take a measurement, process it, and (on a sensor event) transmit it over the networks. For example, security systems have a requirement to take a reading at an interval varying between 10 seconds and 15 minutes. IEEE 802.15.4 security sensors have been designed for long operational life. Figure 2.4 illustrates the superiority of this approach compared with a Bluetooth-based approach. Benefits of C2WSNs include those identified in Sections 2.2.1 to 2.2.3, as advocated in [2.29].

2.4.1 Home Control

Home control applications provide control, conservation, convenience, and safety, as follows (see Figure 2.5) [2.29]:

- Sensing applications facilitate flexible management of lighting, heating, and cooling systems from anywhere in the home.
- Sensing applications automate control of multiple home systems to improve conservation, convenience, and safety.
- Sensing applications capture highly detailed electric, water, and gas utility usage data.

---

[5]ZigBee can also be employed in C1WSns for mesh networking support; principal applications to date, however, tend to fall in the C2WSN category.

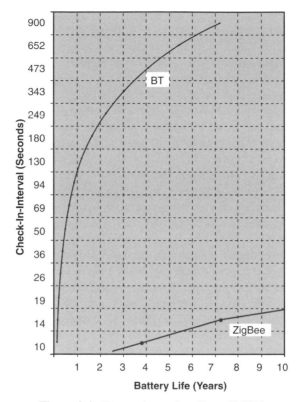

Figure 2.4 Battery longevity. (From [2.32].)

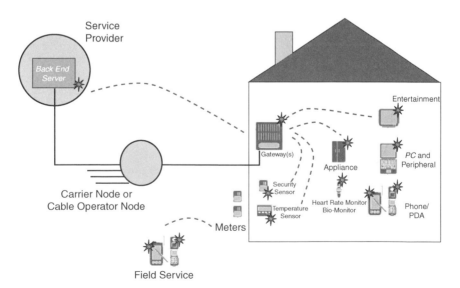

Figure 2.5 Home control applications.

- Sensing applications embed intelligence to optimize consumption of natural resources.
- Sensing applications enable the installation, upgrading, and networking of a home control system without wires.
- Sensing applications enable one to configure and run multiple systems from a single remote control.
- Sensing applications support the straightforward installation of wireless sensors to monitor a wide variety of conditions.
- Sensing applications facilitate the reception of automatic notification upon detection of unusual events.

Body-worn medical sensors (e.g., heartbeat sensors) are also emerging. These are battery-operated devices with network beacons occurring either every few seconds that could be worn by home-resident elderly or people with other medical conditions. These sensors have two ongoing processes: heartbeat time logging and transmission of heart rate and other information (instantaneous and average heart rate, body temperature, and battery voltage) [2.32].

2.4.2 Building Automation

Wireless lighting control can easily be accomplished with C2WSNs in general and ZigBee technology in particular (e.g., dimmable ballasts, controllable light switches, customizable lighting schemes, energy savings on bright days). Hotel energy management is another task that can easily be accomplished with C2WSNs in general and ZigBee technology in particular. Energy is a major operating expense for a hotel; centralized HVAC management allow hotel operators to make sure that empty rooms are not cooled. Asset management is yet another application for C2WSNs. For example, within each container, sensors form a WSN; multiple containers in a ship form an extended WSN to report sensor data. Sensors provide increased security through on-truck and on-ship tamper detection. Faster container processing can be achieved because manifest data and sensor data are known before a ship docks at port [2.26]. Building automation applications provide control, conservation, flexibility, and safety, as follows [2.29]:

- Sensing applications integrate and centralize management of lighting, heating, cooling, and security (e.g., see Figure 2.6).
- Sensing applications automate control of multiple systems to improve conservation, flexibility, and security.
- Sensing applications reduce energy expenses through optimized HVAC management.
- Sensing applications enable one to allocate utility costs equitably based on actual consumption.
- Sensing applications enable the rapid reconfiguring of lighting systems to create adaptable workspaces.

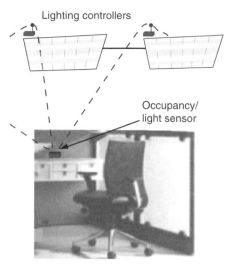

Figure 2.6   Indoor lighting control.

- Sensing applications enable the extension and upgrading of building infrastructure with minimal effort.
- Sensing applications enable one to network and integrate data from multiple access control points.
- Sensing applications enable one to deploy wireless monitoring networks to enhance perimeter protection.

As noted, there is interest in the environmental control for buildings, including energy scavenging. The ultimate preference would be to use microsensor technology that utilizes ultralow-power radio systems and compact packaging; such control can be achieved with multimodal wireless sensing and communication technology [2.30]. Recent focus has been in two areas: airflow measurement technology and the use of sensor networks for controlling indoor temperature. With multisensor single-actuator control of temperature, one can use information from a WSN to control multiple spaces in a building; as a result, one can reduce energy consumption and improve comfort at the same time. This is achieved by replacing the single sensor typical of many building environments with a sensor network that has at least one sensor in each space. The performance improvement is achieved without changing the actuation, making the strategy ideal for retrofits in existing buildings [2.30].

It is advantageous to use WSNs to control systems that are designed to produce a temperature gradient indoors; these systems, now common in many commercial buildings, are called *underfloor air distribution* (UFAD) *systems*. UFAD systems are commonly controlled with a single temperature sensor; traditionally such functions have been localized in a single point. Studies show that one can improve energy performance significantly by using a sensor network with two or more sensors in each space to control such a system [2.30]. WSNs facilitate the distribution of functions over a wide physical space within the building, leading to improved operation. In a cabled design, the cost of installing cabling for the sensors typically represents 50 to 90% of the total cost of the system; therefore, WSNs have the potential of greatly reducing the overall cost. MEMS-based WNs are expected to reduce the cost further. In the future, sensors may be embedded directly in products such as ceiling tiles and furniture, enabling improved control of the indoor environment.

A WSN used for building energy monitoring and control can improve living conditions for the building's occupants, resulting in improved thermal comfort, improved air quality, health, safety, and productivity; at the same time, it can reduce the energy budget needed to condition the space (first-order estimations indicate that such technology could reduce source energy consumption in the United States by 2 quads,[6] which translates to $55 billion per year and 35 million metric tons of reduced carbon emissions [2.30]). Lighting energy accounts for approximately 50% of commercial building electricity consumption [2.31]. In many buildings, much of this energy use is a result of lighting that is turned on unnecessarily because of inadequate control mechanisms; this results from the fact that traditional switches are expensive to install and difficult to adapt to changing requirements; typically, a few switches control many light fixtures, and occupants cannot control the lighting in individual workspaces. Although there are wireless lighting switches on the market today, most have been developed for the residential market; a higher level of flexibility is required in commercial buildings. WSNs systems consist of wireless motes with relays that can turn lights off and on. These systems need to be designed to be compatible with existing lighting systems and will not require replacement of existing lighting ballasts or existing switches [2.31].

Wireless lighting control systems can be used for retrofit applications as well as new construction. These C2WSNs utilize wireless motes installed in individual lighting fixtures in conjunction with a remote wireless switch capable of controlling the light fixtures. There is interest in developing integrated sensor or wireless communication and energy source WNs that [2.30]:

1. Support multiple sensing of temperature, light, sound, flow, and localization (called multimodal sensing)
2. Support a seamless wireless network interface
3. Support an integrated energy source that allows the node to be self-contained and to operate independently for at least 10 years
4. Support building control applications software

Research on this at present is sponsored through the NSF program "XYZ on a Chip: Integrated Wireless Sensor Networks for the Control of the Indoor Environment in Buildings" [2.30].

---

[6]A quad is a quadrillion British thermal units (Btu).

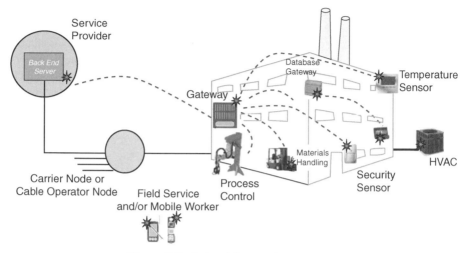

Figure 2.7   Industrial control applications.

2.4.3   Industrial Automation

Industrial automation applications provide control, conservation, efficiency, and safety, as follows (see Figure 2.7) [2.29]:

- Sensing applications extend existing manufacturing and process control systems reliably.
- Sensing applications improve asset management by continuous monitoring of critical equipment.
- Sensing applications reduce energy costs through optimized manufacturing processes.
- Sensing applications help identify inefficient operation or poorly performing equipment.
- Sensing applications help automate data acquisition from remote sensors to reduce user intervention.
- Sensing applications provide detailed data to improve preventive maintenance programs.
- Sensing applications help deploy monitoring networks to enhance employee and public safety.
- Sensing applications help streamlining data collection for improved compliance reporting.

Specific applications for industrial and commercial spaces include [2.32]:

- Warehouses, fleet management, factories, supermarkets, office complexes
- Gas, water, and electric meters

- Smoke, CO, and H_2O detectors
- Refrigeration cage or appliance
- Equipment management services and preventive maintenance
- Security services (including peel-n'-stick security sensors)
- Lighting control
- Assembly line and workflow and inventory
- Materials processing systems (heat, gas flow, cooling, chemical)
- Gateway or field service links to sensors and equipment (monitored to support preventive maintenance, status changes, diagnostics, energy use, etc.)
- Remote monitoring from corporate headquarters of assets, billing, and energy management

According to some observers, RFID tags are poised to become the most far-reaching wireless technology since the cell phone [2.33]. Worldwide revenues from RFID tags was expected to jump to $2.8 billion in 2009. During this period, the technology will appear in many industries, with a significant impact on the efficiency of business processes. In the near term, the largest RFID segment is cartons and supply chains; the second-largest market for RFIDs is consumer products, although this market is sensitive to privacy concerns. Some C2WSNs (e.g., supported with RFID technology) has applications for livestock and domestic pets; humans; carton and supply chain uses; pharmaceuticals; large freight containers; package tracking; consumer products; security, banking, purchasing and access control; and others [2.34]. For example, Airbus's A380 airplane is equipped with about 10,000 RFID chips; the plane has passive RFID chips on removable parts such as passenger seats and plane components. The benefits of RFID tagging of airplane parts include reducing the time it takes to generate aircraft-inspection reports and optimizing maintenance operations.

2.4.4 Medical Applications

A number of hospitals and medical centers are exploring applications of WSN technology to a range of medical applications, including pre-hospital and in-hospital emergency care, disaster response, and stroke patient rehabilitation. WSNs have the potential to affect the delivery and study of resuscitative care by allowing vital signs to be collected and integrated automatically into the patient care record and used for real-time triage, correlation with hospital records, and long-term observation [2.35,2.53]. WSNs permit home monitoring for chronic and elderly patients, facilitating long-term care and trend analysis; this in turn can sometimes reduce the length of hospital stays. WSNs also permit collection of long-term medical information that populates databases of clinical data; this enables longitudinal studies across populations and allows physicians to study the effects of medical intervention programs [2.40]. These WSNs tend to be of the C2WSN category.

Vital sign data, such as pulse oximetry, are poorly integrated with pre-hospital and hospital-based patient care records. Harvard University and others have developed a small, wearable wireless pulse oximeter and two-lead electrocardiogram (EKG). These devices collect heart rate, oxygen saturation, and EKG data and relay it over a short-range (100-m) wireless network to any number of receiving devices, including PDAs, laptops, or ambulance-based terminals. The data can be displayed in real time and integrated into the developing pre-hospital patient care record. The sensor devices themselves can be programmed to process the vital sign data, for example, to raise an alert condition when vital signs fall outside normal parameters; any adverse change in patient status can then be signaled to a nearby EMT or paramedic [2.35–2.42].

In collaboration with the Motion Analysis Laboratory at the Spaulding Rehabilitation Hospital, Harvard University has also developed a tiny wearable device for monitoring the limb movements and muscle activity of stroke patients during rehabilitation exercise. These devices, consisting of three-axis accelerometer, gyroscope, and electromyogram sensors, allow researchers to capture a rich data set of motion data for studying the effect of various rehabilitation exercises on this patient population [2.35–2.42].

In addition to the hardware platform, Harvard University developed a scalable software infrastructure called CodeBlue, for wireless medical devices. CodeBlue is designed to provide routing, naming, discovery, and security for wireless medical sensors, PDAs, PCs, and other devices that may be used to monitor and treat patients in a number of medical settings (see Figure 2.8). CodeBlue is designed to scale across a different network densities, ranging from sparse clinic and hospital deployments to very dense ad hoc deployments at a mass casualty site. Part of the CodeBlue system includes a system for tracking the location of individual patient devices indoors and outdoors using radio signal information [2.35–2.42].

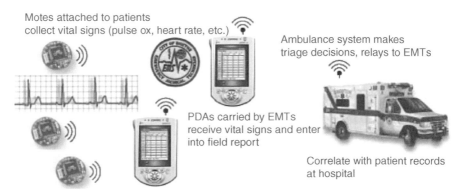

Motes attached to patients collect vital signs (pulse ox, heart rate, etc.)

Ambulance system makes triage decisions, relays to EMTs

PDAs carried by EMTs receive vital signs and enter into field report

Correlate with patient records at hospital

Figure 2.8 Use of CodeBlue for emergency response: PDA displaying real-time vital signs of multiple patients. (Courtesy of Harvard University and Boston University School of Medicine.)

2.5 EXAMPLES OF CATEGORY 1 WSN APPLICATIONS

In this section we discuss a number of WSN applications that either fall in the C1WSN category or have a strong research–scientific focus. The applications discussed below are just a few examples of the many possible applications that exist or are evolving. The ability to deploy WSNs that interconnect in an effective manner with unattended WNs is expected to have a significant bearing on the efficacy of military and civil applications such as, but not limited to, combat field surveillance, security, and disaster management. These WSNs process data assembled from multiple sensors in order to monitor events in an area of interest. For example, in a disaster management event, a large number of sensors can be dropped by a helicopter; networking these sensors can assist rescue operations by locating survivors, identifying risky areas, and making the rescue crew more aware of the overall situation and improving overall safety. Some WSNs have camera-enabled sensors; one can have aboveground full-color visible-light cameras as well as belowground infrared cameras. The use of WSNs will limit the need for military personnel involvement in dangerous reconnaissance missions. Security applications include intrusion detection and criminal hunting [2.43]. Some examples of WSN applications are [2.6]:

1. Military sensor networks to detect and gain as much information as possible about enemy movements, explosions, and other phenomena of interest
2. Law enforcement and national security applications for inimical agent tracking or nefarious substance monitoring (e.g., see Figure 2.9)

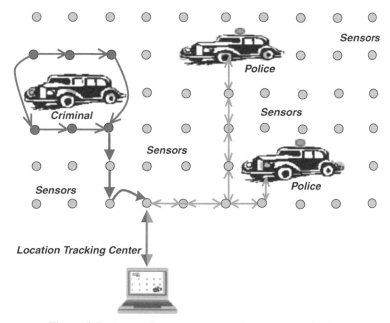

Figure 2.9 Law enforcement–national security application.

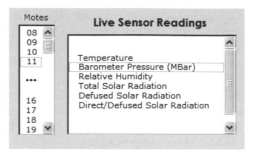

Figure 2.10 Typical real-time administrative access to distributed motes.

3. Sensor networks to detect and characterize chemical, biological, radiological, nuclear, and explosive (CBRNE) attacks and material

4. Sensor networks to detect and monitor environmental changes in plains, forests, oceans, and so on

5. Wireless traffic sensor networks to monitor vehicle traffic on highways or in congested parts of a city

6. Wireless surveillance sensor networks for providing security in shopping malls, parking garages, and other facilities

7. Wireless parking lot sensor networks to determine which spots are occupied and which are free

8. Borders monitoring with sensors and satellite uplinks

Figure 2.10 depicts the typical real-time administrative access to distributed WNs (motes) in an open-space sensor field. Real-time monitoring and sensor interrogation is typically supported. A number of illustrative examples are described in the subsections that follow. These examples just scratch the surface of the plethora of possible applications.

2.5.1   Sensor and Robots

Two technologies appear poised for a degree of convergence: mobile robotics and wireless sensor networks. Some researchers expect that mobile robotics will use WSNs to achieve ubiquitous computing environments. For example, Intel envisions mobile robots acting as gateways into wireless sensor networks, such as into the Smart Dust networks of wireless motes. These robots embody sensing, actuation, and basic (miniaturized) robotics functions. The field of mobile robotics deals with mechanical aspects (the wheels, motors, grasping arms, or physical layout) as well as with the logic aspects (the microprocessors, the software, and the telemetry). Two questions of interest are [2.15]:

- Can a mobile robot act as a gateway into a wireless sensor network?
- Can sensor networks take advantage of a robot's mobility and intelligence?

To affect this convergence, inexpensive standards-based hardware, open-source operating systems, and off-the-shelf connectivity modules are required (e.g., Intel XScale microprocessors and Intel Centrino mobile technology).

One major issue with a mobile robot acting as a gateway is the communication between the robot and the sensor network. Some propose that a sensor network can be equipped with IEEE 802.11 capabilities to bridge the gap between robotics and wireless networks. For example, Intel recently demonstrated how a few motes equipped with 802.11 wireless capabilities can be added to a sensor network to act as wireless hubs [2.15]. Other motes in the network then utilize each other as links to reach the 802.11-equipped hubs; the hubs forward the data packets to the main 802.11-capable gateway, which is usually a PC or laptop. Using some motes as hubs reduces the number of hops that any one data packet has to make to reach the main gateway, and also reduces power consumption across the sensor network. As an example, Intel recently installed small sensors in a vineyard in Oregon to monitor microclimates. The sensors measured temperature, humidity, and other factors to monitor the growing cycle of the grapes, then transmitted the data from sensor to sensor until the data reached a gateway. At the gateway, the data were interpreted and used to help prevent frostbite, mold, and other agricultural problems [2.15].

Intel, Carnegie Mellon University, University of Southern California, University of Pennsylvania, Northwestern, Georgia Tech, NASA, DARPA (the Defense Advanced Research Projects Agency), and NIST (National Institute of Standards and Technology) are just some of the institutions researching this topic. The Robotics Engineering Task Force (RETF; modeled after the Internet Engineering Task Force) has the goal of enabling government and university researchers to work collaboratively to establish standard software protocols and interfaces for robotics systems. The most pressing issue for the RETF is developing standards for commanding and controlling mobile robots.

Other examples of WSN applications include preventive maintenance for equipment in a semiconductor manufacturing fab, and sensor networks for theme parks. Both applications leverage the concept of heterogeneous WSNs, and both solve important business problems in their domains [2.15]. At[7] Intel's semiconductor fabs, thousands of sensors track vibrations coming from various pieces of equipment to determine if the machines are about to fail. There is an established science that enables managers to determine the particular signature that a well-functioning machine should have. Typically, employees in the fab must gather the sensor data manually from each node—a costly and time-consuming process that is carried out periodically, on a schedule determined by the expected failure rate of the equipment. Going forward, networking the sensors could make the process more efficient and cost-effective. Intel reportedly plans to make use of the mote technology to build an application that acquires data automatically; prototypes have already been built (see Figure 2.11). Intel is also exploring the deployment of heterogeneous sensor networks in theme parks. Such networks

---

[7]The paragraphs that follow are based on Intel sources.

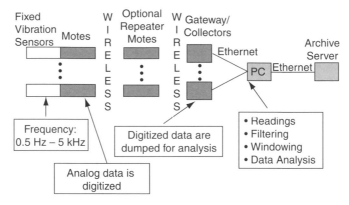

Figure 2.11   Intel fab environment with WSNs.

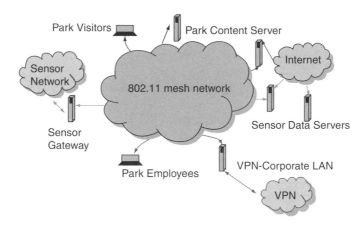

Figure 2.12   Theme park WSN example.

could be used for multiple purposes. One potential use is monitoring the quality of water in tanks (see Figure 2.12); currently, such monitoring is done manually; a WSN can make the process more accurate and efficient. Another potential use of the network is to provide Internet access to park visitors. Visitors can use the wireless network to reserve a space at a particular park attraction or to learn more about an exhibit. The wireless network could improve park management as well. Sensors could track attendance at park exhibits and rides, and management could use the network to access office applications from various stations throughout the park.

2.5.2   Reconfigurable Sensor Networks

Military applications require support for tactical and surveillance arrangements that employ reconfigurable sensor WNs that are capable of forming networks on the fly, assembling themselves without central control, and being deployed incrementally. Reconfigurable "smart" WNs are self-aware, self-configurable, and autonomous. Self-organizing WSNs utilize mechanisms that allow newly deployed WNs to establish connectivity (to build up a network topology) spontaneously. Also, these networks have mechanisms for managing WN mobility (if any), WN reconfiguration, and WN failure (if and when that happens) [2.44].

2.5.3 Highway Monitoring

Transportation (traffic flow) is a sector that is expected to benefit from increased monitoring and surveillance. A specific example follows. (Traffic in the United States is growing at three times the rate of population growth and causing an estimated $75 billion lost annually due to traffic congestion.) Traffic Pulse Technology is an example of a WSN developed by Traffic.com [2.2,2.45]. The[8] goal of this system (which uses stationary WNs; see Figure 2.13) is to collect data through a sensor network, process and store the data in a data center, and distribute those data through a variety of applications. Traffic Pulse is targeted for open-air environments; it

Figure 2.13   Typical highway traffic-sensing installation. (Courtesy of Traffic.com.)

---

[8]The rest of this subsection is based on [2.45].

provides real-time collection of data (e.g., to check temperature or monitor pollution levels). The system is installed along major highways; the digital sensor network gathers lane-by-lane data on travel speeds, lane occupancy, and vehicle counts. These basic data elements make it possible to calculate average speeds and travel times. The data are then transmitted to the data center for reformatting. The network monitors roadway conditions continuously on a 24/7 basis and provides updates to the data center in real time. The system collects key traffic information, including vehicle speeds, counts (volume), and roadway density, transmitting the data over a wireless network to a data center every 60 seconds.

In each major city, Traffic.com maintains a traffic pulse operations center that collects and reports on real-time event, construction, and incident data. This information supplements the data collected from the sensors. Each center produces the information through a wide range of methods: video, aircraft, mobile units, and monitoring of emergency and maintenance services frequencies. Applications include the following:

1. Private traffic information providers in the United States: The company's real-time and archived data offer valuable tools for a variety of commercial and governmental applications.
2. Telematics: For mobile professionals and others, the company's traffic information complements in-vehicle navigation devices, informing drivers not only how to get from point A to point B but how long it will take to get there, or even direct them to an alternative route.

2.5.4　Military Applications

A number of companies have developed WSNs that include customizable, sensor-laden, networked nodes and both mobile and Internet-hosted user interfaces [2.2,2.46]. For example, Rockwell Scientific's wireless sensing network development system allows examination of issues relative to design, deployment, and use of microsensor networks. Wireless distributed microsensor networks consist of a collection of communicating nodes, where each node incorporates (1) one or more sensors for measuring the environment, (2) computing capability to process sensor data into "high-value" information and to accomplish local control, and (3) a radio to communicate information to and from neighboring nodes and eventually to external users. The company[9] has developed new prototype development platforms for experimenting with microsensor networks under a number of government- and industry-sponsored programs (see Figure 2.14). The baseline prototype wireless sensing unit is based on an open, modular design using widely available commercial-off-the-shelf (COTS) technology. These nodes combine sensors (such as mechanical vibration, acoustic, and magnetic) with a commercial digital cordless telephone radio and an embedded commercial RISC microprocessor in a small package.

---

[9]The rest of this subsection and Figure 2.14 are based on Rockwell Scientific sources [2.46].

Figure 2.14 Military examples. (Courtesy of Rockwell Scientific.)

Condition-Based Monitoring Again as an illustrative example, Rockwell Scientific is developing WSNs specifically tailored to the requirements for monitoring complex machinery and processes. Their WSNs have been deployed on board U.S. Navy ships as part of a developmental program with the Office of Naval Research. Exploratory studies have also been done for use of WSNs on aircraft, rotorcraft, and spacecraft as part of an overall integrated vehicle health management system. Machinery maintenance has evolved from run to fail (no maintenance) to scheduled maintenance (e.g., change oil every three months) to condition-based maintenance (CBM). All three techniques are in current use. The economic trade-off is between the cost of the CBM equipment and the staffing resources expended to determine the machine's health and the cost of unexpected, as opposed to scheduled, repair and process downtime. With the emphasis of industry in the last couple of decades on just-in-time processes, unexpected machinery failure can be costly. The successful application of machinery monitoring programs can optimize the use of machinery and keep manufacturing costs in check by making the process more efficient [2.46]. The costs associated with CBM can be allocated into equipment, installation, and labor costs in collecting and analyzing the machine health data. WSNs are positioned to minimize all three costs and, in particular, to eliminate the staffing costs, which often are the largest. With the continuing advances in data processing hardware and RF transceiver hardware (cell phone markets drive this), the technology is now becoming available to install compact monitoring systems on machinery that avoid the installation expense of data cabling through RF link technology; these systems provide a mechanism for data acquisition and analysis on the monitoring unit itself. The primary challenge faced by WSNs for machinery and process monitoring is related to the quality of the information produced by both the individual sensors and the distributed sensor network. Nodes located on individual components must not only be able to provide information on the present state of the component (e.g., a bearing or gearbox), but also provide an indicator of the remaining useful life of the component [2.46].

The approach taken at Rockwell Scientific has been to mount two parallel efforts. Existing diagnostic routines and expert systems are being ported to WSN hardware with modifications for autonomous data collection and analysis. The firm is also involved in developing advanced diagnostics algorithms for machinery vibration monitoring that provide advances over present systems. The main thrust in this area is to generalize diagnostic algorithms so that they do not depend on detailed knowledge of the machinery on which they are installed. Data-processing algorithms that determine critical machine parameters, such as the shaft speed or the number of rolling elements in a bearing, have been developed. The company is also developing the ability for distributed collections of WSN nodes located on machine components and/or throughout a process to provide information on the overall machine and/or process on which they are deployed. This is a primary advantage of a distributed sensing system in that it enables inferences from individual component data to be used to provide diagnostics for aspects of the system that are not being sensed directly. For example, monitoring bearing vibrations or motor currents can provide information not only on bearing health but also on the inception and severity of pump cavitation. Pump cavitation, in turn, can provide information on the state of valves located throughout a pumping process.

The dynamically reconfigurable nature of WSNs is being exploited by Rockwell Scientific in an application of WSNs to space vehicle status monitoring in collaboration with the Boeing Company. WSNs are deployed throughout space vehicles to perform a variety of missions during the different phases of the space flight. For example, during the launch phase, WSN nodes located on various critical components of the spacecraft can monitor vibration levels for out-of-compliance signals. During flight and reentry, the WSN monitor structural disturbances caused by the significant temperature gradients encountered as different portions of the vehicle are alternately exposed and shadowed from the sun and atmosphere. This is accomplished via coherent collection and processing of vibration and strain data. Upon landing, critical components will once again be monitored for out-of-compliance signals. These data are used to determine those components needing postflight maintenance or replacement, enabling faster turnaround for the space vehicle, thereby lowering costs [2.46].

Military Surveillance   For military users, an application focus of WSN technology has been area and theater monitoring. WSNs can replace single high-cost sensor assets with large arrays of distributed sensors for both security and surveillance applications. The WSN nodes are smaller and more capable than sensor assets presently in the inventory; the added feature of robust, self-organizing networking makes WSNs deployable by untrained troops in essentially any situation. Distributed sensing has the additional advantages of being able to provide redundant and hence highly reliable information on threats as well as the ability to localize threats by both coherent and incoherent processing among the distributed sensor nodes. WSNs can be used in traditional sensor network applications for large-area and perimeter monitoring and will ultimately enable every platoon, squad, and soldier to deploy WSNs to accomplish a number of mission and self-protection goals. Rockwell Scientific has been working with the U.S. Marine Corps and U.S. Army to test and refine WSN performance in desert, forest, and urban terrain.

For the urban terrain, WSNs are expected to improve troop safety as they clear and monitor intersections, buildings, and rooftops by providing continuous vigilance for unknown troop and vehicle activity. The primary challenge facing WSNs is accurate identification of the signal being sensed; one needs to develop state-of-the-art vibration, acoustic, and magnetic signal classification algorithms to accomplish this goal. Currently, WSNs run vibration detection algorithms based on energy thresholding; although this is a simple technique, it is subject to false alarms, leading to a desire for more sophisticated spectral signature algorithms. Low-power algorithms to classify a detected event as an impulsive event (e.g., either a footstep or gunshot) or vehicle (e.g., wheeled or tracked, light or heavy) have also been demonstrated.

The inclusion of multiple sensors on each node enables fusion of different sensed phenomenologies, leading to higher-quality information and decreased false alarm rates. Algorithms for fusing the seismic, acoustic, and magnetic sensors on a single node are being developed. Algorithms utilizing the advantages of a network of spatially separate nodes span a range of cooperative behaviors, each of which trades off detection quality versus energy consumption. Examples of cooperative fusion range from high-level decision corroboration (e.g., voting), to feature fusion, to full coherent beam formation. The examples discussed above are simply representative of many efforts under way at many companies involved in theater technology.

Borders Monitoring   At press time Boeing Co. had secured a contract from the Department of Homeland Security to implement SBInet, the Secure Borders Initiative, along the northern and southern U.S. borders. The program was announced by DHS in 2005, and contracts were awarded in late 2006. The SBInet portion of the Secure Borders Initiative is the development of a technological infrastructure that facilitates the use of a variety of sensors and detection devices, and which enables that data to be forwarded to remote operations centers via Ku-band satellite uplinks.

2.5.5   Civil and Environmental Engineering Applications

Sensors can be used for civil engineering applications. Research has been under way in recent years to develop sensor technology that is applicable for buildings, bridges, and other structures. The goal is to develop "smart structures" that are able to self-diagnose potential problems and self-prioritize requisite repairs [2.47]. This technology is attractive for earthquake-active zones. Although routine mild tremors may not cause visible damage, they can give rise to hidden cracks that could eventually fail during a higher-magnitude quake. Furthermore, after a mild earthquake, a building's true structural condition may not be ostensibly visible without some "below-the-skin" measurement. Smart Dust motes, tiny and inexpensive sensors developed by UC–Berkeley engineers, are promising in this regard (see Figure 2.15). The battery-powered matchbox-sized WNs operating on TinyOS are designed to sense a number of factors, ranging from light and temperature (for energy-saving applications) to dynamic response (for civil engineering analysis) [2.47].

Figure 2.15   Motes. [Courtesy of Steve Glaser and David Pescovitz, Center for Information Technology Research in the Interest of Society (CITRIS) program, UC–Berkeley.]

Up to the present, wired seismic accelerometers (devices able to measure movement) have been used; however, these devices are expensive (several thousands of dollars each) and are difficult to install. This predicament limits the density of sensor deployment, which in turn limits the planner's view of a building's structural integrity. As a result, a safety-impacting structural problem does not become visible until the entire building is affected. On the other hand, if sensors that cost a few hundreds of dollars and that can be installed relatively easily and quickly become available, one arrives at a situation where dense packs of sensors can be deployed to surround all critical beams and columns. This arrangement is able to provide detailed structural data. UC Berkeley's Richmond Field Station seismic laboratory is pursuing research in this area [2.47]. Data from the Smart Dust motes is expected to increase the accuracy of finite element analyses, a method of computer modeling where mathematical equations represent a structure's behavior under given conditions.

2.5.6   Wildfire Instrumentation

Collecting real-time data from wildfires is important for life safety considerations and allows predictive analysis of evolving fire behavior. One way to collect such data is to deploy sensors in the wildfire environment. FireBugs are small wireless sensors (motes) based on TinyOS that self-organize into networks for collecting real-time data in wildfire environments [2.48]. The FireBug system combines state-of-the-art sensor hardware running TinyOS with standard off-the-shelf World Wide Web and database technology, allowing rapid deployment of sensors and behavior monitoring.

2.5.7   Habitat Monitoring

As an illustrative example, in the recent past, the Intel Research Laboratory at Berkeley undertook a project with the College of the Atlantic in Bar Harbor and UC–Berkeley to deploy wireless sensor networks on Great Duck Island in Maine. These networks monitor the microclimates in and around nesting burrows used by Leach's storm petrel. The goal was to develop a habitat-monitoring kit that enables researchers worldwide to engage in nonintrusive and nondisruptive monitoring of sensitive wildlife and habitats [2.49]. About three dozen motes were deployed on the island. Each mote has a microcontroller, a low-power radio, memory, and batteries. Sensor motes monitor the nesting habitat of Leach's storm petrel on the island and relay their readings into a satellite link that allows researchers to download real-time environmental data over the Internet. For habitat monitoring the planner needed sensors that can take readings for temperature, humidity, barometric pressure, and midrange infrared. Motes sample and relay their sensor readings periodically to computer base stations on the island [2.49].

2.5.8 Nanoscopic Sensor Applications

There is keen interest in WSNs for biological sensing. In particular, there is interest in the "labs on a chip" concept, including new methodologies supported by nano-techniques. An example follows from work done at UC–Berkeley (work is funded by the Defense Advanced Research Projects Agency). In particular, a nanoscopic microscale confocal imaging array (micro-CIA) is a device that merges MEMSs (micro electromechanical systems), ultrasmall lasers, lenses, and plumbing. These devices are fabricated by micromachining silicon or polymers. Using this technology, one can detect biowarfare pathogens and can use it as a diagnostic tool in medicine. A single nanoscopic micro-CIA is essentially a massively scaled down confocal microscope. Confocal microscopes work by shining a laser at a molecule that has been tagged with a fluorescent die. The laser "excites" the fluorescent molecule so that it emits a specific color of light. To create a clear image of the sample, only one tiny point is illuminated at any moment. The laser scans the sample many times a second, imaging each tiny point, and a complete three-dimensional image is built. A series of rapidly moving mirrors inside the nanoscopic micro-CIA enables the beam to scan over the sample bit by bit just as with a full-sized confocal microscope. Meanwhile, sensors act as eyes, looking for fluorescence and feeding those data to the computer controlling the nanoscopic micro-CIA. By analyzing that information, the computer can detect and identify a single biomolecule automatically. At this juncture, the latest microoptical scanning design operates in only two dimensions; still, this is adequate to detect biowarfare agents. This work is expected to lead to the development of a very sensitive wristwatch biomonitor that soldiers can wear; through a wireless radio link, physicians can then keep tabs on each soldier's physiology on a cellular and molecular level and can identify any substance that a person might encounter [2.50].

2.6 ANOTHER TAXONOMY OF WSN TECHNOLOGY

In this section we provide another taxonomy of WSNs; we include it here to provide another perspective on the taxonomy that we have adopted in this book. The taxonomy discussed, summarized from [2.52], is based on physical placement of the various sensors and the connectivity of these nodes to nodes in the wired infrastructure; the network configuration determines the amount of routing intelligence that needs to be supported in the sensor nodes. Specifically, key factors used in the classification process under discussion are the size of the system, the number of sensors used, the average (and/or maximum) distance (in hops) of the sensors to the wired

TABLE 2.2   Classification Factors

Network Configuration Factor	Implication
Size of overall system (total *number* of sensor nodes)	Determines the effort needed to configure the system for the particular application.
Distance of sensors to wired infrastructure (in hops)	Determines the amount of intelligence needed in a sensor for routing information to specific high-processing nodes.
Deployment distribution of sensor nodes (deterministic or nondeterministic)	Determines the design. In deterministic deployment distributions, the administrator has control over the placement of WNs and he or she can perform remedial operations in case of faults (determinism generally decreases as the number of WNs increases). In nondeterministic deployment distributions, the deployment may be random (e.g., dropped from an helicopter) or time dependent (time of day, or failure over time due to power drainage).

infrastructure, and the distribution of the sensor nodes (see Table 2.2). This taxonomy is somewhat similar, but not identical, to the one utilized in the book.

Three types of WSN system (technology) that have been described in [2.52] are:

1. Nonpropagating WSN systems
2. Deterministic routing WSN systems
   a. Aggregating
   b. Nonaggregating systems
3. Self-configurable and self-organizing WSN systems
   a. Aggregating
   b. Nonaggregating systems

In nonpropagating WSN systems, WNs are not responsible to support dynamic routing of packets to end systems. This follows because the wired infrastructure is the basic connecting component in this case and WNs are generally in close proximity (one hop) to the wired infrastructure. WNs collect and report their sensor measurements to nodes connected to the wired network, which, in turn, route the information to the end system. These systems are generally manually configurable and are highly deterministic in deployment distribution. Environmental sensors deployed in buildings or within a physically restricted area belong to this category.

In deterministic routing WSN systems, the wired and wireless infrastructures both play an active role in routing packets. For packets to reach the wired infrastructure in these environments, the WNs have to route or forward packets through a number of wireless hops. However, the routes to the wired infrastructure are deterministic and can be configured manually. In home networking systems, the WNs are in prespecified positions and route information through predetermined routes. The number of nodes in such a system is usually relatively small.

In aggregating systems, the information received from "downstream" WNs can be aggregated and forwarded. Intermediary nodes in the network have the ability to fuse the information received from downstream sources. In nonaggregating systems, the information gathered by every source node is independent and is transmitted separately. Weather monitoring systems are examples of aggregating WSNs, and toll-badge-reading systems are examples of nonaggregating WSNs. (Nonpropagating systems are not generally viewed under this partitioning since the nodes are often just one hop away from the wired node, and hence the in-network aggregation issue is moot. If present, the aggregation functionality is typically performed in the wired infrastructure or at the gateway connecting the wireless network to the wired infrastructure; hence, such systems do not require specialized aggregating functionality to be embedded into the WSN itself.)

The class of self-configurable systems relates to WSNs where WNs need to self-organize themselves (initially or as time goes by) into a connected network. Many self-configurable systems are nondeterministic in topological deployment; the number of nodes in these systems can be from hundreds to hundreds of thousands (when the number of nodes exceeds a few hundreds, even deterministic deployment systems need to be self-configurable). In these environments, specific gateway WNs have connectivity to the wired infrastructure for transferring information to the end systems. A large-scale security network is an example of a deterministic system that belongs to the category of self-configurable systems; a target-tracking system is an example of a self-configurable WSN. In self-configurable WSNs, the nodes may also aggregate data [2.60].

2.7 CONCLUSION

In this chapter we provided a sample of possible WSN applications. The reader should be able to envision dozens of other potential applications, as they appear to be almost unlimited. Basically, wherever one wants to instrument, observe, and react to events and phenomena in a specified environment, one can use WSNs; the environment can be the physical world, a biological system, or an IT framework.

REFERENCES

[2.1] "Cougar: The Sensor Network Is the Database," Cornell University, Ithaca, NY, http://www.cs.cornell.edu/database/cougar/.

[2.2] "Sensor Network Applications," Computer Science Instructional Facility, University of California–Davis, http://wwwcsif.cs.ucdavis.edu/~yick/sensor/Sensorapplication.htm.

[2.3] "Distributed Surveillance Sensor Network," http://www.spawar.navy.mil/robots/undersea/dssn/dssn.html.

[2.4] S. Cowen, S. Briest, J. Dombrowski, "Underwater Docking of Autonomous Undersea Vehicles Using Optical Terminal Guidance," *IEEE Oceans 1997* (Annual Joint Conference of the IEEE Ocean Engineering Society and the Marine Technology Society), Halifax, Nova Scotia, Canada, Oct. 1997.

[2.5] "Sensors and Sensor Networks," Program Solicitation, NSF 03-512, Mar. 6, 2003, National Science Foundation, Directorate for Engineering, http://www.nsf.gov/ cgi-bin/getpub?gpg; NSF Publications Clearinghouse, pubs@nsf.gov.

[2.6] "Smart Sensor Networks," National Institute of Standards and Technology, Gaithersburg, MD, http://w3.antd.nist.gov/wahn_ssn.shtml.

[2.7] N. Bulusu, S. Jha, Eds., *Wireless Sensor Networks: A Systems Perspective*, Artech House, Norwood, MA, 2004.

[2.8] R. R. Brooks, S. S. Iyengar, *Multi-sensor Fusion: Fundamentals and Applications with Software*, Prentice Hall, Upper Saddle River, NJ, 1998.

[2.9] K. Whitehouse, X. Jiang, "Calamari: A Sensor Field Localization System," http://www.cs.berkeley.edu/~kamin/calamari/.

[2.10] K. Whitehouse, "The Design of Calamari: An Ad-Hoc Localization System for Sensor Networks," Master's thesis, University of California–Berkeley, 2002.

[2.11] K. Whitehouse, F. Jiang, A. Woo, C. Karlof, D. Culler, "Sensor Field Localization: A Deployment and Empirical Analysis," Technical Report, University of California– Berkeley, Apr. 9, 2004.

[2.12] K. Whitehouse, C. Sharp, E. Brewer, D. Culler, "Hood: A Neighborhood Abstraction for Sensor Networks," *Proceedings of the ACM International Conference on Mobile Systems, Applications, and Services* (MobiSys'04), Boston, MA, June 2004.

[2.13] W. S. Conner, J. Heidemann, L. Krishnamurthy, X. Wang, M. Yarvis, "Workplace Applications of Sensor Networks," in *Wireless Sensor Networks: A Systems Perspective*, N. Bulusu and S. Jha, Eds., Artech House, Norwood, MA, 2004.

[2.14] L. Krishnamurthy, "PSFQ: A Reliable Transport Protocol for Wireless Sensor Networks," *Proceedings of the 1st ACM International Workshop on Wireless Sensor Networks and Applications* (WSNA'02), in conjunction with *ACM MobiCom 2002*, Atlanta, GA, Sept. 2002.

[2.15] Promotional materials, Intel, Santa Clara, CA, www.intel.com, 2003.

[2.16] T. Clouqueur et al., "Sensor Deployment Strategy for Target Detection," *Proceedings of the 1st Workshop on Sensor Networks and Applications* (WSNA'02), Atlanta, GA, Sept. 2002.

[2.17] I. F. Akyildiz et al., "A Survey of Sensor Networks," *IEEE Communications*, Aug. 2002, pp. 102 ff.

[2.18] C.-Y. Chong, S. P. Kumar, "Sensor Networks: Evolution, Opportunities, and Challenges," *Proceedings of the IEEE*, Vol. 91, No. 8, Aug. 2003, pp. 1247ff.

[2.19] M. Martonos, "Habitat Monitoring Using Zebranet: Design and Experience," in *Wireless Sensor Networks: A Systems Perspective*, N. Bulusu and S. Jha, Eds., Artech House, Norwood, MA, 2004.

[2.20] K. A. Delin, "SensorWebs in the Wild," in *Wireless Sensor Networks: A Systems Perspective*, N. Bulusu and S. Jha, Eds., Artech House, Norwood, MA, 2004.

[2.21] W. M. Merrill, L. Girod, B. Schiffer, D. McIntire, G. Rava, K. Sohrabi, F. Newberg, J. Elson, W. Kaiser, "Defense Systems: Self-Healing Land Mines," in *Wireless Sensor Networks: A Systems Perspective*, N. Bulusu and S. Jha, Eds., Artech House, Norwood, MA, 2004.

[2.22] J. Cox, "Wireless Sensor Networks Grabbing Greater Attention," *NetworkWorld*, Sept. 27, 2004, pp. 9ff.

[2.23] M. Hatler, *Wireless Sensor Networks: Mass Market Opportunities*, One World Inc., San Diego, California, February 22, 2004.

[2.24] M. J. Thompson, "Invisible Computing Is Hard to Miss," *Technology Review*, Feb. 2005, p. 86.

[2.25] "Short Distance Wireless Landscape: Will ZigBee Bring the Jetsons' Home to Life?" In-Stat, San Jose, CA, 2004.

[2.26] B. Heile, "Emerging Standards: Where Do ZigBee/UWB Fit?" ZigBee Alliance, Bishop Ranch, CA, June 8, 2004.

[2.27] E. H. Callaway, Jr., *Wireless Sensor Networks Architectures and Protocols*, Auerbach Publications, New York, 2003.

[2.28] J. A. Gutierrez, E. H. Callaway, Jr., R. L. Barrett, Jr., *Enabling Wireless Sensors with IEEE 802.15.4, Low-Rate Wireless Personal Area Networks*, IEEE Press, Piscataway, NJ, 2003.

[2.29] ZigBee Alliance, Bishop Ranch, CA, http://www.zigbee.org/.

[2.30] C. Federspiel, E. Arens, D. Auslander, C. Lin, S. Tang, D. Wang, "XYZ on a Chip: Integrated Wireless Sensor Networks for the Control of the Indoor Environment in Buildings," Center for the Built Environment, a group of industry and government leaders teamed up with faculty and researchers at the University of California–Berkeley, http://www.cbe.berkeley.edu/research/briefs-wirelessxyz.htm.

[2.31] K. Maclay, "UCB Center Wins Funding to Develop Wireless Lighting Controls," press release, Berkeley Wireless Research Center, University of California–Berkeley, Sept. 24, 2004.

[2.32] J. Adams, "Designing with 802.15.4 and ZigBee," Industrial Wireless Applications Summit, San Diego, CA, Mar. 9, 2004.

[2.33] In-Stat, San Jose, CA, http://www.in-stat.com.

[2.34] "RFID Tags and Chips: Changing the World for Less Than the Price of a Cup of Coffee," In-Stat/MDR Report IN0402440WT, In-Stat, San Jose, CA, Jan. 2005.

[2.35] "CodeBlue: Wireless Sensor Networks for Medical Care," Division of Engineering and Applied Sciences, Harvard University, Boston University School of Management, Boston Medical Center, Spaulding Rehabilitation Hospital, Dec. 17, 2004, http://www.eecs.harvard.edu/~mdw/proj/codeblue/.

[2.36] K. Lorincz, D. J. Malan, T. R. F. Fulford-Jones, A. Nawoj, A. Clavel, V. Shnayder, G. Mainland, M. Welsh, S. Moulton, "Sensor Networks for Emergency Response: Challenges and Opportunities," *IEEE Pervasive Computing*, Special Issue on Pervasive Computing for First Response, Oct.–Dec. 2004.

[2.37] W. Tollefsen, M. Pepe, D. Myung, M. Gaynor, M. Welsh, S. Moulton, "iRevive, a Pre-Hospital Mobile Database for Emergency Medical Services," *International Journal of Healthcare Technology and Management*, Summer 2004.

[2.38] T. R. F. Fulford-Jones, G. Y. Wei, M. Welsh, "A Portable, Low-Power, Wireless Two-Lead EKG System," *Proceedings of the 26th IEEE Engineering Medicine and Biology Society (EMBS) Annual International Conference*, San Francisco, CA, Sept. 2004.

[2.39] D. Malan, T. Fulford-Jones, M. Welsh, S. Moulton, "CodeBlue: An Ad Hoc Sensor Network Infrastructure for Emergency Medical Care," presented at the International Workshop on Wearable and Implantable Body Sensor Networks, Apr. 2004.

[2.40] M. Welsh, D. Malan, B. Duncan, T. Fulford-Jones, S. Moulton, "Wireless Sensor Networks for Emergency Medical Care," presented at the GE Global Research Conference, Harvard University and Boston University School of Medicine, Boston, MA, Mar. 8, 2004.

[2.41] D. Myung, B. Duncan, D. Malan, M. Welsh, M. Gaynor, S. Moulton, "Vital Dust: Wireless Sensors and a Sensor Network for Real-Time Patient Monitoring," presented at the 8th Annual New England Regional Trauma Conference, Burlington, MA, Nov. 2003.

[2.42] M. Welsh, D. Myung, M. Gaynor, S. Moulton, "Resuscitation Monitoring with a Wireless Sensor Network," American Heart Association, Resuscitation Science Symposium, Supplement to *Circulation: Journal of the American Heart Association*, Oct. 28, 2003.

[2.43] K. Akkaya, M. Younis, "A Survey on Routing Protocols for Wireless Sensor Networks," Department of Computer Science and Electrical Engineering, University of Maryland, Baltimore County, Baltimore, MD, Aug. 18, 2003.

[2.44] A. Lim, Institute for Reconfigurable Smart Components, "DARPA SENSIT Project, Self-Organizing Sensor Networks," Department of Computer Science and Engineering, Auburn University, Auburn, AL, 2005.

[2.45] Mobility Technologies, Wayne, PA, http://www.mobilitytechnologies.com/ntdc/.

[2.46] Rockwell Scientific, Thousand Oaks, CA, http://wins.rockwellscientific.com/.

[2.47] S. D. Glaser, Center for Information Technology Research in the Interest of Society, University of California–Berkeley, Berkeley, http://www.coe.berkeley.edu/labnotes/1101smartbuildings.html.

[2.48] Center for Information Technology Research in the Interest of Society (CITRIS), "Design and Construction of a Wildfire Instrumentation System Using Networked Sensors," work developed as a part of Adaptive Real-Time Geoscience and Environmental Data Analysis, Modeling and Visualization grant from the National Science Foundation Information Technology Research Initiative, http://firebug.sourceforge.net/.

[2.49] Intel Research Laboratory at Berkeley, "Habitat Monitoring on Great Duck Island," Intel, Santa Clara, CA, and University of California–Berkeley, http://www.greatduck-island.net/.

[2.50] "Nano-Microscope Spots Single Molecules," Lab Notes: Research from the Berkeley College of Engineering, University of California–Berkeley, www.coe.berkeley.edu/labnotes/1101micro.html.

[2.51] K. Whitehouse, C. Karlof, D. Culler, "Getting Ad-Hoc Signal Strength Localization to Work," Technical Report, University of California–Berkeley, May 26, 2004.

[2.52] L. Subramanian, R. H. Katz, "An Architecture for Building Self-Configurable Systems," Department of Electrical Engineering and Computer Sciences, University of California–Berkeley, http://www.eecs.berkeley.edu/.

[2.53] M. Welsh, Technical materials, Harvard University, Cambridge, MA, mdw@eecs.harward.edu.

[2.54] P. Desai, "Wireless Sensor Networks: A Survey," Teaching materials, University of Florida, Gainesville, FL, www.ufl.edu.

[2.55] UC–Davis, "Wireless Sensor Network News," Department of Electrical and Computer Engineering, University of California–Davis, http://www.ece.ucdavis.edu.

[2.56] Nature Publishing Group, "Beating the Lights," http://www.nature.com, Dec. 8, 2004.

[2.57] W. Knight, "Smart Bullet Reports Back Wirelessly," NewScientist.com News Service, May 2004.

[2.58] NFC Forum, Wakefield, MA, info@nfc-forum.org, http://www.nfc-forum.org.

[2.59] K. L. Vantran, "WolfPack Proves Strength in Numbers," U.S. Department of Defense, American Forces Press Service, DefenseLink, http://www.defenselink.mil, Aug. 2003.

[2.60] E. Ekici, Y. Gu, D. Bozdag, "Mobility-Based Communications in Wireless Sensor Networks," *IEEE Communications*, July 2006, pp. 56ff.

3

BASIC WIRELESS SENSOR
TECHNOLOGY

3.1 INTRODUCTION

In this chapter we look at basic sensor node systems technology at several levels.
First, we focus on the sensor node technology itself (Section 3.2), providing a
survey of sensor technology, including a taxonomy that classifies devices in
families, such as large sensors (e.g., radar sensors), microsensors (tiny sensors),
nanosensors, tag-reading sensors, and other sensors (Section 3.3). As already
noted, WSNs are characterized by the fact that they need to operate in
resource-constrained environments; in turn, this fact imposes strict design guidelines and limitations on the WNs; to this end, we address sensor functionality and
components, including the sensing and actuation unit, processing unit, communication unit, power unit, and other application-dependent units. Second, we look
at fundamental networking and topological issues (Section 3.4). Building on the
introduction provided herein, these issues are revisited in more detail in
subsequent chapters. Finally, we look at some current research trends in sensor
technology (Section 3.5).

The terms *sensor node*, *wireless node* (WN), *Smart Dust*, *mote*, and *COTS*
(commercial off-the-shelf) *mote* are used somewhat interchangeably in the industry;
the most general terms used here are *sensor node* and *WN*.

Wireless Sensor Networks: Technology, Protocols, and Applications, by Kazem Sohraby, Daniel Minoli,
and Taieb Znati

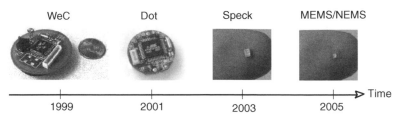

Figure 3.1 Progression of sensor technology (motes) over time (partial sample). (WeC and Dot motes: from Seth Hollar, Kris Pister, and James McClurkin, UC–Berkeley; speck motes: from SpeckNet Consortium/Scottish Higher Education Funding Council; MEMS/NEMS: authors' synthesis.)

3.2 SENSOR NODE TECHNOLOGY

3.2.1 Overview

As we saw in earlier chapters, a WSN consists of a group of dispersed sensors (motes) that have the responsibility of covering a geographic area (the sensor field) in terms of some measured parameter (also known as the measurand); alternatively, a sensor supports a point-to-point link in which the "reader" end is attached to a wireline network (e.g., a stationary tag reader sensing a mobile tag). Sensor nodes have wireless communication capabilities and some logic for signal processing, topology management (if and where applicable), and transmission handling (including digital encoding and possibly encryption and/or forward error correction). Figure 3.1 depicts the progression of sensor technology over time during the past few years. WSNs that combine physical sensing of parameters such as temperature, light, or seismic events with computation and networking capabilities are expected to become ubiquitous in the future [3.3]. Successful development of low-cost robust miniaturized sensors and detection equipment (such as mass spectrometers and chromatographs) will be of benefit; design of such systems is now being encouraged by U.S. research agencies (e.g., the National Science Foundation) [3.5]. Some sensor applications also support e-money purchases at point-of-sale locations such as from soft-drink machines, kiosks, gas stations, and checkout counters.

At the design level a WSN sits at the confluence of research in disciplines such as database query processing, networking, algorithms, and distributed systems [3.3]; hence, a lot of thought and engineering go into the development of both WNs and WSNs. The basic functionality of a WN generally depends on the application, but the following requirements are typical [3.4]:

1. Determine the value of a parameter at a given location. For example, in an environment-oriented WSN, one might need to know the temperature, atmospheric pressure, amount of sunlight, and the relative humidity at a number of locations. This example shows that a given WN may be connected to different types of sensors, each with a different sampling rate and range of allowed values.

2. Detect the occurrence of events of interest and estimate the parameters of the events. For example, in a traffic-oriented WSN, one would like to detect a vehicle moving through an intersection and estimate the speed and direction of the vehicle.

3. Classify an object that has been detected. For example, is a vehicle in a traffic sensor network a car, a minivan, a light truck, a bus?

4. Track an object. For example, in a military WSN, track an enemy tank as it moves through the geographic area covered by the network.

Naturally, the data collected must be transmitted to the appropriate data-consumption entity in a timely fashion. In many cases there are real-time or near-real-time requirements; for example, the detection of an intruder should be communicated to the police in real time so that relevant action can be taken promptly.

As noted in Chapter 1, sensors are either passive or active devices. Passive sensors in single-element form include, among others, seismic-, acoustic-, strain-, humidity-, and temperature-measuring devices. Passive sensors in array form include optical- (visible, infrared 1 μm, infrared 10 μm) and biochemical-measuring devices. Arrays are geometrically regular clusters of WNs (e.g., following some topographical grid arrangement). Passive sensors tend to be low-energy devices. Active sensors include radar and sonar; these tend to be high-energy systems.

Sensing principles include, but are not limited to, mechanical, chemical, thermal, electrical, chromatographic, magnetic, biological, fluidic, optical, ultrasonic, and mass sensing. WNs may be exposed to hostile environments; the environment may include high temperatures, high vibration or noise levels, or corrosive chemicals. WNs may be incorporated in mobile robotic systems; they could also be integral to manufacturing systems. As discussed in Chapter 1, *embedded sensing* refers to the synergistic incorporation of microsensors in structures or environments; embedded sensing enables spatially and temporally dense monitoring of the system under consideration (e.g., an environment, a building, a battlefield). In biological systems, the sensors themselves must not affect the system or organism adversely [3.5]. The technology for sensing and control includes electric and magnetic field sensors; radio-wave frequency sensors; optical-, electrooptic-, and infrared sensors; radars; lasers; location and navigation sensors; seismic and pressure-wave sensors; environmental parameter sensors (e.g., wind, humidity, heat); and biochemical national security–oriented sensors. Typical sensor parameters (measurands) include:

- *Physical measurement.* Examples include two-axis magnetometers; light and ultraviolet intensity (photo resistor); radiation levels, radio, and microwave; humidity, temperature (thermistor), atmospheric pressure, fog, and dust; sound and acoustics; two-axis accelerometers, shock wave, seismic, physical pressure, and motion; video and image (visible or infrared); and location (GPS) and locomotion measurements.

- *Chemical and biological measurements.* Examples include the presence or concentration of a substance or agent at specified concentration levels (there are no less than 50 biological agents of interest [3.9]).

• *Event measurement.* Examples include determination of the occurrence of human-made or natural events, including cyber-level events; tracking of internal and external events.

Small, low-cost, robust, reliable, and sensitive sensors are needed to enable the realization of practical and economical sensor networks. Although a large number measurands are of interest for WSN applications, commercially available sensors exist for many of these measurands; one prominent exception is that a wide range of appropriate chemical sensors is not yet broadly available [3.8].

Sensor nodes come in a variety of hardware configurations: from nodes connected to a LAN and attached to permanent power sources, to nodes communicating via wireless multihop RF radio powered by small batteries [3.3]. The trend is toward very large scale integration (VLSI), integrated optoelectronics, and nanotechnology; in particular, work is under way in earnest in the biochemical arena. The goal of recent research and engineering is to build cubic millimeter (mm^3)–scale advanced WNs and motes. As shown in Figure 3.1, motes developed in the early 2000s were on the order of a cubic inch (this is approximately 16,387 mm^3). By 2007, researchers expect to have 1-mm^3 nodes able to operate in a functional network (e.g., SpeckNet research [3.1]).

3.2.2 Hardware and Software

Related to WN design, the following functionality typically needs to be supported: intrinsic node functionality; signal processing, including digital signal processing (e.g., FFT/DCT), compression, forward error correction, and encryption; control and actuation; clustering and in-network computation; self-assembly; communication; routing and forwarding; and connectivity management. To support this functionality, the hardware components of a WN include the sensing and actuation unit (single element or array), the processing unit, the communication unit, the power unit, and other application-dependent units. Figure 3.2 (which builds on Figure 1.3) shows hardware and software components of a typical sensing node.

As we noted in Chapter 1, the following are important sensor-node issues (refer to Table 1.1): sensor type, sensor power consumption, operating environment, computational and sensing capabilities, signal-processing capabilities, connectivity, and telemetry and control of remote devices. Clearly, the sensor node architecture, scope, and complexity depend on the application. Table 2.1 identified over 200 applications, many of which probably have their own sensor technology.

Sensors, particularly Smart Dust and COTS motes [3.2], have four basic hardware subsystems:

1. *Power.* An appropriate energy infrastructure or supply is necessary to support operation from a few hours to months or years (depending on the application).

2. *Computational logic and storage.* These are used to handle onboard data processing and manipulation, transient and short-term storage, encryption, forward

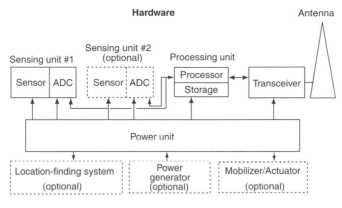

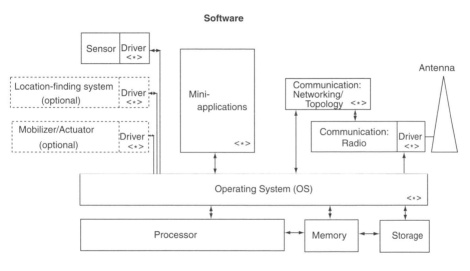

Figure 3.2 Hardware and software components of WNs.

error correction (FEC), digital modulation, and digital transmission. WNs have computational requirements typically ranging from an 8-bit microcontroller to a 64-bit microprocessor. Storage requirements typically range from 0.01 to 100 gigabytes (GB).

3. *Sensor transducer(s).* The interface between the environment and the WN is the sensor. Basic environmental sensors include, but are not limited to, acceleration, humidity, light, magnetic flux, temperature, pressure, and sound.

4. *Communication.* WNs must have the ability to communicate either in C1WSN arrangements (mesh-based systems with multihop radio connectivity among or between WNs, utilizing dynamic routing in both the wireless and wireline portions

of the network), and/or in C2WSN arrangements (point-to-point or multipoint-to-point systems generally with single-hop radio connectivity to WNs, utilizing static routing over the wireless network with only one route from the WNs to the companion terrestrial or wireline forwarding node). Researchers have developed many protocols specifically for WSNs. Transmission range, transmission impairments, modulation techniques, routing, and network topologies are issues of interest. Distances range from a few meters to a few kilometers; lower-layer communication protocols tend to be of the IEEE 802.11/802.15/802.16 class, although other methods have also been used. Throughput ranges from 10 to 256 kbps in most applications (some of the video-based application may require more bandwidth).

Sensors typically have five basic software subsystems:

1. *Operating system (OS) microcode* (also called *middleware*). This is the board-common microcode that is used by all high-level node-resident software modules to support various functions. As is generally the case, the purpose of an operating system is to shield the software from the machine-level functionality of the microprocessor. It is desirable to have *open-source operating systems* designed specifically for *WSNs*; these OSs typically utilize an architecture that enables rapid implementation while minimizing code size. TinyOS is one such example of a commonly used OS.

2. *Sensor drivers.* These are the software modules that manage basic functions of the sensor transceivers; sensors may possibly be of the modular/plug-in type, and depending on the type and sophistication, the appropriate configuration and settings must be uploaded into the sensor (drivers shield the application software from the machine-level functionality of the sensor or other peripheral).

3. *Communication processors.* This code manages the communication functions, including routing, packet buffering and forwarding, topology maintenance, medium access control (e.g., contention mechanisms, direct-sequence spread-spectrum mechanisms), encryption, and FEC, to list a few (e.g., see Figure 3.3).

4. *Communication drivers* (encoding and the physical layer). These software modules manage the minutia of the radio channel transmission link, including clocking and synchronization, signal encoding, bit recovery, bit counting, signal levels, and modulation.

5. *Data processing mini-apps.* These are numerical, data-processing, signal-value storage and manipulations, or other basic applications that are supported at the node level for in-network processing.

3.3 SENSOR TAXONOMY

Because of the variety of sensor types (sensor systems) that exist, a taxonomy is useful. The taxonomy in Table 3.1 is, in effect, an elaboration of Table 1.1. This

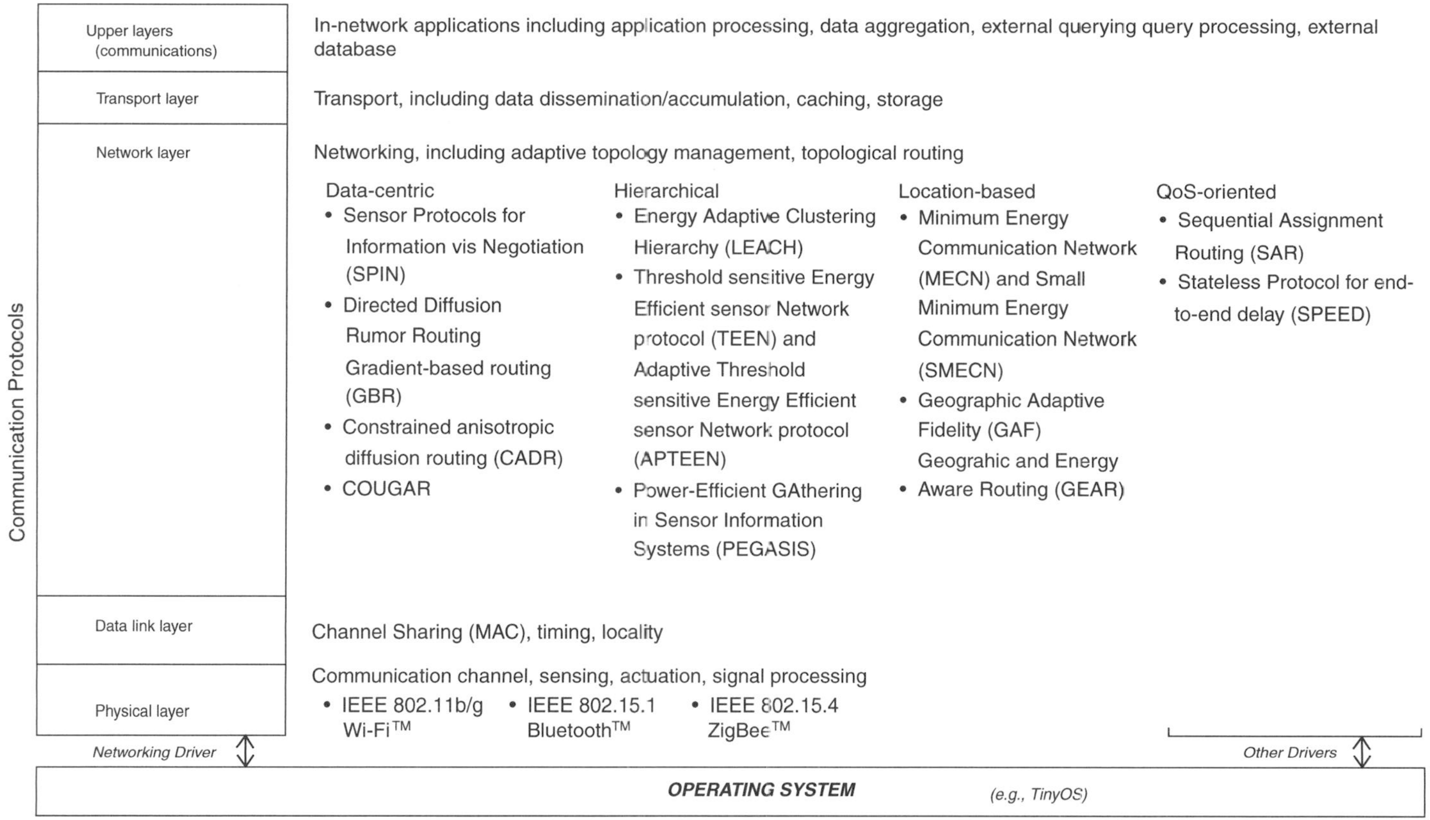

Communication Protocols

Upper layers (communications)	In-network applications including application processing, data aggregation, external querying query processing, external database
Transport layer	Transport, including data dissemination/accumulation, caching, storage
Network layer	Networking, including adaptive topology management, topological routing

Network layer

Data-centric
- Sensor Protocols for Information vis Negotiation (SPIN)
- Directed Diffusion Rumor Routing
- Gradient-based routing (GBR)
- Constrained anisotropic diffusion routing (CADR)
- COUGAR

Hierarchical
- Energy Adaptive Clustering Hierarchy (LEACH)
- Threshold sensitive Energy Efficient sensor Network protocol (TEEN) and Adaptive Threshold sensitive Energy Efficient sensor Network protocol (APTEEN)
- Power-Efficient GAthering in Sensor Information Systems (PEGASIS)

Location-based
- Minimum Energy Communication Network (MECN) and Small Minimum Energy Communication Network (SMECN)
- Geographic Adaptive Fidelity (GAF)
- Geograhic and Energy Aware Routing (GEAR)

QoS-oriented
- Sequential Assignment Routing (SAR)
- Stateless Protocol for end-to-end delay (SPEED)

Data link layer	Channel Sharing (MAC), timing, locality
Physical layer	Communication channel, sensing, actuation, signal processing • IEEE 802.11b/g Wi-Fi™ • IEEE 802.15.1 Bluetooth™ • IEEE 802.15.4 ZigBee™

Networking Driver ⇕

OPERATING SYSTEM (e.g., TinyOS)

Other Drivers ⇕

Figure 3.3 Some of the networking protocols supported by WNs.

81

TABLE 3.1 Basic Taxonomy of Sensor Nodes

Size of Sensor	Mobility of Sensor	Power of Sensor	Computation Logic; Storage Capability of Sensor	Sensor Mode	Communication Apparatus; Lower-Layer Protocols	Communication Apparatus; Upper-Layer Protocols
Very large (10^3 mm^3)	Fully mobile at deployment; fully mobile postdeployment	Self-replenishable, continuous	High-end processor (e.g., 64-bit micro); high-end storage (e.g., 100 GB)	High-end multimodal; physics	Multihop/mesh; hops in 10^1–10^2 m; IEEE MAC	Dynamic routing; data-centric
Large (10^2 mm^3)	Fully mobile at deployment; semimobile postdeployment	Self-replenishable, sporadic	Midrange processor (e.g., 16- or 32-bit micro); high-end storage	High-end multimodal; chemistry–biology	Multihop/mesh; hops in 10^2 to 10^4 m; IEEE MAC	Dynamic routing; hierarchical
Medium (10^1 mm^3)	Fully mobile at deployment; immobile postdeployment	Battery, 10^1 hours	Low-end processor (e.g., 8-bit micro); high-end storage	High-end multimodal; physics–chemistry–biology	Multihop/mesh; hops in 10^4 or more meters; IEEE MAC	Dynamic routing; location-based
Small (10^0 mm^3)	Semimobile at deployment; fully mobile postdeployment	Battery, 10^2 hours	High-end processor (e.g., 64-bit micro); midrange storage (e.g., 1 GB)	Midrange multimodal; physics	Multihop/mesh; hops in 10^1 to 10^2 m; special MAC	Dynamic routing; QOS-based

Scale	Mobility	Power	Processor/storage	Function	Network	
Very small (10^{-1} mm^3)	Semimobile at deployment; semimobile postdeployment	Battery, 10^3 hours	Midrange processor (e.g., 16- or 32-bit micro); midrange storage	Midrange multimodal; chemistry–biology	Multihop/mesh; hops in 10^2 to 10^4 m; special MAC	Static routing (single hop)
Ultrasmall (10^{-2} mm^3)	Semimobile at deployment; immobile postdeployment	Battery, 10^4 hours	Low-end processor (e.g., 8-bit micro); Midrange storage	Midrange multimodal; physics–chemistry–biology	Multihop/mesh; hops in 10^4 or more meters; special MAC	
Microscale (10^{-3} mm^3)	Immobile at deployment; fully mobile postdeployment	Battery, 10^5 hours	High-end processor (e.g., 64-bit micro); low-end storage (e.g., 0.01 GB)	Single function; physics	Single hop; hops in 10^1 to 10^2 m; IEEE MAC	
Nanoscale (<10^{-4} mm^3)	Immobile at deployment; semimobile postdeployment		Midrange processor (e.g., 16- or 32-bit micro); low-end storage	Single function; chemistry–biology	Single hop; hops in 10^2 to 10^4 m; IEEE MAC	
	Immobile at deployment; immobile postdeployment		Low-end processor (e.g., 8-bit micro); low-end storage	Single function; physics–chemistry–biology	Single hop; hops in 10^4 or more meters; IEEE MAC; Single hop; special MAC	

TABLE 3.2   Reduced-Complexity Taxonomy of Sensor Nodes

Size of Sensor, s	Mobility of Sensor, m	Power of Sensor, p	Computation Logic and Storage Capability of Sensor, cp	Sensor Mode, md	Communication Apparatus or Protocols of Sensor, cm
1 Large	1 Mobile	1 Self-replenishable	1 High-end processor and storage	1 Multimodal, physics	1 Multihop/mesh with dynamic routing
2 Small	2 Static	2 Battery, hours–days	2 Midrange processor and storage	2 Multimodal, chemistry/biology	2 Single hop with static routing
3 Microscopic		3 Battery, weeks–months	3 Low-end processor and storage	3 Single function, physics	
4 Nanoscopic		4 Battery, years		4 Single function, chemistry–biology	

taxonomy is somewhat daunting since there are $8 \times 9 \times 7 \times 9 \times 9 \times 10 \times 5 = 2{,}041{,}200$ cases or combinations. However, the classification "buckets" are reasonable, and a large majority of the combinatorial combinations are, in fact, valid. To reduce the scope of the taxonomy, we suggest the use of the modified classification shown in Table 3.2; here one has only $4 \times 2 \times 4 \times 3 \times 4 \times 2 = 768$ cases or combinations. For example, a s(2)m(2)p(3)cp(2)md(1)cm(1) WN is a system that is small, static, battery-powered, has multiple measurands, and supports multihop networking.

3.4   WN OPERATING ENVIRONMENT

As we saw in Chapter 1, networking implies a need to support physical and logical connectivity. In WSNs, physical connectivity is supported over a wireless radio link of one or more hops, at a distance of tens, hundreds, or thousand of meters. Logical connectivity has the goal of supporting topology maintenance and multihop routing (when present). The design and engineering of WNs clearly needs to take into account all the issues described in Section 3.2 as well as in this section.

Sensor nodes have to deal with the following resource constraints [3.3] (see also Table 3.3):

- *Power consumption.* Almost invariably, WNs have a limited supply of operating energy; it follows that energy conservation is a key system design consideration.
- *Communication.* The wireless network usually has limited bandwidth; the networks may be forced to utilize a noisy channel; and the communication channel may be relegated to an unprotected frequency band. The implications

TABLE 3.3   Design Constraints or Requirements for WSNs and WNs

WSN/WN Requirement	Motivation
Collaborative data processing	A factor that distinguishes WSNs from simple ad hoc networks is that the goal in WSNs is detection or estimation of specified events, not just communications. One needs to provide scalable, fault-tolerant, flexible data access and intelligent data reduction [3.3]. This drives the overall architecture because detection and estimation often require fusing data from multiple sensors; data fusion requires the transmission of data and control messages. Quantification of sensor data, including limits of detection, calibration, interferences, sampling, and verification of accuracy, also needs to be taken into account [3.5].
Constrained energy use	In many applications the WNs are deployed in remote areas; in these cases, the lifetime of a node may be determined by the battery life; this in turn requires a minimization of energy consumption.
Large topology support	Networks of 10,000 or even 100,000 nodes are envisioned for some applications. Fortunately, most WSNs/WNs are stationary (aside from the deployment of sensors on the ocean surface or the use of mobile, unmanned, robotic sensors in military operations).
Querying capabilities	A data-consumption entity may need to query an individual node or group of nodes for information collected in the region. Because it may not be feasible to transmit a large amount of the data across a network, various local sink nodes need to collect the data from a given area and create summary messages to reply to the query.
Self-organization	It is typically a requirement that WSNs be able to self-organize: Given the large number of nodes and their potential placement in hostile locations, manual configuration is typically not feasible. Also, nodes may fail (from lack of energy or from physical destruction), and new nodes may join the network: the network must be able to reconfigure itself so that it can continue to operate properly and support reliable connectivity.

Source: Adapted from [3.4].

are limited reliability, poor quality of service (e.g., high latency, high variance, high frame loss), and security exposure (e.g., denial of service, jamming, interference, high bit-error rates).

- *Computation.* WNs typically have limited computing power and memory resources. The implications are restrictions on the types of data-processing algorithms that can run on a sensor node. This also limits the scope and volume of intermediate results that can be stored in the WNs. Research aims at developing a distributed data management layer that scales with the growth of sensor interconnectivity and computational power on the sensors; the goal is to deploy mechanisms that reside directly on the sensor nodes and create the abstraction of a single processing node without centralizing data or computation.

- *Uncertainty in measured parameters.* Signals that have been often have various detected or collected degrees of intrinsic uncertainty. Desired data may be commingled with noise and/or interference from the environment. Node malfunction could collect and/or forward inaccurate data. Node placement (particularly in ad hoc networks without mobility) may impair operation and bias individual readings.

Some of the intrinsic factors that the design constraints or requirements that WSNs and WNs need to take into account include the following:

- WNs may be deployed in a dense manner (close proximity), implying communication complexity (e.g., in support of packet forwarding and topology management)
- For military and/or national security applications, WNs need to support rapid deployment; the deployment must be supportable in an ad hoc fashion; and the environment is expected to be highly dynamic.
- WNs may be prone to failure. Unattended, untethered, self-powered low-duty-cycle systems are typical, yet some WSNs require sensing systems that are long-lived and environmentally resilient.
- As just noted, WNs are limited in power, computational capacity, and memory. Communication circuitry and antennas are the primary elements that use up most of the energy.
- The topology that the WNs need to maintain may change very frequently. Communication links may be expensive (not only from an electromagnetic spectrum perspective, but also in terms of the operational support of the requisite infrastructure); the bandwidth may be limited; and as just noted, the power availability at the sensor may be limited and/or expensive in reference to supporting a high-capacity, high-range link (i.e., to feed a high-power antenna).
- WNs may not have global addresses because of the potentially large number of sensors and overhead needed to support such global addresses (IPv6 could be applicable in this context).
- WNs require special routing and data dissemination mechanisms (e.g., data-centric, hierarchical, and/or location-based routing).
- WNs often require in-network processing, even while the data are being routed. One wants to be able to perform data processing in the network in the proximity of the source of the data, and then forward only summarized,

aggregated, fused, and/or synthesized results. Typical functionality involves signal processing, data aggregation, data fusion, and data analysis. There is also an interest in database management, including querying mechanisms and data storage and warehousing.

- Arrays of ultralow-power wireless nodes may be incorporated in reconfigurable networks with high-speed connectivity to processing centers for decision and responsive action [3.5].

3.5 WN TRENDS

For WSNs to achieve wide-scale deployment, the size, cost, and power consumption of the nodes must decrease considerably and the intelligence of the WNs must increase [3.6]. To meet evolving functional requirements of the various user communities, it will be necessary for sensor systems to leverage and incorporate advances in adjacent technologies, such as nanofabrication, biosystems, massively distributed networks, ubiquitous computing, broadband wireless communications, and information and decision systems [3.5].

Evolving requirements for new WSNs and WNs include, among others: (1) the ability to respond to new toxic chemicals, explosives, and biological agents; (2) enhanced sensitivity, selectivity, speed, robustness, and fewer false alarms; and (3) the ability to function, perhaps autonomously, in unusual, extreme, and complex environments. These needs can be addressed by the design and synthesis of functionalized receptors and materials, resulting in next-generation devices. The materials may be of varying porosity, enabling them to detect single toxic compounds in complex mixtures or physical configurations that have surfaces with microchannels for microfluidic discrimination. Advanced biological, chemical, and materials research can be brought to bear on this challenge, including the design of functional nano- and mesoscale complex structures (e.g., quantum dots, nanowires, gels). Robustness under anticipated manufacturing schemes is also required [3.5].

Miniaturization, manufacturability, and cost are also critical issues. Integration of sensors, processors, energy sources, and the communications network interface on a chip would facilitate the exchange of sensor data and critical information with the outside world. Information extraction may involve detection of events or objects of interest, estimation of key parameters, and human-in-the-loop or closed-loop adaptive feedback [3.5]. Some of the goals (e.g., as defined by the PicoRadio effort at UC–Berkeley [3.7]) are to develop mesoscale low-cost (i.e., <50 cents) transceivers for ubiquitous wireless data acquisition that minimize power or energy dissipation [i.e., minimize energy (<5 nJ/(correct) bit)] for an energy-limited source and minimize power (i.e., $<100\,\mu W$ for a power-limited source, enabling energy scavenging) by using the following strategies: self-configuring networks, fluid trade-off between communication and computation, an integrated system-on-a-chip (SOC) approach, and aggressive low-energy architectures and circuits.

Standardization is important. As the definition of sockets has made the use of communication services on the Internet independent of the underlying protocol stack, communication medium, and even operating system, the application interface one needs for WSNs should be an abstraction that is offered to any sensor network application and supported by any sensor network platform [3.7]. Research and engineering activity now under way seeks to advance fundamental knowledge in new sensor technologies, including sensors for toxic chemicals, explosives, and biological agents; sensor networking systems in a distributed environment; the integration of sensors into commercial systems; and the interpretation and use of sensor data in decision-making processes [3.5]. Table 3.4 provides a partial list of near-term research efforts as sponsored by U.S. government agencies.

Of late, one has seen targeted efforts to develop chemical sensors for sensor networks, particularly for monitoring soil contamination and for habitat monitoring. Specifically, one needs an array of miniaturized chemical sensors to monitor the flow of contaminants accurately (e.g., see Figure 3.4). Optimally, one is interested in developing microscale liquid chromatography systems [3.8]. According to published reports, the U.S. Department of Homeland Security (DHS) is coordinating an effort for the end-of-decade deployment of a nationwide sensor network to provide a real-time early-warning system for a plethora of chemical, biological, and nuclear threats across the United States. Planners at DHS are working on developing capabilities to deal with multifaceted threats targeted at airports, subways, and buildings; they are also looking at issues related to water sources, animal herds, and flocks of birds that could spread contaminants or harmful biological agents. This type of technology is currently under development [3.9].

National research laboratories have been working on core issues in materials, sensors, networks, and electronics, and have already established field trials of prototype networks. The multifaceted nature of the global threat has led researchers to consider a system that consists of a suite of different types of sensors. Researchers are planning to use MEMSs and nanotechnology for low-cost, high-reliability, and high-accuracy biological and chemical sensors. In one approach, researchers are studying hybrid sensors that use surface-chemical detection as a first trigger, which could then use technology on the same device for more time-consuming techniques, such as DNA testing. Other researchers are studying the use of infrared or ultraviolet spectrum analysis as well as biometric sensors that mimic human cells to create test reactions. Further into the future, MEMS technology is seen as having promise for creating miniature benchtop labs on a chip. Sensors could use polymer- or gel-coated silicon devices to trap targeted chemicals, then send the agents through fluidic channels to on-chip arrays of surface-acoustic-wave detectors. A follow-on device would integrate the fluidics, surface acoustic waves, and support electronics on a single device [3.9]. Other research teams are exploring nanotechnology to deliver new sensor materials (e.g., researchers at the Pacific Northwest National Laboratory have

TABLE 3.4   Partial List of Near-Term Research Efforts as Sponsored by U.S. Government Agencies

Designs, materials, and concepts for new sensors and sensing systems	Examples include novel sensing materials and devices; the design of solid and liquid surfaces with molecular recognition, long lifetime, and regenerability of the sensing site; biomimetic sensors, including hybrids consisting of proteins, enzyme fragments and components, bioorganometallics, or other biocatalysts that can be linked to surfaces; bioMEMS; sensors for toxic agents (biological, chemical, radiation); sensors for operation in harsh environments; wireless sensors; chip-based systems incorporating multiple sensors, computation, actuation, and wireless interfaces; sensor systems capable of remote activation and interrogation; sensor power sources; novel optical imaging concepts; novel techniques for metrology at the nanoscale; new modeling and simulation tools; new techniques for on-sensor self-calibration and self-test; enhanced specificity to maximize accuracy and minimize false alarms; and new methods for sensor fabrication, manufacture, and encapsulation.
Arrayed sensor networks and networking	This area includes: Enabling networking technologies for distributed wireless and wired sensor networks Scalable and robust architectures Design Automated tasking Querying techniques Adaptive management and control of sensor nodes Design trade-offs and performance optimization in resource-constrained sensor networks Design of ultralow-power processing nodes for local information management Investigation of localized versus distributed versus centralized processing of sensor data Common building blocks and interfaces for sensor networking Strategies for using heterogeneous sensor and network nodes to enhance performance and reduce false alarms Security and authentication for resource-constrained sensor networks Embedded and hybrid systems Application-specific network and system services, including data-centric routing, attribute-based addressing, location management, and service discovery Energy-efficient media access, error control, and traffic management protocols

(Continued)

TABLE 3.4   (*Continued*)

	Mobile sensor networks
	Scalable reconfigurability and self-organization
Interpretation, decision, and action based on sensor data	Examples include decision theory for intelligent use of sensed information; detection and identification of false alarms; feedback theory; development of new statistical algorithms, sampling theories, and supervisory control systems tailored to needs; concepts for optimal sensor locations for effective process and system control; mathematical hybrid system tools for monitoring distributed networks of large arrays of sensors and actuators; handheld diagnostic kits; and pattern recognition and state estimation. System-level sensor applications include biomedical health monitoring, diagnostic, and therapeutic systems; image-guided surgery; health monitoring systems for civil structures; crisis management sensor systems; surveillance technology; robotics; mobile sensors; tracking and monitoring of mobile units (endangered species, inventory control, transportation); and sensor assessment (reliability, verification, validation).

Source: National Science Foundation materials [3.5].

developed nanosized preconcentrators for nerve agents, botulism, and other toxins) [3.9].

Sandia has been testing handheld sensors designed to detect chemical-weapons agents on the battlefield with high sensitivity; the detection window is 2 minutes or less. The lab has been asked to explore adding networking and GPS capability

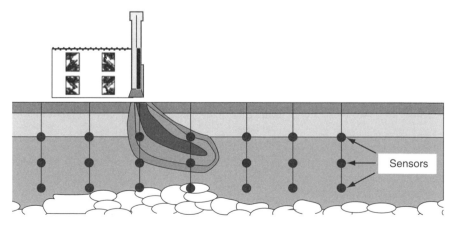

Figure 3.4   Sensor array for chemical contamination analysis.

to those sensors so that they could be mounted on military vehicles, creating a mobile battlefield sensor network. The expectation is that by the turn of the present decade, a bio smoke alarm detector will be ready for commercial deployment [3.9].

On the networking front, researchers are considering peer-to-peer network with multilevel security and quality-of-service guarantees, spanning terrestrial wireless, wireline, and satellite links. The underlying network architecture for a national sensor network has been studied at Oak Ridge National Labs. The aim is to use off-the-shelf technology as much as possible and to leverage existing infrastructure, such as the 30,000 cellular towers and 100,000 cellular base stations in the United States today. However, developing quality-of-service guarantees and multilevel security for a hybrid wired, wireless, and satellite network is a challenge [3.9]. Several pilot sensor network projects are being field tested, including systems developed by Los Alamos and UC–Berkeley researchers to safeguard crops. Trial sensor networks are also in place in Boston subways, at the San Francisco airport, and on the Miami docks. The Washington subway recently went operational with a chemical-sensor system developed by Sandia and Argonne National Laboratories in Chicago [3.9].

3.6 CONCLUSION

In this chapter we looked at basic sensor node technology along with a taxonomy of sensor types. Some current trends were also discussed.

REFERENCES

[3.1] D. Roman, "Scottish Universities Plan Speckled Computing Net," *EE Times*, Oct. 27, 2003.

[3.2] *The Scientist and Engineer's Guide to TinyOS Programming*, University of California–Berkeley, http://tinyos.org. This book was developed as an open source, freely available manuscript on the TinyOS Documentation Project.

[3.3] "Cougar: The Sensor Network Is the Database," Cornell University, Ithaca, NY, http://www.cs.cornell.edu/database/cougar/.

[3.4] "Smart Sensor Networks," National Institute of Standards and Technology, Gaithersburg, MD, http://w3.antd.nist.gov/wahn_ssn.shtml.

[3.5] Sensors and Sensor Networks, Program Solicitation, NSF 03-512, Mar. 6, 2003, National Science Foundation, Directorate for Engineering, http://www.nsf.gov/cgi-bin/getpub?gpg; also, NSF Publications Clearinghouse, pubs@nsf.gov.

[3.6] J. M. Rabaey, M. J. Ammer, J. L. da Silva Jr., D. Patel, S. Roundy, "PicoRadio Supports Ad Hoc Ultra-low Power Wireless Networking," *Computer*, July 2000; wireless sensor network research at the Berkeley Wireless Research Center, http://bwrc.eecs.berkeley.edu/Research/Pico_Radio/.

[3.7] J. M. Rabaey, "Ultra Low-Power Computation and Communication Enables Ambient Intelligence," presented at the Smart Objects Conference, Grenoble, France, Apr. 2003.

[3.8] Center for Embedded Networked Sensing, University of California–Los Angeles, http://www.cens.ucla.edu/portal/micro_nano_sensor_tech/.

[3.9] R. Merritt, "Planned U.S. Sensor Network Targets Terror Threats," *EE Times*, July 14, 2003.

4

WIRELESS TRANSMISSION
TECHNOLOGY AND SYSTEMS

4.1 INTRODUCTION

In this chapter we look at radio-channel-related issues. It should immediately be noted that to maximize the opportunity for widespread and cost-effective deployment of WSN, one needs to make use of existing and/or emerging commercial off-the-shelf (COTS) wireless communications and infrastructures rather than having to develop an entirely new, specially designed apparatus. WSNs can use a number of wireless COTS technologies, such as Bluetooth/Personal Area Networks (PANs), ZigBee, wireless LANs (WLAN)/hotspots, broadband wireless access (BWA)/ WiMax, and 3G.

Given this pragmatic perspective, we focus here less on the science of radio transmission per se as a discrete system component and more on an integrated system-level view of the field. In other words, we explore the use of the just-named technologies as a plug-and-play system integration opportunity more than looking at the fundamentals of modulation, transmission, encoding, radio impairments, and so on. Stated differently, the developer of WSN systems should not be required to have a deep understanding of radio science (beyond basic issues such as power, range and coverage, bandwidth, performance, security, and a few other factors), but rather, which off-the-shelf wireless systems already defined by various standards bodies (e.g., Bluetooth, Wi-Fi, WiMax, ZigBee/IEEE 802.15.4) can be used by way of employing and/or integrating preconfigured chipsets and ICs (integrated circuits), antennas, drivers, and protocol machinery.

Wireless Sensor Networks: Technology, Protocols, and Applications, by Kazem Sohraby, Daniel Minoli, and Taieb Znati

Consistent with this perspective, in this chapter we look at some macro-level issues, while the chapters that follow provide more in-depth technical information. In Section 4.2 we provide a basic primer on radio technology; Appendix A provides some additional details related to modulation. In Section 4.3 we survey off-the-shelf technologies (IEEE family) that can be used by WSNs. Chapter 5 will expand on these concepts.

4.2 RADIO TECHNOLOGY PRIMER

This section comprises a terse primer on radio technology.

The electromagnetic spectrum provides an unguided medium (channel) for point-to-point and/or broadcast radio transmission. Radio transmission is usually (frequency)-bandlimited by design. The analog bandwidth of the channel (the slice of electromagnetic frequency domain used) determines how much information (analog or digital) can be transmitted over the channel. A transmission channel in general, and a radio-based channel in particular, is never perfect because it is subjected to external (and even internal) noise sources; noise has a tendency to degrade, disrupt, or otherwise affect the quality of an intelligence-bearing signal. A lot of radio-transmission engineering has to do with how to deal with the noise problem; the goal is nearly always to optimize the signal-to-noise ratio, subject to specified constraints (e.g., bandwidth requirements, cost, reliability, power consumption, equipment and antenna size).

4.2.1 Propagation and Propagation Impairments

Issues of interest in radio design include, among others, propagation, impairments, environment (i.e., indoors–outdoors, unobstructed–obstructed, benign–hostile, etc.), sensitivity, antenna design, channel bandwidth (analog and/or digital), and frequency of operation. Many design factors (e.g., propagation, attenuation, impairments) are related parametrically to the frequency band in use. In particular, directionality becomes more of an issue at higher frequency ranges; also, generally, bandwidth increases as one moves to higher-frequency bands (given that larger portions of the spectrum are in principle available). For the purpose of this primer, we focus on operation at 2.4 GHz. However, as noted, the commercial WSN developer need not worry about all of these issues at a fundamental level if he or she employs off-the-shelf technology (beyond basic considerations about distance, antenna type, bit-error rate, bandwidth, and power requirements[1]).

The most basic model of radio-wave propagation typically found in WSN environments involves the *direct* or *free-space wave* (see Figure 4.1). In this model, radio waves emanate from a point source of radio energy, traveling in all directions

---

[1]What we mean is that many of the relevant issues have already been studied, addressed, traded off, and optimized by the developers of the particular standard in question.

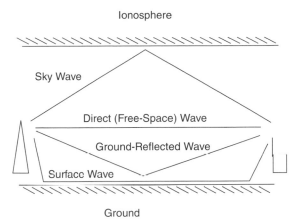

Figure 4.1 Radio propagation modes. [*Note:* For WSNs the direct (free-space) wave is the most common.]

in a straight line, filling the entire spherical volume of space with radio energy that varies in strength with a $1/(\text{distance})^2$ rule (or 20 dB per tenfold increase in distance) [4.1]; attenuation in environments that are not free space (e.g., waters, coaxial cable, heavily wooded areas, confined rooms or structures) is considerably more severe.

Three basic physical mechanisms affect radio propagation [4.2]:

1. *Reflection.* A propagating wave impinges on an object that is large compared to the wavelength. (e.g., the surface of the Earth, buildings, walls).
2. *Diffraction.* A radio path between the transmitter and receiver is obstructed by a surface with sharp irregular edges; waves bend around the obstacle, even when line of sight (LOS) does not exist.
3. *Scattering.* Objects smaller than the wavelength of the propagating wave are encountered along the way (e.g., foliage, street signs, lampposts).

These phenomena cause radio signal distortions and signal fading as described in Table 4.1. Signal strength fluctuations caused by the fact that the composite signal received comprises a number of components from the various sources of reflections from different directions as well as scattered and/or diffracted signal components affect both mobile and stationary receivers, whether the receivers are indoors or outdoors. In this phenomenon, called *multipath*, signal fluctuations can be as much as 30 to 40 dB. The intrinsic electromagnetic (radio) signal strength attenuation caused by these phenomena is called a *large-scale effect*; signal-strength fluctuations related with the motion of the broadcasting or receiving antenna are called *small-scale effects*.

Reflection, diffraction, and scattering all give rise to additional radio propagation paths beyond the direct line-of-sight path between the radio transmitter and receiver;

TABLE 4.1   Basic Phenomena Affecting Signals

Phenomenon	Description
Reflection	A phenomenon that occurs when a propagating electromagnetic wave impinges upon an object that is large compared to the wavelength of the propagating wave. Reflections occur from the surface of the Earth and from buildings and walls.
Diffraction	A phenomenon that occurs when the radio path between the transmitter and receiver is obstructed by a surface that has sharp irregularities (edges). The secondary waves resulting from the obstructing surface are present throughout the space and even behind the obstacle, giving rise to a bending of waves around the obstacle, even when a line-of-sight path does not exist between transmitter and receiver. At high frequencies, diffraction, like reflection, depends on the geometry of the object as well as the amplitude, phase, and polarization of the incident wave at the point of diffraction.
Scattering	A phenomenon that occurs when the medium through which the wave travels consists of objects with dimensions that are small compared to the wavelength and where the number of obstacles per unit volume is large. Scattered waves are produced by rough surfaces, small objects, or by other irregularities in the channel. In practice, foliage, street signs, and lampposts induce scattering in a mobile communications system.

Source: Adapted from [4.3].

multipath arises when more than one path is available for radio signal propagation [4.3]. Metallic materials as well as dielectrics (or electrical insulators) cause reflections. When multiple signal propagation paths exist, the actual signal level received is the vector sum of all the signals incident from any direction or angle of arrival. Some signals will aid (constructively reinforce) the direct path; others will subtract (destructively interfere with or vector-cancel out) from the direct signal path (see Table 4.2).

TABLE 4.2   Multipath Types

Type	Description
Specular multipath	Arises from discrete, coherent reflections from smooth metal surfaces. Can cause complete signal outages and radio dead spots within a building; the problem is especially difficult in underpasses, tunnels, stairwells, and small enclosed rooms.
Diffuse multipath	Arises from diffuse scatterers and sources of diffraction (the visible glint of sunlight off a choppy sea is an example of diffuse multipath). It gives rise to a background noise level of interference.

TABLE 4.3   Fade Factors

Type	Description
Large-scale fades	Attenuation: in free space, power decreases as a function of $1/d^2$ ($d =$ distance from the transmitting antenna)
	Shadows: signals blocked by obstructing structures
Small-scale fades	Rapid changes in signal strength over a small area or time interval due to multipath
	Random frequency modulation due to varying Doppler shifts on different multipath signals
	Time dispersion (echoes) caused by multipath propagation delays:
	Multipath propagation yields signal paths of different paths with different times of arrival at the receiver
	Spreads (smears) the signal; can cause intersymbol interference and limits the maximum symbol rate (signals related to previous bit or symbol interfere with the next symbol)
	Typical values of delay spread: open spaces, $<0.2\ \mu s$; suburban spaces, $0.5\ \mu s$; urban spaces, $3\ \mu s$
	Frequency-selective fading and Rayleigh fading:
	Combination of direct and out-of-phase reflected waves at the receiver yields attenuated signals.
	Addressed via antenna diversity (use two antennas a quarter-wavelength separated to combine received signals) and/or equalization (subtract delayed and attenuated images of the direct signal from the received signal—should be done adaptively to determine what these subtractions should be, since they change as the mobile devices moves around).

The impact of mobility on transmission characteristics is fairly difficult to model exactly. Channel performance varies with user location and time, and the radio propagation pattern is complex. One needs to deal with multipath scattering from nearby objects, shadowing from dominant objects, and attenuation effects from various physical phenomena. All of these factors result in rapid fluctuations of received power; even when the device mobile is stationary, the signals received may fade, due to movement of surrounding objects [4.2]. Table 4.3 highlights some of the issues.

Figure 4.2 describes pictorially issues related to outdoor propagation. For indoor propagation applications, the signal decays much faster: walls, floors, and furniture attenuate or scatter radio signals; also, the coverage is restricted to the local environment by walls and the like. The path loss formula is [4.2]

$$\text{path loss} = \text{unit loss} + 10n \log(d) = kF + IW$$

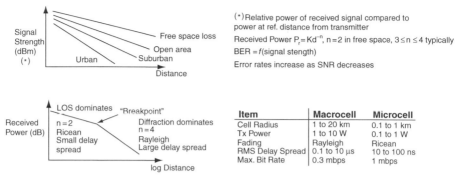

Figure 4.2   Outdoor radio propagation. (Based in part on [4.2].)

where unit loss = power loss (dB) at a 1-m distance (30 dB)

n = power-delay index

d = distance between transmitter and receiver

k = number of floors that the signal traverses

F = loss per floor

I = number of walls that the signal traverses

W = loss per wall

Additional contributing factors include the following [4.2]:

- People moving around (additional multipath-induced attenuation of up to 10 dB)
- Buildings with few metal and hard partitions: root-mean-square (rms) delay spread of 30 to 60 ns (equaling several Mbps without equalization)
- Buildings with metal or open aisles: rms delay spread of up to 300 ns (hundreds of kbps without equalization)
- Between floors:
  - Concrete or steel flooring yields less attenuation than that of steel plate flooring
  - Metallic-tinted windows yield greater attenuation
  - 15 dB for first-floor separation, 6 to 10 dB for the next four floors, 1 to 2 dB for each additional floor of separation

The indoor signal strength received depends on the office plan, construction materials, density of personnel, furniture, and so on (e.g., wall losses, 10 to 15 dB; floor losses, 12 to 27 dB; delay spread, varies between 15 and 100 ns, requiring sophisticated equalization techniques to achieve acceptable bit-error rates). Table 4.4 depicts signal attenuation values for signals typically used in networking and telecom applications. A drawback of higher-frequency bands (e.g., 5 GHz for IEEE 802.11a applications) compared to lower-frequency bands (e.g., 2.4 GHz for

TABLE 4.4   Signal Attenuation Due to Typical Obstacles

Wall Type	Frequency	Transmission Loss (dB)
Exterior wood frame wall	800 MHz	4–7
	5–6 GHz	9–18
Brick, exterior	4–6 GHz	14
Concrete block, interior	2.4 GHz	5
	5 GHz	5–10
Gypsum board, interior	2.4 GHz	3
	5 GHz	5
Wooden floors	5 GHz	9
Concrete floors	900 MHz	13

IEEE 802.11b/g applications) is the shorter wavelength of the signal at the higher band. It turns out that short-wavelength signals have more difficulty propagating through physical obstructions encountered in an office (walls, floors, and furniture) than do those at longer wavelengths.

There are few "RF-friendly" buildings that are free of multipath reflections, reflections from internal partitions, absorption from various office materials, diffraction around sharp corners, and scattering from wall, ceiling, or floor surfaces. Radio-wave propagation inside smooth-walled metal buildings can be so problematic that radio dead spots can exist to the point where the signal is virtually nonexistent. The dead spots arise because of almost perfect, lossless reflections from smooth metal walls, ceilings, or fixtures that interfere with signals radiated directly. The dead spots exist in three-dimensional space within the building, and motions of only a few inches can alter reception from a state of no signal to a state of full signal. Proper functioning of the radio communication link requires that multipath be minimized or eliminated [4.1]. Figure 4.3 depicts a simple example of indoor multipath.

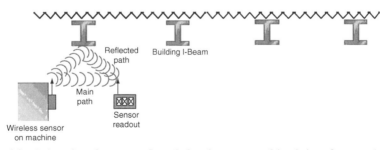

Figure 4.3   Indoor interference: reflected signal creates multipath interference. A factory building I-beam reflects transmission of wireless sensor data to a digital readout. The readout receives both the main and reflected signals that interfere with each other, disrupting the display. The readout can be moved out of the reflected signal path. Typically, movement of only a few inches is all that is required for better signal reception. If the readout cannot be moved, repositioning the antenna will have the same effect. With interference eliminated, the readout works normally (From [4.4].).

Error bursts are an outcome of fades in radio channels. Doppler-induced frequency or phase shifts due to motion can also cause loss of synchronization. Errors increase as the bit period approaches the delay spread. The typical acceptable BER for data communications is 10^{-6}; in some wireless situations this goal may not be met on a consistent basis. Strategies for overcoming errors include antenna diversity, forward error correction techniques, and traditional automatic repeat request (retransmission protocol for blocks in error) [4.2]. The outdoor-to-indoor penetration or *building loss* depends on building materials, orientation, layout, height, percentage of windows, and transmission frequency. The strength of the signal received increases with increasing building height; the penetration loss decreases with increasing frequency (e.g., 6 dB loss through windows).

In an industrial environment, care is needed when placing sensors in order to minimize interference. One needs to keep WNs away from other sources of radio-frequency interference (RFI), such as brush-type electrical motors, other radio transmitters or transceivers, or unshielded computer equipment and/or cables. Sensors that must be located near such devices should connect to the transceiver via a short piece of shielded cable so that they can stay as far away as possible from the source of the RFI. In a factory environment, large iron and steel structures may create multipath problems. As noted, multipath propagation occurs when nearby metal reflects the radio signal in the same way that a mirror reflects light. The receiver detects multiple signals simultaneously—the original and the reflections—and cannot decode any of them. Moving the receiving or transmitting antenna just a few inches is sometimes enough to fix this problem [4.4]. More generally, RF multipath problems can be mitigated in a number of ways [4.1]:

1. *Radio system design:* redundant paths for each receiver, if possible
2. *Antenna system design:* dual diversity antennas used at each receiver
3. *Signal/waveform design:* spread-spectrum radio design with the highest feasible chip rate
4. *Building/environment design:* not much can be done in this area unless RF-friendly greenfield buildings are constructed

Interference can also be caused by other legitimate or illegitimate users of a given frequency band. Interference can occur when a user starts to broadcast signal in a band while in proximity to other transmitters and/or receivers. (The scope of proximity depends on the frequency band, the power utilized by transmitting entities, and the modulation scheme, among other factors.) In the United States, most frequency bands are assigned by the Federal Communications Commission (FCC) to a specific (private) user or organization; (a few) other frequency bands can be utilized by anyone. In the former, while interference can be caused by accidental spillage of signal and/or malicious injection (e.g., jamming), the source of the interference can be stopped legally (or militarily). In the latter case there is no recourse because the band is open to anyone and coexistence is managed by "good citizenship." In the latter case, unfortunately there is also intraband interference between various technologies (e.g., IEEE 802.15.1 Bluetooth technology can interfere with IEEE 802.11b/g-based systems). As just stated, most frequency bands require a license from the FCC (the license is needed for transmission, but generally not for reception). The license is granted (usually for a fee) to a specific user and/or organization. Use of the industrial, scientific and medical (ISM) band (at 2.4 GHz—more exactly, 2.412 to 2.484 GHz), and of the Unlicensed Network Information Infrastructure (U-NII) band (at 5 GHz) does not require a license. However, there still are technical guidelines that must be followed in terms of the radiated power, radiation pattern, and so on.

The current frequency-assignment system under which the FCC operates was formulated in the 1920s; under this system, different radio bands are assigned to different services and licenses are then required to operate inside those bands [4.45]. In recent years there has been interest on the part of the FCC to explore innovative ways to open new spectrum to commercial unlicensed use. Examples include the release of new spectrum in the 5-GHz U-NII band in 2003 as well as the opening up of 7.5 GHz of bandwidth for ultrawideband (UWB) signaling in the region between 3.1 and 10.6 GHz. Although the power levels allowed for UWB are extremely low—a roof of -41 dBm—the move marked the first time the FCC had allowed unlicensed use across otherwise licensed bands [4.45]. Cognitive radio (CR) technology is a new way to look at this issue (this topic is discussed later in the chapter).

4.2.2 Modulation

Modulation is the overlay of an intelligent signal over an underlying carrying signal, which is then transmitted over the medium in question (be it a cable, wireless, or fiber-optic medium). Baseband applications are those applications where the coded signal is carried directly over a medium without having to overlay it onto a carrier signal. Non-baseband systems use modulation; baseband systems do not. In traditional environments modulation allows transmission over long distances (e.g., tens to hundreds of miles); baseband systems usually are limited to the carriage of information over a fraction of a mile. Traditional wired LAN systems are baseband systems: The signal is encoded by some appropriate mechanism (e.g., Manchester encoding) and then transmitted over unshielded twisted-pair cable. Analog radio and TV transmission use modulation.

Three types of modulation typically used in radio applications are amplitude modulation (AM), frequency modulation (FM), and phase modulation (PM). In AM, the amplitude of the carrying signal is modulated (summed over or superimposed) by (the amplitude of) the incoming intelligence-bearing signal. In FM, the frequency of the carrying signal is modulated (summed over or superimposed) by (the frequency of) the incoming intelligence-bearing signal. In PM, the phase of the carrying signal is modulated (summed over or superimposed) by (the phase of) the incoming intelligence-bearing signal.

In an AM environment, when the incoming intelligence-bearing signal is digital (a sequence of 0 and 1 values), the modulation process is called *amplitude shift keying* (ASK). In an FM environment, when the incoming intelligence-bearing signal is digital, the modulation process is called *frequency shift keying* (FSK). In a PM environment, when the incoming intelligence-bearing signal is digital, the modulation process is called *phase shift keying* (PSK). When the incoming signal is interpreted as a sequence of n bits at a time (e.g., 00, 01, 10, 11; or 000, 001, 010, 011) and a combination of PSK and ASK techniques are used, the modulation process is called *quadrature amplitude modulation* (QAM). *Note:* In a digital environment the concept of the carrying signal, which was so prominent in the analog context, degenerates and the process is seen as giving rise to a sequence of discrete states (implemented in amplitude, frequency, or phase values). In all of these cases the modulation is said to be digital. In these situations the design goal is to maximize channel throughput by making the incoming digital signal pulse as dense as possible (time axis as small as possible) or by finding a way to encode groups of incoming bits over a single signal change (also known as *baud*).

The maximum digital capacity C of a single-carrier system with spectral bandwidth W is defined by *Shannon's equation:*

$$C = W \log_2(1 + S/N)$$

S is the signal power received and N is the noise power (the channel here is assumed to be an additive white Gaussian noise channel). In a typical environment, the log term usually ranges from 1 to 10, depending on the modulation technique (the signal-to-noise ratio is usually between -1 and 20). Figure 4.4 depicts some typical

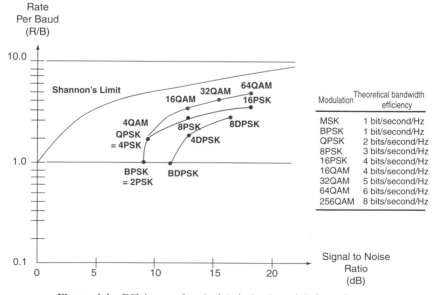

Figure 4.4   Efficiency of typical (wireless) modulation schemes.

digital modulation schemes in terms of their efficiency as measured by the bit rate per baud (in this context, 1 baud equates to 1 hertz). Typically, in wireless communication it is desirable to maximize the bandwidth efficiency; in traditional wireless communication, sophisticated (high-complexity) modulation methods are used to maximize the link throughput. For example, 64-point quadrature amplitude modulation (64 QAM) is used in WLANs operating with the IEEE 802.11a to achieve 54 Mbps throughput in a 20-MHz channel. High efficiency, however, comes with a price: first, the circuit complexity goes up considerably; second, the power consumption increases when one targets a high channel throughput. As might be expected, high throughput and efficiency are also desirable in WSNs; however, a trade-off between efficiency and power must be accepted: Schemes that support high efficiency require complex designs (read "high-count transistor chipsets") and fairly high power consumption. Research has shown that advanced modulation results in degraded energy efficiency for systems operating with short packets and/or a low duty cycle [4.5].

Spread-spectrum modulation techniques have a higher effective signal-to-noise ratio than narrowband techniques, but require more channel bandwidth. Direct-sequence spread spectrum (DSSS) is one of the two common spread-spectrum techniques (it being used, for example, in commercial implementation of the WLAN standards, including ZigBee). Frequency hopping spread spectrum (FHSS) is the other technique (it is used in the Bluetooth environment for PANs). In DSSS, the incoming data stream is hashed by a pseudorandom sequence that generates a sequence of output microbits or chips that are distributed across the underlying broadband channel. To the casual eye, these distributed microbits appear like noise. Fairly complex digital signal processing functions are needed to recover the original signal; processing must occur at the chip rate, and timing synchronization of all the nodes in the system must be within a fraction of the chip interval (which is the reciprocal of the chip rate).

Compared to DSSS systems, FHSS uses relatively low complexity baseband hardware. The synchronization mechanism is also less complex; however, agile frequency hopping requires fast signaling settling. There are prima facie advantages in the use of FHSS for WSNs (e.g., improved multipath performance can be achieved with FHSS); however, the requirement for low-power operation and the widebandnature of operation gives rise to practical engineering challenges.

Appendix A provides some additional information related to modulation.

4.3 AVAILABLE WIRELESS TECHNOLOGIES

As we noted in passing in Chapter 3, two frequency bands are typically used by WNs: the ISM band and the U-NII band. As we just described, indoor and outdoor interference arises from both natural sources and/or phenomena (e.g., loss or attenuation, absorption, fading, multipath) as well as from other users in proximity utilizing these "unprotected bands." A WSN will experience interference whether it uses one of the IEEE PAN/LAN/MAN technologies or even some other generic radio technology. For example, as noted above, other devices, such as Bluetooth-based PDAs and cellular phones, which share operating frequencies with wireless sensors on both the ISM and UNII bands, can affect confined spaces (open spaces are subject to other issues). Microwave ovens, which operate at 2.45 GHz, may overwhelm many wireless technologies in the 2.4-GHz ISM band. On a manufacturing floor, improperly filtered electric motors may generate enough electrical noise to make wireless transmissions unreliable. Even the physical placement of a transmitter can cause a significant loss of signal [4.4].

Nonetheless, IEEE PAN/LAN/MAN technologies are broadly implemented technologies and are probably the ones utilized in the majority of (commercial) WSNs on a going-forward basis. Protocols determine the physical encoding of signal transmitted as well as the data link layer framing of the information; channel-sharing and data- and event-handling procedures are also specified by the protocol. There are several wireless protocols; the most widely used are (1) the IEEE 802.15.1 (also known as Bluetooth); (2) the IEEE 802.11a/b/g/n series of wireless LANs; (3) the IEEE 802.15.4 (ZigBee); (4) the MAN-scope IEEE 802.16 (also known as WiMax); and (5) radio-frequency identification (RFID) tagging. Each standard possesses different benefits and limitations. Figure 4.5 depicts graphically some of the features of these protocols see also Table 4.5).

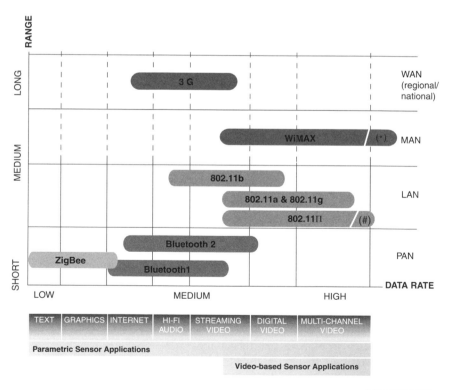

Figure 4.5 Graphical comparison of available protocols.

TABLE 4.5 Wireless Protocol Comparison

| Property | IEEE Standard | | |
	802.11	802.15.1/Bluetooth	802.15.4/ZigBee
Range (m)	~100	~10 to 100	~10
Data throughput (Mbps)	~2 to 54	~1 to 3	~0.25
Power consumption	Medium	Low	Ultralow
Battery life measured in:	Minutes to hours	Hours to days	Days to years
Size relationship	Large	Smaller	Smallest
Cost/complexity ratio	>6	1	0.2

Source: [4.4].

The IEEE 802.15.4 standard supports a maximum data rate of 250 kbps, with rates as low as 20 kbps (slower than most telephone modems); however, it has the lowest power requirement of the group. ZigBee devices are designed to run several years on a single set of batteries, making them ideal candidates for unattended or difficult-to-reach locations. Bluetooth is a short-range communication protocol widely used in cellular-type phones and PDAs (has a range of about 10 m, or a maximum of 100 m with power boost); it operates in the 2.4-GHz ISM band and has a bandwidth of approximately 1 to 3 Mbps. IEEE 802.11a/b/g/n is a collection of related technologies that operate in the 2.4-GHz ISM band, the 5-GHz ISM band, and the 5-GHz U-NII bands; it provides the highest power and longest range of the common unlicensed wireless technologies. Transmission data rates can reach 54 Mbps (twice as much with the latest IEEE 802.11n protocol). Typically, hardware implementation of some or all of 802.11 protocols comes preinstalled on most new laptop computers; the technology is often also available for PDAs and cellular phones. RFID is the one form of wireless sensing that requires no power in the tag; it is a passive technology used for labeling and tracking. The RFID tag is the sensor; the sensor responds when power is beamed to it through the reading device. Current RFID tags can hold only 96 bits of information, but newer tags that support 128 and 256 bits are becoming available [4.4]. Most RFID tags have *integrated circuits* (ICs), microelectronic semiconductor devices with a large number of interconnected transistors and other components. Although the topic of RFIDs is not covered further in this book, a glossary of basic terms is included in Table 4.6 for completeness [4.41].

The subsections that follow provide additional details on these standardized wireless technologies. We partition the discussion into campus and MAN/WAN application spaces.

4.3.1 Campus Applications

Campus sensor communications can occur over Bluetooth, wireless LAN (WLAN), ZigBee, or WiMax/hotspot systems.

TABLE 4.6   RFID Glossary[a]

Active tag	An RFID tag that comes with a battery that is used to power the microchip's circuitry and transmit a signal to a reader. Active tags can be read from 100 ft or more away, but they are expensive—more than $20 each. Tags are used for tracking expensive items over long ranges. For instance, the U.S. military uses active tags to track containers of supplies arriving in ports.
Automatic identification	(a.k.a. automatic data capture) A method of collecting data and entering them directly into computer systems without human involvement. Technologies normally considered part of auto-ID include bar codes, biometrics, RFID, and voice recognition.
Backscatter	A method of communication between tags and readers. RFID tags using backscatter technology reflect back to the reader a portion of the radio waves that reach them. The signal reflected is modulated to transmit data. Tags using backscatter technology can be either passive or active, but either way, they are more expensive than tags that use inductive coupling.
Chipless RFID tag	An RFID tag that does not depend on an integrated microchip. Instead, the tag uses materials that reflect back a portion of the radio waves beamed at them. A computer takes a snapshot of the waves beamed back and uses it like a fingerprint to identify the object with the tag. Companies are experimenting with embedding RF reflecting fibers in paper to prevent unauthorized photocopying of certain documents. But chipless tags are not useful in the supply chain because even though they are inexpensive, they cannot communicate a unique serial number that can be stored in a database.
Closed-loop systems	RFID tracking systems set up within a company. Since the item being tracked never leaves the company's control, the company does not need to worry about using technology based on open standards.
Contactless smart card	A credit card or loyalty card that contains an RFID chip to transmit information to a reader without having to be swiped through a reader. Such cards can speed checkout, providing consumers with more convenience.
EEPROM (electrically erasable programmable read-only memory)	A nonvolatile storage device on microchips. Usually, bytes can be erased and reprogrammed individually. RFID tags that use EEPROM are more expensive than factory-programmed tags, but they offer more flexibility because the end user can write an ID number to the tag at the time the tag is going to be used.
Electromagnetic compatibility (EMC)	The ability of a system or product to function properly in an environment where other electromagnetic devices are used and not itself be a source of electromagnetic interference.
Electromagnetic interference (EMI)	Interference caused when the radio waves of one device distort the waves of another. Cells phones, wireless computers, and even robots in factories can produce radio waves that interfere with RFID tags.

TABLE 4.6 (*Continued*)

Electronic article surveillance (EAS)	Simple electronic tags that can be turned on or off. When an item is purchased (or borrowed from a library), the tag is turned off. When someone passes a gate area holding an item with a tag that has not been turned off, an alarm sounds. EAS tags are embedded in the packaging of most pharmaceuticals.
Electronic product code (EPC)	A 96-bit code created by the auto-ID center that will one day replace bar codes. The EPC has digits to identify the manufacturer, product category, and the individual item. It is backed by the Uniform Code Council and the European Article Numbering Association the two main bodies that oversee bar code standards.
Error-correcting code	A code stored on an RFID tag to enable a reader to determine the value of missing or garbled bits of data. It is needed because a reader might misinterpret some data from the tag and think that a Rolex watch is actually a pair of socks.
Error-correcting mode	A mode of data transmission between the tag and the reader in which errors or missing data are corrected automatically.
Error-correcting protocol	A set of rules used by readers to interpret data correctly from the tag.
Excite	A reader is said to "excite" a passive tag when the reader transmits RF energy to wake up the tag and enable it to transmit back.
Factory programming	The process of writing the identification number into a silicon microchip at the time the chip is made, as is necessary for some read-only tags.
Field programming	Tags that use EEPROM, or nonvolatile memory, can be programmed after being shipped from the factory.
GTAG (global tag)	A standardization initiative of the Uniform Code Council and the European Article Numbering Association for asset tracking and logistics based on RFID. The GTAG initiative is supported by Philips Semiconductors, Intermec, and Gemplus, three major RFID tag makers.
High-frequency tags	Tags that operate typically at 13.56 MHz. They can be read from about 10 ft away and transmit data faster, but they consume more power than do low-frequency tags.
Inductive coupling	A method of transmitting data between tags and readers in which the antenna from the reader picks up changes in a tag's antenna.
Low-frequency tags	Tags that typically operate at 125 kHz. The main disadvantages of low-frequency tags are that they have to be read from within 3 ft and the rate of data transfer is slow. But they are less expensive than high-frequency tags and less subject to interference.
Memory	The amount of data that can be stored on a tag.
Microwave tags	Radio-frequency tags that operate at 5.8 GHz. They have very high transfer rates and can be read from away as far as 30 ft, but they use a lot of power and are expensive.
Multiple-access schemes	Methods of increasing the amount of data that can be transmitted wirelessly within the same frequency spectrum. RFID readers use time-division multiple access (TDMA), meaning that they read tags at different times to avoid interfering with one another.

(*Continued*)

TABLE 4.6 (*Continued*)

Nominal range	The read range at which a tag can be read reliably.
Null spot	Area in the reader field that does not receive radio waves. This is essentially the reader's blind spot. It is a phenomenon common to ultrahigh-frequency systems.
Object name service (ONS)	An auto-ID center–designed system for looking up unique electronic product codes and pointing computers to information about the item associated with the code. ONS is similar to the domain name service, which points computers to sites on the Internet.
Passive tag	An RFID tag without a battery. When radio waves from the reader reach the chip's antenna, it creates a magnetic field. The tag draws power from the field and is able to send back information stored on the chip. At this juncture simple passive tags cost from about 50 cents to several dollars.
Patch antenna	A small square antenna made from a solid piece of metal or foil.
Power level	The amount of RF energy radiated from a reader or an active tag. The higher the power output, the longer the read range, but most governments regulate power levels to avoid interference with other devices.
Programming	Writing data to an RFID tag.
Proximity sensor	A device that detects the presence of an object and signals another device. Proximity sensors are often used on manufacturing lines to alert robots or routing devices on a conveyor to the presence of an object.
Read	The process of turning radio waves from a tag into bits of information that can be used by computer systems.
Read range	The distance from which a reader can communicate with a tag. Active tags have a longer read range than passive tags because they use a battery to transmit signals to the reader. With passive tags, the read range is influenced by frequency, reader output power, antenna design, and method of powering up the tag. Low-frequency tags use inductive coupling (see above), which requires the tag to be within a few feet of the reader.
Read rate	The maximum rate at which data can be read from a tag, expressed in bits or bytes per second.
Reader (also called an interrogator)	The reader communicates with an RFID tag via radio waves and passes the information in digital form to a computer system.
Reader field	The area of coverage. Tags outside the reader field do not receive radio waves and cannot be read.
Read-only tag	A tag that contains data that cannot be changed unless the microchip is reprogrammed electronically.
Read–write tag	An RFID tag that can store new information on its microchip. San Francisco International Airport uses a read–write tag for security. When a bag is scanned for explosives, the information on the tag is changed to indicate that it has been checked. The tag is scanned again before it is loaded on a plane. Read–write tags are more expensive than read-only tags and therefore are of limited use for supply chain tracking.

TABLE 4.6 *(Continued)*

RFID tag	A microchip attached to an antenna that picks up signals from and sends signals to a reader. The tag contains a unique serial number but may have other information, such as a customer's account number. Tags come in many forms, such as smart labels that are stuck on boxes, smart cards and keychain wands for paying for things, and a box that you stick on your windshield to enable you to pay tolls without stopping. RFID tags can be active tags, passive tags, or semipassive tags.
RFID tags' frequency	RFID tags use low, high, ultrahigh, and microwave frequencies. Each frequency has advantages and disadvantages that make them more suitable for some applications than for others.
Scanner	An electronic device that can send and receive radio waves. When combined with a digital signal processor that turns the waves into bits of information, the scanner is called a reader or interrogator.
Semipassive tag	Similar to active tags, but the battery is used to run the microchip's circuitry but not to communicate with the reader. Some semipassive tags sleep until they are woken up by a signal from the reader, which conserves battery life. Semipassive tags cost $1 or more.
Sensor	A device that responds to a physical stimulus and produces an electronic signal. Sensors are increasingly being combined with RFID tags to detect the presence of a stimulus at an identifiable location.
Silent commerce	This term covers all business solutions enabled by tagging, tracking, sensing, and other technologies, including RFID, which make everyday objects intelligent and interactive. When combined with continuous and pervasive Internet connectivity, they form a new infrastructure that enables companies to collect data and deliver services without human interaction.
Smart label	A label that contains an RFID tag. It is considered "smart" because it can store information, such as a unique serial number, and communicate with a reader.
Tag antenna	The antenna is the conductive element that enables the tag to send and receive data. Passive tags usually have a coiled antenna that couples with the coiled antenna of the reader to form a magnetic field. The tag draws power from this field.
Time-division multiple access (TDMA)	A method of solving the problem of the signals of two readers colliding. Algorithms are used to make sure that readers attempt to read tags at different times.
Transponder	A radio transmitter–receiver that is activated when it receives a predetermined signal. RFID tags are sometimes referred to as transponders.
Ultrahigh frequency (UHF) tag	Typically, tags that operate between 866 and 930 MHz. They can send information faster and farther than can high- and low-frequency tags. UHF tags are also more expensive than low-frequency tags, and they use more power.

(Continued)

TABLE 4.6   *(Continued)*

Uniform Code Council (UCC)	The nonprofit organization that oversees the Uniform Product Code, the bar code standard used in North America.
Uniform Product Code (UPC)	The bar code standard used in North America. It is administered by the Uniform Code Council.
Write rate	The rate at which information is transferred to a tag, written into the tag's memory and verified as being correct.

Source: [4.41].
[a]RFID is a method of identifying unique items using radio waves. Typically, a reader communicates with a tag, which holds digital information in a microchip; however, there are chipless forms of RFID tags that use material to reflect back a portion of the radio waves beamed at them.

Bluetooth   Bluetooth is a specification for short-range RF-based connectivity for portable personal devices. It is a short-range wireless data exchange protocol designed for a small variety of tasks, such as synchronization, voice headsets, cell modem calls, and mouse and keyboard input. The specification began as a de facto industry standard; more recently, IEEE Project 802.15.1 developed a wireless PAN standard based on the Bluetooth v1.1 Foundation Specifications. The IEEE 802.15.1 standard was published in 2002. Bluetooth is directed principally to the support of personal communication devices such as telephones, printers, headsets, and PC keyboards and mice. The technology has restricted performance characteristics by design; hence, its applicability to WSN is rather limited in most cases. For these same environments, ZigBee is probably a better solution; however, given the popularity and longevity of the standard, it is given some coverage here.

As part of its effort, the IEEE has reviewed and provided a standard adaptation of the Bluetooth Specification v1.1 Foundation media access control (MAC) (L2CAP, LMP, and baseband) and the physical layer (PHY) (radio). Also specified is a clause on service access points (SAPs), which includes a LLC–MAC interface for the ISO/IEC 8802-2 LLC. A normative annex that provides a protocol implementation conformance statement (PICS) proforma has been developed. Also specified is an informative high-level behavioral ITU-T Z.100 specification and description language (SDL) model for an integrated Bluetooth MAC sublayer [4.6].

The Bluetooth specification defines a low-power, low-cost technology that provides a standardized platform for eliminating cables between mobile devices and facilitating connections between products. The system uses omnidirectional radio waves that can transmit through walls and other nonmetal barriers. Unlike other wireless standards, the Bluetooth wireless specification includes both link layer and application layer definitions for product developers. Radios that comply with the Bluetooth wireless specification operate in the unlicensed, 2.4-GHz ISM radio spectrum, ensuring communication compatibility worldwide.

Bluetooth radios use a spread-spectrum, frequency-hopping, full-duplex signal. While point-to-point connections are supported, the specification allows up to seven simultaneous connections to be established and maintained by a single radio [4.7]. AFH (adaptive frequency hopping), available with newer versions, allows for more graceful coexistence with IEEE 802.11 WLAN systems. The signal hops among 79 frequencies at 1-MHz intervals to give an acceptable degree of interference immunity between multiple Bluetooth devices and between a Bluetooth device and a WLAN device (at least in the case where not all the available frequencies are used by the WLAN—this is probably the case in a SOHO environment, where only one or two access points are used at a location). To minimize interference with other protocols that use the same band, the protocol can changes channels up to 1600 times per second. If there is interference from other devices, the transmission does not stop, but its speed is downgraded.

Bluetooth version 1.2 allowed a maximum data rate of 1 Mbps; this results in an effective throughput of about 723 kbps. In late 2004, a new version of Bluetooth known as Bluetooth version 2 was ratified; among other features it included enhanced data rate (EDR). With EDR the maximum data rate is able to reach 3 Mbps (throughput of 2.1 Mbps) within a range of 10 m (up to 100 m with a power boost). Older and newer Bluetooth devices can work together with no special effort [4.8]. Because a device such as a telephone headset can transmit the same information faster with Bluetooth 2.0 + EDR, it uses less energy, since the radio is on for shorter periods of time. The data rate is improved by more efficient coding of the data sent across the air; this also means that for the same amount of data, the radio will be active less of the time, thus reducing the power consumption [4.7]. Newer Bluetooth devices are efficient at using small amounts of power when not actively transmitting: for example, the headset is able to burst two to three times more data in a transmission and is able to sleep longer between transmissions. Noteworthy features of Bluetooth core specification version 2.0 + EDR include:

- Three times faster transmission speed than that of preexisting technology
- Lower power consumption through a reduced duty cycle
- Simplification of multilink applications due to increased available bandwidth
- Backwardly compatible to earlier versions
- Improved bit-error-rate performance

Hardware developers were shifting from Bluetooth 1.1 to Bluetooth 1.2 in the recent past; Bluetooth 2.0 products were being introduced at press time. To be exact, version 2.0 devices have a higher power consumption; however, the fact that the transmission rate is three times faster (thereby reducing the transmission burst times) effectively reduces consumption to half that of 1.x devices. Devices are able to establish a trusted relationship; a device that wants to communicate only with a trusted device can authenticate the identity of the other device cryptographically. Trusted devices may also encrypt the data that they exchange over the air.

A Bluetooth device playing the role of "master" can communicate with up to seven devices playing the role of "slave" (groups of up to eight devices are called *piconets*). At any given instant in time, data can be transferred between the master and one slave; but the master switches rapidly from slave to slave in a round-robin fashion. (Simultaneous transmission from the master to multiple slaves is possible but is not used much in practice.) The Bluetooth specification also makes it possible to connect two or more piconets to form a *scatternet*, with some devices acting as a bridge by simultaneously implementing the master role in one piconet and the slave role in another piconet.

The Bluetooth SIG recently established a road map for future improvements to Bluetooth. Priorities for 2005 included quality of service (QoS), security, and power consumption; priorities for 2006 were to include multicast, additional security, and long-range performance. The Bluetooth SIG is also working with developers of UWB to ensure backward compatibility with the new standard. UWB is a short-distance wireless protocol capable of transmitting up to 100 Mbps of data a distance of about 10 m; Bluetooth is only capable of 1 to 3 Mbps over the same distance. It is conceivable that Bluetooth could be supplanted by this faster technology, so the Bluetooth SIG is working to make sure that UWB is backwardly compatible with current Bluetooth devices (at present, two groups are competing for their technology to be ratified as the UWB standard). Depending on the usage cases, technologies such as ZigBee and UWB can be either complementary or overlapping [4.7]. It is hypothetically possible that Bluetooth wireless technology and UWB could converge, but work and agreements will need to take place to make this happen. The immediate problems for UWB—the two competing standards and the lack of the international regulatory approval—need to be resolved for the idea of convergence to be interesting for Bluetooth wireless technology.

WLAN   The following are areas where advances in wireless LAN (WLAN) is taking place:

1. Higher WLAN speeds to support an adequate number of users in high-density environments and also voice over IP (VoIP) users. The transition to an IEEE 802.11g and/or 802.11n environment is a basic necessity in a high-density and/or high-bandwidth context.
2. Support of QoS over the wireless (and also core intranet) infrastructure. The deployment of IEEE 802.11e QoS-supporting technology is another basic necessity.
3. Secure communications is highly desirable. The deployment of IEEE 802.11i security capabilities is yet another requirement.
4. Roaming between access points, floors, and subnets is needed, as is a handoff to a cellular service when corporate WLAN service is no longer available or generally, for WN mobility situations. The deployment of IEEE 802.11r roaming capabilities addresses this requirement (capabilities not expected to be available and/or implemented until sometime in the future). Roaming also brings up the question of whether a traditional IP solution is adequate or if one needs to utilize Mobile IP (MIP) (IETF RFC 3344) [4.9]; this is a fairly complex issue.

The IEEE 802.11b and 802.11g specifications postulate a partitioning of the spectrum into 14 overlapping staggered channels whose center frequencies are

5 MHz apart; within this partitioning of the ISM spectrum, channels 1, 6, and 11 (and if available in the regulatory domain, channel 14) do not overlap. These channels (or other sets with similar gaps) can be used so that multiple networks can operate in close proximity without interfering with each other (see Figure 4.6).

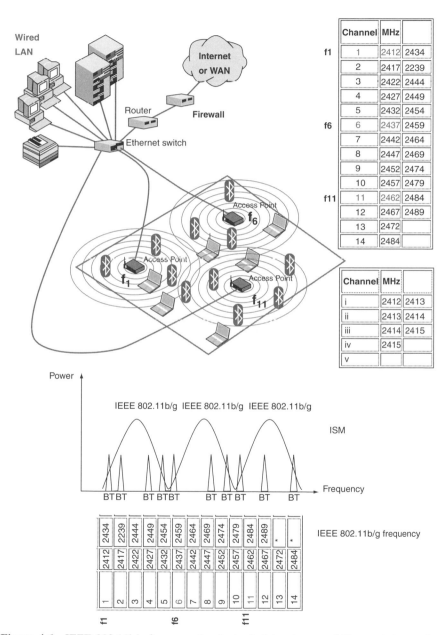

Figure 4.6 IEEE 802.11b/g frequency bands, typical topology, and bluetooth interaction.

TABLE 4.7   IEEE WLAN-Relevant Frequencies in Various Parts of the World

Channel	MHz	U.S.	Canada	Europe (ETSI)	Spain	France	Japan
1	2412	×	×	×		×	×
2	2417	×	×	×		×	×
3	2422	×	×	×		×	×
4	2427	×	×	×		×	×
5	2432	×	×	×		×	×
6	2437	×	×	×		×	×
7	2442	×	×	×		×	×
8	2447	×	×	×		×	×
9	2452	×	×	×		×	×
10	2457	×	×	×	×	×	×
11	2462	×	×	×	×	×	×
12	2467			×		×	×
13	2472			×		×	×
14[a]	2484						

[a]Channel 14, where available, is restricted to 802.11b operation.

The spectral mask for 802.11b requires that the signal be at least 30 dB down from its peak energy at ±11 MHz from the center frequency and at least 50 dB down from its peak energy at ±22 MHz from the center frequency. Note that if the transmitter is sufficiently powerful, the signal can be relatively strong even beyond the ±22-MHz point (e.g., a powerful transmitter on channel 6 can easily overwhelm a weaker transmitter on channel 11); in most situations, however, the signal in a given channel is sufficiently attenuated to interfere only minimally with a transmitter on any other channel.

The channels that are available for use in a particular country differ according to the regulations of that country. Table 4.7 identifies IEEE-relevant frequencies in various parts of the world. In the United States, for example, FCC regulations allow only channels 1 to 11 to be used. Channels 10 and 11 are the only channels that work in all parts of the world, because Spain has not licensed channels 1 to 9 for 802.11b operation.

The UNII band used in the IEEE 802.11a context is in the range 5.15 to 5.85 GHz. The 802.11a standard uses 300 MHz of bandwidth; the spectrum is divided into three *domains*, each having restrictions imposed on the maximum output power allowed. The first 100 MHz in the lower-frequency portion is restricted to a maximum power output of 50 mW; the second 100 MHz has a higher maximum, 250 mW; and the third, 100 MHz, intended primarily for outdoor applications, has a maximum power output of 1.0 W. It is generally recognized that the higher-frequency UNII band is limited intrinsically to shorter ranges than the ISM band, due to higher path loss, limiting the utility of 802.11a relative to that of 802.11b/g in the WSN context, except for within-building applications. In particular, there is an increase of excess path loss with frequency. Table 4.8 provides a comparison

TABLE 4.8   A Comparison of IEEE 802.11b/g and IEEE 802.11a

	802.11b/802.11g	802.11a
Available bandwidth	83.5 MHz	300 MHz
Unlicensed frequencies of operation	2.4–2.4835 GHz	5.15–5.35 GHz, 5.725–5.825 GHz
Number of non-overlapping channels	3 (indoor–outdoor)	4 (indoor–outdoor)
Data rate per channel	1, 2, 5.5, 11, 54 Mbps	6, 9, 12, 18, 24, 36, 48, 54 Mbps
Modulation	DSSS	OFDM

between IEEE 802.11b/g and IEEE 802.11a. The IEEE 802.11a protocol uses a complex digital modulation method: specifically, orthogonal frequency-division multiplexing (OFDM); this digital modulation method requires more linearity in amplifiers because of the higher peak-to-average power ratio of the OFDM signal transmitted. In addition, better phase noise performance is required because of the closely spaced overlapping carriers. These issues tend to add to the implementation cost of 802.11a products. Although IEEE 802.11a was approved in the late 1990s, new product development has proceeded much more slowly than with 802.11b/g, due to the cost and complexity of implementation.

Frequency-division multiplexing (FDM) is a multiplexing technology that transmits multiple signals from or for different users simultaneously over a single transmission path, such as a cable or wireless system (commercial FM radio is an example). Each signal occupies its own unique frequency range (carrier), which is modulated by the data (text, voice, video, etc.). The OFDM spread-spectrum technique distributes the data over a large number of carriers that are spaced apart at precise frequencies. This spacing provides the orthogonality, which prevents the demodulators from seeing frequencies other than their own. The benefits of OFDM are high spectral efficiency, resiliency to RF interference, and lower multipath distortion. This is useful because in a typical terrestrial broadcasting scenario there are multipath channels (i.e., the signal transmitted arrives at the receiver using various paths of different length). Since multiple versions of the signal interfere with each other [intersymbol interference (ISI)] it becomes difficult to extract the original information. OFDM is the modulation technique used for digital television in Europe, Japan, and Australia [4.10].

As stated previously, a drawback of 5 GHz is that higher-frequency signals experience more difficulties propagating through physical obstructions encountered in an office (walls, floors, and furniture) than do those at 2.4 GHz. There is an intrinsic degradation in throughput as the distance between the transmitter and receiver increases. See Figure 4.7 for a comparison of the two standards or bands with regard to propagation or performance and distance. An advantage of 802.11a is its ability to deal with delay spread and multipath reflection effects: The slower symbol rate and placement of significant guard time around each symbol reduces

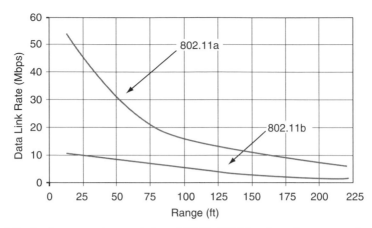

Figure 4.7 Performance characteristics of IEEE 802.11a: throughput comparison versus distance (indoor applications).

the ISI caused by multipath interference; by contrast, 802.11b networks are generally range limited by multipath interference rather than the loss of signal strength over distance.

Now-emerging multiple-input, multiple-output (MIMO) systems use multiple antennas to transmit and receive radio signals. MIMO methods increase the throughput and quality of the signals received. IEEE 802.11n uses MIMO techniques. For example, MIMO–OFDM will allow service providers to deploy a broadband wireless access (BWA) system that has non-line-of-sight (NLOS) functionality. Specifically, MIMO–OFDM takes advantage of the multipath properties of environments using base station antennas that do not have LOS. As noted, in multipath environments the original signal and the individual echoes each arrive at the receiver antenna at slightly different times, causing the echoes to interfere with one another, thus degrading signal quality. The MIMO system uses multiple antennas to transmit data simultaneously in small segments to the receiver, which can process the data flows and put them back together. This process, called *spatial multiplexing*, increases the data-transmission speed proportionally by a factor equal to the number of antennas transmitting. In addition, since all data are transmitted both in the same frequency band and with separate spatial signatures, this technique utilizes the spectrum fairly efficiently [4.10].

ZigBee   In this section we provide a brief description of ZigBee. ZigBee is the only standards-based technology designed to address the unique needs of low-cost, low-power WSNs for remote monitoring, home control, and building automation network applications in the industrial and consumer markets [4.11]. The wireless systems discussed in previous subsections provide high data rates at the expense of power consumption, application complexity, and cost. However, there are many wireless monitoring and control applications for industrial and home markets that require longer battery life, lower data rates, and less complexity than those made available by existing wireless standards. For commercial success one needs a standards-based wireless technology that has performance characteristics that closely meet the requirements for reliability, security, low power, and low cost [4.12], [4.40].

For such wireless applications a targeted standard has been developed by the IEEE [4.13]. The IEEE 802.15 Task Group 4 was chartered to investigate a low-data-rate solution with multimonth to multiyear battery life and very low complexity. The standard is intended to operate in an unlicensed international frequency band. Potential applications fot this standard are home automation, wireless sensors, interactive toys, smart badges, and remote controls. The scope of the task group has been to define the physical layer (PHY) and the media access control (MAC) [4.14]. This standards-based interoperable wireless technology is optimized to address the specific needs of low-data-rate wireless control and sensor-based networks [4.12]. Functionality defined by the ZigBee Alliance is used at the upper layers.

The ZigBee Alliance ratified the first ZigBee specification in 2004, making the development and deployment of power-efficient, cost-effective, low-data-rate monitoring, control, and sensing networks a reality. ZigBee/IEEE 802.15.4 is expected to become the leading wireless technology for a plethora of uses, ranging from building automation to industrial and residential applications. Developers were anticipating ZigBee-compliant consumer products as quickly as early 2005 [4.11]. A graphical representation of the areas of responsibility between the IEEE standard, ZigBee Alliance, and user is presented in Figure 4.8; the definition of the application profiles is organized by the ZigBee Alliance [4.11], [4.42].

Hotspot/WiMax   In recent years service providers have deployed IEEE 802.11b/11g-based hotspot services to support Internet access and VoIP applications [4.15]. Furthermore, there is interest in delivering metro-wide Internet/VoIP services using WiMax (IEEE 802.16-based) connectivity. Since WiMax is newer, we focus here on this technology (see Table 4.9 for a technical comparison of WiMax to Wi-Fi [4.16]).

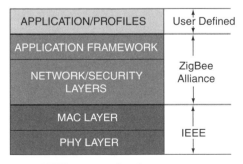

IEEE 802.15.4 Stack

Figure 4.8   ZigBee protocol stack.

TABLE 4.9   Comparative Overview of Wi-Fi and Mobile WiMax Technology

	Wi-Fi Based on 802.11	WiMax Based on 802.16e-2005
Spectrum	Unlicensed, 2.4 GHz (ISM band) and 5.8 GHz (UNII band)	Licensed less than 6 GHz
Range and coverage	Typically less than 100 meters	PMP, NLOS typically 1 to 10 km depending on frequency and terrain characteristics. Point-to-point, LOS up to 50 km
Applications	Indoor WLAN, fixed and nomadic usage model	Outdoor WMAN, fixed, nomadic, portable, and mobile applications
Peak downstream data rate	Up to 54 Mbps in 20 MHz channel BW	Up to 50 Mbps in 10 MHz channel BW
QoS	802.11e provides better QoS support than 802.11a,b,g but only set traffic priorities (up to 8). It is not deterministic, so it is possible for one connection or traffic type to override and starve another connection	Provides guaranteed service levels for specific types of traffic on a connection-by-connection basis. 802.16 uses priority, committed, and peak information rates and can meet specific latency and jitter requirements for specific types of traffic. Can also regenerate network clocks over the air
Privacy and security	WEP uses a repetitive key and is easily defeated. This has been upgraded to WPA and WPA2 with 802.11i	802.16 has two data encryption modes: mandatory 56-bit DES and 128 AES. Also supports device base station, subscriber station, and user authentication. Has secure key exchange and is 802.1x compliant
Latency	CSMA/CA approach for scheduling increases latency with multiple connections. Latency is not deterministic and therefore, adversely affects QoS	Uses a grant-request mechanism as opposed to CSMA/CA; this eliminates delays with multiple users sharing the same channel. This is necessary to support latency sensitive traffic such as VoIP
System gain	System gain is limited by transmit power limits in the unlicensed 2.4 and 5 GHz bands, thus limiting the range capability of 802.11. Support for MIMO in 802.11n will improve this somewhat. Lack of support for subchannelization limits uplink system gain with battery-operated laptops as subscriber stations	Licensed frequency bands permits higher base station Tx power. Subchannelization provides increased system gain in the uplink direction. Adaptive antenna systems including MIMO, beamforming, space-time coding (STC), and spatial multiplexing (SM) also enhance system gain and range
Support for battery-operated handsets	Subchannelization is not supported so subscriber station Tx power must be sufficient to transmit full channel. This is satisfactory for a laptop with a large battery or	Uplink subchannelization reduces Tx power requirements for battery-operated subscriber devices. Various sleep mode options are available to conserve battery life

TABLE 4.9   (*Continued*)

	access to AC power, but not acceptable for mobile handhelds, PDAs, etc. Also no sleep mode	
Multipath immunity	OFDM with a FFT size of 64 provides some immunity to multipath	S-OFDMA with FFT size of 512 to 2048 FFT for channel BWs from 5 to 20 MHz
Interference immunity	802.11 does not have support for transmit power control (TPC) or dynamic channel selection (DCS). Some of these issues are addressed with 802.11h	Aided by transmit power control, subchannelization and support for adaptive antenna systems

Source: WiMax Forum.

The IEEE 802.16 Working Group has developed a point-to-multipoint (PMP) broadband wireless access standard for systems in the frequency ranges 10 to 66 GHz and sub-11 GHz. This technology is targeted to metropolitan area environments. The IEEE 802.16 standard covers both the MAC and PHY layers. A number of PHY considerations were taken into account for the target environment. At higher frequencies, line of sight (LOS) is a must. This requirement eases the effect of multipath, allowing for wide channels, typically greater than 10 MHz in bandwidth. This gives the IEEE 802.16 protocol the ability to provide very high capacity links on both the uplink and downlink. For sub-11 GHz, non-line-of-sight (NLOS) capability is a requirement. The original IEEE 802.16 MAC was enhanced to accommodate different PHYs and services, which address the needs of different metropolitan environments. The standard is designed to accommodate either time-division duplexing (TDD) or frequency-division duplexing (FDD) deployments, allowing for both full- and half-duplex terminals in the FDD case [4.16]. IEEE 802.16a has a LOS radius of 50 km and an NLOS of 10 km or thereabouts, depending on the types of obstacles in the topography. WiMax is the marketing name of the IEEE 802.16 standard.

The MAC was designed specifically for the PMP wireless access environment. It supports higher layer or transport protocols, such as ATM, Ethernet, and IP, and is designed to accommodate easily future protocols that have not yet been developed. The MAC is designed for high bit rates (up to 268 Mbps each way) and operates on a broadband physical layer, while delivering ATM-compatible QoS, UGS (unsolicited grant service), rtPS (real-time polling service), nrtPS (non-real-time polling service), and best effort services. The frame structure allows terminals to be dynamically assigned uplink and downlink burst profiles according to their link conditions. This allows a trade-off between capacity and robustness in real time and provides roughly a two-fold increase in capacity on average compared to nonadaptive systems while maintaining appropriate link availability. The 802.16 MAC uses a variable-length protocol data unit (PDU) along with a number of other concepts that greatly increase the efficiency of the standard. Multiple MAC PDUs may be concatenated into a single burst to save PHY overhead. Additionally, multiple service data units (SDUs) for the same service may be concatenated into a single MAC PDU, saving on MAC header overhead. Fragmentation allows large SDUs to be sent across frame boundaries to guarantee the QoS of competing services. Payload header suppression can be used to reduce the overhead caused by the redundant portions of SDU headers. The MAC uses a self-correcting bandwidth request–grant scheme that eliminates the overhead and delay of acknowledgments while allowing better QoS handling than that of traditional acknowledgment schemes. Terminals have a variety of options for requesting bandwidth, depending on the QoS and traffic parameters of their services. Terminals can be polled individually or in groups; they can steal bandwidth already allocated to make requests for more; they can signal the need to be polled, and they can piggyback requests for bandwidth [4.16].

A typical WiMax network consists of a base station supported by a tower- or building-mounted antenna. The base station connects to the appropriate terrestrial network (PSTN, Internet, etc.) Applications include, but are not limited to, point-to-point communication between stations, point-to-multipoint communication between the base station and clients, backhaul services for Wi-Fi (802.11) hotspots, broadband Internet services to home users, private-line services for users in remote locations, and metro-wide WSN applications.

4.3.2 MAN/WAN Applications

MAN/WAN sensor communications can occur over WiMax/hotspots or 3G systems. After a brief discussion of a brand-new (but speculative) technology, cognitive radios (CRs), in the remainder of the section we focus on the evolution of cellular networks in terms of the desire to provide a lateral data channel that supports any number of applications, including WSNs.

Cognitive Radios and IEEE 802.22 With the plethora of wireless services that are becoming available, stakeholders believe that the limiting factor at this time is the scarcity of radio spectrum. Studies have shown that most of this spectrum scarcity is concentrated in the unlicensed bands; this is where the major advancements in spectrum use have taken place (e.g., Wi-Fi, cordless phones). Licensed bands, however, typically experience considerable underutilization. CR-based approaches represent a new paradigm in wireless communications that aims at utilizing the large amount of underused spectrum in an intelligent way while not interfering with other incumbent devices in frequency bands already licensed for specific uses [4.43].

The IEEE 802.22 wireless regional area network (WRAN) standard is the first worldwide project to employ CR concepts for dynamically sharing spectrum with television broadcast signals. IEEE 802.22 seeks to develop a standard for a cognitive radio-based PHY–MAC–air interface for use by license-exempt devices on a noninterfering basis in spectrum allocated to the television broadcast service. This standard specifies the air interface, including the MAC and PHY, of fixed point-to-multipoint wireless regional area networks operating in the VHF–UHF TV broadcast bands between 54 and 862 MHz. This standard is intended to enable deployment of interoperable IEEE 802 multivendor wireless regional area network products, to facilitate competition in broadband access by providing alternatives to wireline broadband access and extending the deployability of such systems into diverse geographic areas, including sparsely populated rural areas, while preventing harmful interference to incumbent licensed services in the TV broadcast bands [4.44].

There is a large untapped market for broadband wireless access in rural and other unserved or underserved areas where wired infrastructure cannot be deployed economically. Products based on this standard will be able to serve those markets and increase the efficiency of spectrum utilization in spectrum currently allocated to, but unused by, the TV broadcast service. WRAN supports an approach for operation over large, potentially sparsely populated areas (e.g., rural areas), taking advantage of the favorable propagation characteristics in the VHF and low-UHF TV bands. The unique requirements of operating on a strict noninterference basis in spectrum assigned to, but unused by, the incumbent licensed services requires a new approach using purpose-designed cognitive radio techniques that will permeate both the PHY and MAC layers [4.44]. In principle, this wireless service can also be used to support metro-area WSNs.

Cognitive radio—where a device can sense its environment and location and then alter its power, frequency, modulation, and other parameters so as to dynamically reuse available spectrum—is now just emerging. CR can, in theory, allow multidimensional reuse of spectrum in space, frequency, and time, obliterating the spectrum and bandwidth limitations that have slowed broadband wireless development in the United States and elsewhere. This new technology is in a way similar to *software-defined radio* (SDR). With SDR the software embedded in a radio cell phone, for example, can define the parameters under which the phone should operate in real time as its user moves from place to place; traditional cell phone parameters, by contrast, are relatively fixed in terms of frequency band and protocol. A SDR is a flexible wireless communications device that implements its signal processing entirely in software: Software radios can easily change such features as modulation, bandwidth, and coding, which are fixed in more traditional radios. The basic technology of software radio is now being deployed in military and commercial applications. CR is even more advanced than SDR: CR, as noted, can sense its environment and learn from it [4.45]. The FCC is currently investigating commercial applications, and the Defense Advanced Research Projects Agency is proposing military applications (under the XG—or next-generation communications—program). DARPA's aim is to develop technology that allows multiple users to share spectrum in a way that coexists with, and complements, sharing protocols included in today's Wi-Fi technologies. Work on CR and IEEE 802.22 is currently under way.

3G Cellular Networks   Over the past decade, mobile communications technology has evolved from first-generation (1G) analog voice-only communications to second-generation (2G) digital, voice, and data communications. The demand for more cost-effective and feature-enhanced mobile applications has led to the development of new-generation wireless systems (or simply 3G). State-of-the-art 3G handsets are designed to provide multimegabit Internet access with an "always on" feature and data rates of up to 2.048 Mbps [4.17].

In reference to cellular applications, the core network of traditional cellular systems is typically based on a circuit-switched architecture similar to that utilized in wireline networks. Wireless service providers are now in the process of evolving their core networks to IP technology. Wireless telecommunications started as a subdiscipline of wireline telephony, and the absence of global standards resulted in regional standardization. Two major mobile telecommunications standards have emerged: time-division multiple access/code-division multiple access (TDMA/CDMA) developed by the Telecommunications Industry Association (TIA) in North America, and Global System for Mobile Communications (GSM) developed by the European Telecommunications Standards Institute (ETSI) in Europe. As one moves toward third-generation (3G) wireless services, there is a need to develop standards that are more global in scope [4.18].

In the late 1990s there were discussions on the development of standards for a 3G mobile system with a *core network* based on evolutions of the GSM and an *access network* based on all the radio access technologies (i.e., both frequency- and time-division duplex modes) supported by the plethora of different carriers (in different countries). This project was called the Third Generation Partnership Project (3GPP) [4.19]. Around the turn of this decade, the American National Standards Institute (ANSI) decided to establish the Third Generation Partnership Project 2 (3GPP2), a 3G partnership initiative for evolved ANSI/TIA/Electronics Industry Association (EIA) networks [4.20]. In addition, there also was the establishment of a strategic group called International Mobile Telecommunications-2000 (IMT-2000) within the International Telecommunication Union (ITU) [4.21], which focused its work on defining interfaces between 3G networks evolved from GSM on the one hand and ANSI on the other, with the goal of enabling seamless roaming between 3GPP and 3GPP2 networks. Because of the worldwide ("universal") roaming characteristic, 3GPP started referring to 3G mobile systems as the Universal Mobile Telecommunication System (UMTS) [4.22]. Since then, there has been advocacy for and progress toward an *all-IP UMTS network architecture*. The all-IP UMTS specifications replaced the earlier circuit-switched transport technologies by utilizing packet-switched transport technologies, and introduce multimedia support in the UMTS core network [4.22].

Figure 4.9 depicts some basic industry transition paths to 3G wireless. As implied in the preceding paragraph, currently the 3G world is split into two camps: the cdma2000, which is an evolution of the IS-95 standard, and the wideband code division multiple access (W-CDMA)/time-division synchronous CDMA (TD-SCDMA)/enhanced data rates for GSM evolution (EDGE) camp, whose standards are improvements of GSM, IS-136, and packet data cellular (PDC)—these are all second-generation standards. In the United States, Verizon Wireless and Sprint PCS were the first two carriers to develop 3G networks. The other major carriers have already advanced to the 2.5G technology, with the vision to soon join the 3G community [4.17].

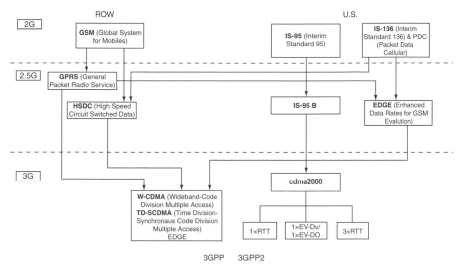

Figure 4.9   Migration path(s) to 3G wireless networks.

The original scope of 3GPP was to produce globally applicable technical specifications and technical reports for a 3G mobile system based on evolved GSM core networks and the radio access technologies that they support [i.e., universal terrestrial radio access (UTRA), both FDD and TDD modes]. The scope was subsequently amended to include maintenance and development of the GSM technical specifications and technical reports, including evolved radio access technologies [e.g., general packet radio service (GPRS) and EDGE] [4.23]. 3GPP and 3GPP2 also address the issue of the limited data throughput capabilities of 2G/2.5G systems, motivating providers to start work on 3G wideband radio technologies that can provide higher data rates (e.g., for Internet access, messaging, location-based services). This work resulted in 3G wireless radio technologies that provide data rates of 144 kbps for vehicular, 384 kbps for pedestrian, and 2 Mbps for indoor environments, and meet the ITU IMT-2000 requirements. Clearly, these channels can be utilized for WSN applications. Now that the radio technology standards to support higher data rates have been developed, the providers are focusing on development of standards for all-IP networks [4.18].

3GPP   The basic characteristics of an all-IP network are end-to-end IP connectivity, distributed control and services, and gateways to legacy networks [4.18]. As noted earlier in the chapter, there are two major protocol suites for supporting VoIP: session initiation protocol (SIP), standardized by the IETF, and H.323, standardized by the ITU. It was decided in 3GPP to use only SIP as the call control protocol between terminals and the mobile network. Interworking with other H.323 terminals (e.g., fixed H.323 hosts) is performed by a dedicated server in the network. New elements in this architecture, compared to a traditional 2G cellular network, are as follows (see also Figure 4.10) [4.22]:

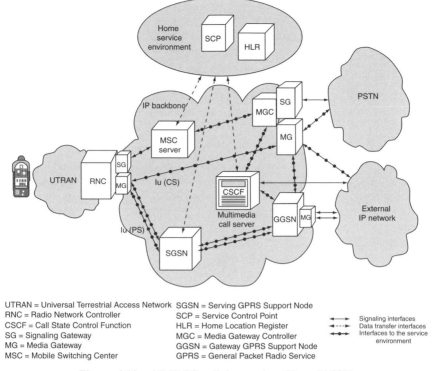

UTRAN = Universal Terrestrial Access Network   SGSN = Serving GPRS Support Node
RNC = Radio Network Controller   SCP = Service Control Point
CSCF = Call State Control Function   HLR = Home Location Register
SG = Signaling Gateway   MGC = Media Gateway Controller
MG = Media Gateway   GGSN = Gateway GPRS Support Node
MSC = Mobile Switching Center   GPRS = General Packet Radio Service

Signaling interfaces
Data transfer interfaces
Interfaces to the service environment

Figure 4.10   All-IP 3G cellular service. (From [4.22].)

1. *Mobile switching center (MSC) server.* The MSC server controls all calls coming from circuit-switched mobile terminals and mobile-terminated calls from a PSTN/GSM network to a circuit-switched terminal. The MSC server interacts with the media gateway control function (MGCF) for calls to and from the PSTN. There is a functional split of the MSC, where the call control and services part is maintained in the MSC server, and the switch is replaced by an IP router [Media Gateway (MG)]. This functional split reduces the deployment cost and guarantees the support of all existing services.

2. *Call state control function (CSCF).* The CSCF is an SIP server that provides or controls multimedia services for packet-switched (IP) terminals, both mobile and fixed.

3. *MG at the Universal Terrestrial Access Network (UTRAN) side.* The MG transforms VoIP packets into UMTS radio frames. The MG is controlled by the MGCF by means of Media Gateway Control Protocol ITU H.248. The media gateway is added to fulfill the second requirement. In Figure 4.10 the MG is drawn at the UTRAN side of the Iu interface, hence the Iu interface between the core network and UTRAN is IP-based. The MG can also be located at the core network side of the Iu interface (without impact on the UTRAN).

4. *MG at the PSTN side*. All calls coming from the PSTN are translated to VoIP calls for transport in the UMTS core network. This MG is controlled by the MGCF using the ITU H.248 protocol.

5. *Signaling gateway* (*SG*). An SG relays all call-related signaling to and from the PSTN and UTRAN on an IP bearer and sends the signaling data to the MGCF. The SG does not perform any translation at the signaling level.

6. *MGCF*. The first task of the MGCF is to control the MGs via H.248. Also, the MGCF performs translation at the call control signaling level between ISDN user part (ISUP) signaling used in the PSTN and SIP signaling used in the UMTS multimedia domain.

7. *Home subscriber server* (*HSS*). The HSS is the extension of the home location register (HLR) database with the subscribers' multimedia profile data.

For the transport of data traffic, UMTS uses the General Packet Radio Service (GPRS) network. For voice calls there are two options: for packet-switched mobile terminals, voice data are transported over the GPRS network using the GPRS tunneling protocol (GTP) on top of IP; all mobility is addressed by the GPRS protocols. For circuit-switched mobile terminals, voice samples are transported over IP between the MGs using the Iu frame protocol; in the latter case there is no tunneling; hence mobility has to be solved in a different way, by media gateway handovers.

An essential architectural principle of the 3GPP framework is to provide separation of service control from connection control. 3GPP started with GPRS as the core packet network and overlaid it with call control and gateway functions required for supporting VoIP and other multimedia services. The functions are provided via IETF-developed protocols to maintain compatibility with the industry direction in all-IP networks. These new networks also provide VoIP capabilities; the same capabilities that support VoIP can also support WSNs. To support VoIP, call control functions are provided by the call state control function (CSCF) (refer to Figure 4.10). The mobile terminal communicates with the CSCF via SIP protocols. The CSCF performs call control functions, service switching functions, address translation functions, and vocoder negotiation functions. For communication to the public-switched telephone network (PSTN) and legacy networks, PSTN gateways are utilized. To support roaming to 2G wireless networks, roaming gateway functions are also provided. The serving GPRS support node (SGSN) uses existing GSM registration and authentication schemes to verify the identity of the data user. This makes the SGSN access-technology-dependent. The GPRS HLR is enhanced for services that use IP protocols. The data terminal makes itself known to the packet network by doing a *GPRS-attach*. The IP address is anchored in the GPRS gateway node, GGSN, during the entire data session. This limits the mobility of the data terminal to within GPRS-based networks. To provide mobility with other networks, a MIP foreign agent can be incorporated in the GGSN [4.18].

3G Release 1999 was the first release of the 3GPP specifications; it was essentially a consolidation of the underlying GSM specifications and the development of the new UTRAN radio access network. The foundations were laid for future high-speed traffic transfer in both circuit- and packet-switched modes. That release was followed over the years by Releases 4, 5, and 6 [4.23]. Release 1999 was an introductory specification on the architecture of the UMTS network. According to Release 1999, UMTS comprises a UTRAN and two core networks [circuit-switched core network (CS-CN) and packet-switched core network (PS-CN)], which link up to services networks such as the PSTN and the Internet. Thus, using both traditional circuit- and modern packet-switched networks, UMTS Release 1999 supports various services, including voice, data (fax, SMS), and Internet access. Later, Release 4 adapted to the same architecture added more services to the UMTS network. The coexistence of two core networks, however, signified many limitations compared to competitive 3G systems, especially in video and multimedia services. Release 5 was a solution to the limitations that came along to modernize the UMTS architecture currently employed in 3G networks around the world. In this final phase, the PS-CN dominates the CS-CN and takes responsibility for telephony services. Systems based on UMTS Release 5 have much lower infrastructure and maintenance costs and provide enhanced services. Release 6 added additional capabilities [4.17].

As seen at the macro level in Figure 4.11, a new component is added to the basic UMTS architecture: the supplementary IP Multimedia Subsystem (IMS). IMS aims at supporting both telephony and multimedia services. IMS's role in UMTS architecture is to interact with both the PSTN and the Internet to provide all types of multimedia services to users. The CSCF element in the IMS infrastructure is responsible for signaling messages between all IMS components in order to control multimedia sessions originated by the user. Consequently, there is a proxy-CSCF (P-CSCF), an interrogating-CSCF (I-CSCF), and a serving-CSCF (S-CSCF), all responsible for particular signaling functions using SIP. The P-CSCF's responsibility

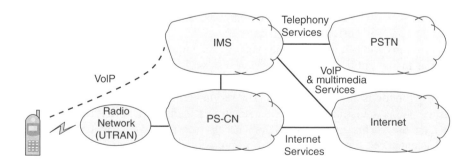

UTRAN = UMTS Terrestrial Access Network
PS-CN = Packet Switched Core Network
IMS = IP Multimedia Subsystem

Figure 4.11   UMTS Release 5 basic architecture.

is to act as the QoS enforcement point and to provide local control for emergency services. I-CSCF is an optional component that interacts with the HSS to find the location of the S-CSCF (it is optional because the P-CSCF can be set up to negotiate directly with the S-CSCF). The S-CSCF controls all the session management functions for the IMS. Depending on the capabilities of the IMS and the capacity requirements, there may be more than one S-CSCF node, and others can eventually be added to the system. The function of the HSS is to handle all user information, such as subscription and location queries. The HSS communicates with the CSCFs via an IP-based protocol called Cx interface; all other IMS components interact with each other via SIP. The media gateway control function (MGCF) is in charge of controlling one or more MGs; the MGCF interacts with the S-CSCF and the transport signaling gateway (T-SGW). MGs are bit processors for end-to-end users; their function is to convert PCM in the PSTN to IP-based formats, and vice versa. Finally, the T-SGW is included in the IMS because of the need to convert signaling system number 7 (SS7) to IP since the PSTN is only SS7-compatible [4.17].

3GPP2 3GPP2 has also undertaken work to enhance the IP architecture for multimedia services (including voice). The approach here is to capitalize on the synergies of Internet technologies and to use a single network for all services. 3GPP2 has created a new packet data architecture building on the CDMA 2G and 3G air interface data services. 3GPP2 has taken advantage of 3G high data rates and existing work in IETF on MIP to enhance the network architecture to provide IP capabilities. One advantage of using IETF protocols is ease in interworking and roaming with other IP networks. The other major advantage is that it can provide private network access (virtual private networking) via a MIP tunnel with IP security [4.18].

In the 3GPP2 architecture, IP connectivity reaches all the way to the base station transceiver (BTS). Both the base station controller (BSC) and BTS are contained in the IP-based radio access network node. This means that the BSC will be a router-based IP node containing some critical radio control functions (e.g., power control, soft handoff frame selection). The remaining control functions, such as call and session control, mobility management, and gateway functions, are moved out to the managed IP network. This allows for a distributed and modular control architecture. Since much of the communication will be between wireless and legacy terminals, gateway functions are provided for roaming to 2G wireless networks and interworking with the PSTN. In the 3GPP2 architecture, the mobile terminal uses mobile-IP-based protocols to identify itself. The packet data serving node (PDSN) contains a MIP foreign agent (FA) functionality. When the mobile terminal attaches to the FA, the FA establishes a mobile IP tunnel to the home agent (HA) and sends a registration message to the HA. The HA accesses the authorization, authentication, and accounting (AAA) server to authenticate the mobile terminal. The IP address of the mobile terminal is now anchored in the HA for the duration of the data session. The data device connected to the mobile terminal can be handed over to any other access device that supports mobile IP. Thus, this approach can provide mobility across different access networks (wireless, wireline, etc.). However, since it essentially uses address translation to provide mobility, it cannot do fast handoff, due to the latency of address updates from distant agents [4.18].

Comparison of Services   The 3GPP and 3GPP2 architectures are different because of the underlying base networks and evolution strategies. In 3GPP, GPRS-based mobility was already defined, so the IP network enhancements were considered on top of GPRS. On the other hand, 3GPP2 needed to develop a mobility mechanism for packet data since one did not exist previously. As noted, 3GPP2 has decided to use MIP as the basis for packet data mobility [4.18].

To illustrate the similarities and differences of the two approaches, mobility needs to be addressed at three levels: air-interface mobility, link-level mobility, and network-level mobility. Air-interface mobility supports cell-to-cell handoff within a radio access network. Link-level mobility maintains a point-to-point protocol (PPP) context across multiple radio access networks. Network-level mobility provides mobility across networks. In both approaches, air-interface mobility is handled in the radio access network. Air-interface mobility is specific to the radio technology, therefore harmonization of the two depends on the harmonization efforts under way for global CDMA. In 3GPP, link-level mobility is handled by GTP; this protocol is used to provide mobility to other 3GPP-defined networks. The 3GPP architecture also provides an option in which an FA may be located in the GGSN. This allows roaming from GPRS-based networks to other IP access networks. In 3GPP2, link-level mobility is provided by defining a tunneling protocol as an extension of MIP. The MIP architecture allows the mobile device to have a point of presence and to roam across any IP network. Registration and authentication in the 3GPP architecture for access and data networks are integrated and utilize the schemes used for wireless. In the 3GPP2 architecture, the registration and authentication for access and data networks are performed separately. For a data network, authentication and registration as defined in MIP are used; hence, the data architecture is access-independent [4.18].

3G Operators   After many delays, 3G networks are now being rolled out. 3G wireless networks offer all the normal mobile telephony services plus high-speed data access. 3G operators may initially limit data access to their own branded data services or at least price open Internet access significantly higher than access to their own traditional data services. The mobile market, however, is competitive, and there are consumer and business requirements for access to the open Internet. In fact, flat-rate bundles for data access services are already available in some markets. This data-channel access can be used to support VoIP services [4.24]. Wireless operators that are looking to continue to displace wireline voice revenues as their business posture need to reduce their overall delivery costs as users move from 2G TDM to 3G VoIP [4.25]. Below we look briefly at the VoIP possibilities because a successful commercial "play" in this space would accelerate the deployment (and ubiquity) of 3G services, thereby indirectly opening up an opportunity for WSN applications.

For example, equipment upgrades can introduce high-speed data capabilities to UMTS networks. Specifically, new technologies now becoming available enable carriers to provide new "blended lifestyle services" via any wireline, wireless, or Wi-Fi/WiMax endpoint by providing a variety of 3GPP IMS functional elements (as discussed previously), including the call session control functions, media resource function controller, policy decision function, and breakout gateway control function. Because this equipment expands the data channel on 3G cellular networks, these upgrades also lay the foundation for operators to introduce VoIP and more advanced multimedia services on their mobile networks (here one can transmit IP-voice datagrams over the data channel). VoIP over 3G gives operators the ability to support a greater number of voice users at a lower cost, in turn helping to ensure that voice services can continue to be delivered profitably. Some researchers estimate that 3G wireless can deliver voice by way of VoIP for a quarter of the cost per minute compared to 2G TDM methods [4.25].

For mobile operators that have invested heavily in 2G and 3G cellular networks, there may be relatively little incentive to offer VoIP services according to observers (their existing networks already deliver better-quality voice services at lower cost than VoIP can achieve today). However, VoIP may look more attractive to those service providers seeking to bypass mobile operators' traditional voice tariffs, particularly if an opportunity to undercut those tariffs using VoIP arises due to significant drops in 3G data pricing. A number of mobile operators have launched unlimited-use data tariffs that could make them vulnerable to customers using VoIP to cut their spend [4.26]. 3G service-provider VoIP offerings could appear in the United States in the 2008 or 2009 time frame. That would come after operators upgrade their 2.5G/3G networks. For example, upgrades to 1xEV DO provide peak data rates of about 1.8 Mbps compared to typical rates of 300 to 400 kbps for the current generation of 1xEV-DO [4.27].

Calculations of the threat to 3G revenues from broadband wireless (WiMax) have focused mainly on data, but as some 3G carriers start to put VoIP in a more central position in their strategies, they could find that this service segment is also affected. The 3G UMTS and CDMA technologies may have been the first to promise both voice and broadband-class data on one network and device, but the emergence of usable VoW has also moved formerly data-only approaches into this space. A potential early limit on VoIP over 3G data access could be the limited upstream capability of the initial 3G services. W-CDMA can deliver up to 384 kbps downstream but only 64 kbps upstream; it is preferable to have data rates exceeding 64 kbps, but if that is all that is available, one can make do for most VoIP services [4.24]. Road maps for data networks such as CDMA EVDO (evolution—data only) and UMTS's data-only strand, TDD,[2] now include VoIP [4.25].

---

[2]UMTS TDD mobile broadband technology is a packet data implementation of the international 3GPP UMTS standard. Unlike W-CDMA, which uses FDD (frequency division), UMTS TDD is designed to work in a single unpaired frequency band. One of the largest benefits of using TDD is that it supports variable asymmetry, meaning that an operator can dictate how much capacity is allocated to downlink versus uplink. As the traffic patterns for data typically heavily favor the downlink, this results in better use of spectrum assets and higher efficiency [4.23].

1. The shift is already visible in the CDMA market, even without taking into account challenges from broadband wireless. New EV-DO equipment aims at peak data rates of 3.1 Mbps and supports VoIP. As such, it could perhaps make a further upgrade to the next CDMA generation, EV-DV (evolution—data and voice) unnecessary. This equipment was expected to start shipping in 2006, and although EV-DO with VoIP will take advantage of the spectral efficiencies of CDMA less well than EV-DV, this will be outweighed by early availability and lower prices [4.25].

2. In the UMTS space, manufacturers have already developed a TDD mobile handset offering VoIP as well as the usual broadband packet-based services, and providers have completed the first successful transmission of a call from a mobile VoIP handset over UMTS TDD and claim that the network is ideal for voice because it features high capacity, low latency, and low power requirements. Their services will be more compelling if they can offer voice, and therefore they will be less likely to opt for a pure IP solution such as 802.16 instead of TDD. TDD-ready handsets are currently becoming commercially available [4.25].

Hotspot/WiMax Operators   For operators considering deployment of broadband wireless access technologies (e.g., WiMax), being able to offer VoIP could strengthen the business case for investing in such networks by moving operators beyond a focus on low-margin Internet access. Fixed/wireline operators have shown interest in use of wireless VoIP in trying to defend against fixed mobile substitution by developing services that combine VoIP over WLAN/hotspot/WiMax with cellular voice elsewhere [4.24,4.26]. Again, a successful VoIP application would drive deployment, which can be advantageous to WSN applications.

Fixed-Mobile Convergence Operators   Recently, there has been interest in fixed-mobile convergence (FMC). Mobile network operators plan to leverage emerging IMS service platforms to deliver "one phone, one number" telephony over both fixed and mobile infrastructure. This means that a mobile handset will use 2G/3G mobile infrastructure when the user is outdoors and VoIP over Wi-Fi when the user is at work or at home. Mobile operators see IMS and FMC as an opportunity to take additional market share from traditional fixed-line operators. However, once high-speed Internet access becomes available on mobile phones, a plethora of VoIP services will follow [4.24].

Most telephone calls originate from inside buildings, where cellular mobile coverage is poorest. As such, residential users are often forced to keep their fixed-line services for use when they are at home; the same applies in office buildings, with the added problem that wireless operators have not been in a position to offer the Centrex or PBX features that enterprises require. In theory, however, that could change with the advent of IMS and FMC [4.24].

To enable converged handsets, FMC relies on broadband Internet access for the fixed portion and WLANs now and WiMax in the future for the mobile portion. WLANs are deployed at a large percentage of enterprises, and home-based Wi-Fi setups are spreading rapidly. Broadband Internet access is also available in thousands of public hotspots. The first round of convergence depends on handsets that support 2G, 3G, and Wi-Fi connections on the same phone. Mobile operators then use an IMS platform to transparently combine regular mobile service on their 2G or 3G mobile network with VoIP services over Wi-Fi and/or fixed broadband access. Because of the fact that the mobile portion of FMC uses the existing mobile number and the existing mobile switching network elements, mobile operators have an advantage [4.24].

Without broadband Internet access, VoIP service providers are less of a threat to mobile operators' FMC services. The business proposition of fixed-mobile convergence is to hit the sweet spot of high convenience and low cost [4.24]. VoIP vendors will be in a better position to provide their own FMC if WiMax delivers on its promise of wireless broadband Internet access; however, widespread WiMax deployment is expected to take a number of years. Instead, the VoIP competitive threat may be enabled by the mobile operators' own data services [4.24]. A successful VoIP penetration could indirectly drive WSN applications by building out the infrastructure.

4.4   CONCLUSION

In this chapter we looked at radio transmission issues. To maximize the opportunity for widespread and cost-effective deployment of WSN, plans are to use existing and/or emerging COTS wireless communications and infrastructures rather than having to develop an entirely new, specially designed apparatus. WSNs can use a number of wireless COTS technologies, such as Bluetooth, ZigBee, WLAN/hotspots, WiMax, and 3G.

APPENDIX A: MODULATION BASICS

Modulation Capsule   We have indicated that WSNs implementers will probably use off-the-shelf radio technology such as ZibBee, WiMax, Wi-Fi, or 3G; this means that they do not necessarily have to worry about the fundamental aspects of radio science and modulation. However, a brief discussion of modulation is in order. Table 4A.1 lists some key terms related to modulation, from various sources, including [4.34], [4.37], and [4.38]. In the context of digital transmission and modulation, the related topic of digital encoding is also of interest.

A basic technique used in radio transmission is phase-shift keying (PSK), mentioned in the body of the chapter (Table 4A.1 lists a number of approaches, but PSK is a fundamental methodology). In PSK the frequency and amplitude of the carrying signal are both kept constant. Phase-coherent PSK utilizes two defined signals: A logic 0 is represented by a π-degree phase shift, and a logic 0 is represented by a 0-degree phase shift; this is, however, a complex situation for the

TABLE 4A.1 Basic Modulation Terminology

(Multiple) phase-shift keying (M-PSK)	(e.g., 8-PSK) In digital transmission, angle modulation in which the phase of the carrier is discretely varied in relation either to a *reference phase* or to *the phase of the immediately preceding signal element*, in accordance with data being transmitted. In a communications system, the representing of characters, such as bits or quaternary digits, by a shift in the phase of an electromagnetic carrier wave with respect to a reference, by an amount corresponding to the symbol being encoded.

For M-ary PSK, M different phases are required, and every n(where $M = 2^n$) bits of the binary bit stream are coded as one signal that is transmitted as $A \sin(\omega t + \theta_j) \, j = 1, \ldots, M$. The output is a baseband representation of the modulated signal. The M-ary number parameter, M, is the number of points in the signal constellation. Baseband M-ary phase-shift keying modulation with a phase offset of Θ maps an integer m between 0 and $M - 1$ to the complex value $\exp(j\Theta + j2\pi m/M)$. The modulator accepts binary representations of integers between 0 and $M - 1$. It modulates each group of K bits, called a binary *word*. The input can be either a vector of length K or a frame-based column vector whose length is an integer multiple of K.

Note 1: BPSK is the same as 2-PSK; QPSK is the same as 4-PSK; 8-ary-PSK is the same as 8-PSK. (Q = quarternary.)

Note 2: For example, when encoding bits, the phase shift could be 0° for encoding a "0" and π for encoding a "1," or the phase shift could be $-\pi/2$ for "0" and $+\pi/2$ for "1", thus making the representations for 0 and 1 a total of π apart.

Note 3: In PSK systems designed so that the carrier can assume only two different phase angles, each change of phase carries one bit of information (i.e., the bit rate equals the modulation rate); if the number of recognizable phase angles is increased to four, 2 bits of information can be encoded into each signal element; similarly, eight phase angles can encode 3 bits in each signal element.

(Noncoherent) differentially detected DPSK (DDPSK)	(e.g., 8-DPSK) Phase-shift keying that is used for digital transmission in which the phase of the carrier is varied discretely in relation to the *phase of the immediately preceding signal element* and in accordance with the data being transmitted.
Binary PSK (BPSK)	*See* (Multiple) phase-shift keying (M-PSK), Note 1.

For binary PSK (BPSK):

$$S_0(t) = A \cos \omega t \qquad \text{represents binary 0}$$
$$S_1(t) = A \cos(\omega t + \pi) \qquad \text{represents binary 1}$$

A BPSK modulator modulates a signal using the binary phase-shift keying method. The output of the modulator is a baseband representation of the signal modulated. The input

TABLE 4A.1 (*Continued*)

	is a discrete-time binary-valued signal. If the input bit is 0 or 1, respectively, the modulated symbol is $\exp(j\Theta)$ or $-\exp(j\Theta)$ respectively, where Θ is the phase offset parameter.
BPSK	*See* Binary PSK.
CDPSK	*See* Coherent(ly detected) DPSK.
Coherent	Pertaining to a fixed phase relationship between corresponding points on an electromagnetic wave. *Note:* A truly coherent wave would be perfectly coherent at all points in space. In practice, however, the region of high coherence may extend over only a finite distance.
Coherent demodulation	Demodulation using a carrier reference that is synchronized in frequency and phase to the carrier used in the modulation process.
Coherent(ly detected) DPSK (CDPSK)	(e.g., 8-CDPSK) Phase-shift keying that is used for digital transmission, in which the phase of the carrier is discretely modulated in *relation to the phase of a reference signal* and in accordance with data to be transmitted, and in which the modulated carrier is of constant amplitude and frequency. *Note:* A phase comparison is made of successive pulses, and information is recovered by examining the phase transitions between the carrier and successive pulses rather than by the absolute phases of the pulses.
DBPSK	*See* Differential binary phase-shift keying.
DDPSK	*See* (Noncoherent) differentially detected DPSK.
Differential(ly encoded) phase-shift keying (DPSK) modulation	A form of PSK in which the reference phase for a given interval is the phase of the signal during the preceding interval.
Differential binary phase-shift keying (DBPSK)	A DBPSK modulator modulates a signal using the differential space binary phase-shift keying method. The output is a baseband representation of the signal modulated. The input is a discrete-time binary-valued signal; the input can be either a scalar or a frame-based column vector.
	• If the first input bit is 0 or 1, respectively, the first symbol modulated is $\exp(j\Theta)$ or $-\exp(j\Theta)$, respectively, where Θ is the phase offset parameter. • If a successive input bit is 0 or 1, respectively, the symbol modulated is the previous modulated symbol multiplied by $\exp(j\Theta)$ or $-\exp(j\Theta)$, respectively.
Differential detection	As an alternative to recovering a coherent reference, some systems just compare the phase in the present interval to the phase in the previous intervals. The signal received in the preceding interval is delayed for one signal interval and is used as a reference to demodulate the signal in the next interval. Assuming that the data have been encoded in terms of phase shift instead of absolute phase positions,

(*Continued*)

TABLE 4A.1   *(Continued)*

	one can decode the data properly. Hence, this technique, referred to as *differential detection*, inherently requires differential encoding. In general, PSK systems require differential encoding since the receivers have no means of determining whether a recovered reference is a sine reference or a cosine reference. Furthermore, the polarity of the recovered reference is ambiguous. Thus, error probabilities for PSK systems are doubled automatically because of the differential encoding process. Differential detection, on the other hand, implies an even greater loss of performance since a noisy reference is used in the demodulation process. Typically, differential detection imposes a penalty of 1 to 2 dB in the SNR.
Differential modulation	Modulation in which the choice of the significant condition for for any signal element is dependent on the significant condition for the preceding signal element.
Differential quaternary phase-shift keying (DQPSK)	A modulator that modulates a signal using the differential quaternary phase-shift keying method. The input contains pairs of binary values. The output is a baseband representation of the signal modulated. The input can be either a vector of length 2 or a frame-based column vector whose length is an even integer. The figure below shows the signal constellation for the DQPSK modulation method when the phase offset parameter Θ is $\pi/4$. The arrows indicate the four possible transitions from each symbol to the next symbol.

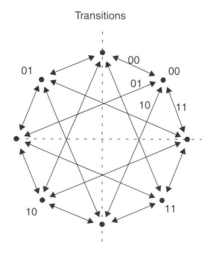

| DQPSK | *See* Differential quaternary phase shift keying. |
| ECC | *See* Forward error-correction coding. |

TABLE 4A.1   (*Continued*)

Forward error-correction (FEC) coding	(a.k.a. error-correction coding) Achieved by adding redundancy, such as parity-check symbols, to a message before transmission.This redundancy provides the corresponding decoder at the receiver with information such that the receiver can detect and correct transmission errors. FEC has potential application whenever digital data move over an imperfect (e.g., noisy) channel, such as satellite communications systems, wireless LANs, WSNs, digital cellular communications, digital video broadcast, and others. Among the most powerful and common FECs today are Reed–Solomon (RS) codes, convolutional codes, and combinations of the two. M. Reed and G. Solomon developed the codes in 1960; the codes are the standard FECs for Intelsat and for digital video broadcasting applications. Convolutional codes are another group of powerful codes that became popular with introduction of the Viterbi decoding algorithm. The concatenation of these two Vcodes, Reed–Solomon–Viterbi (RSV), has for many years represented the state of the art in FEC [4.38].
M-ary differential phase-shift keying (M-DPSK)	The M-DPSK modulator modulates a signal using the M-ary differential phase-shift keying method. The output is a baseband representation of the signal modulated. The M-ary number parameter, M, is the number of possible output symbols that can immediately follow a given output symbol. The input must be a discrete-time signal. The modulator accepts binary representations of integers between 0 and $M-1$. It modulates each group of K bits, called a binary *word*. The input can be either a vector of length K or a frame-based column vector whose length is an integer multiple of K.
M-DPSK	*See* M-ary differential phase-shift keying.
M-PSK	*See* (Multiple) phase-shift keying.
Phase coherent (phase coherence)	The state in which two signals maintain a fixed phase relationship with each other or with a third signal that can serve as a reference for each.
Phase modulation (PM)	Angle modulation in which the phase angle of a carrier is caused to depart from its reference value by an amount proportional to the instantaneous value of the modulating signal.
Phase-shift keying (PSK)	The form of phase modulation in which the modulation function shifts the instantaneous phase of the modulated wave (signal) between predetermined discrete values (e.g., when encoding bits, the phase shift could be $0°$ for encoding a 0 and π for encodinga 1, or the phase shift could be $-\pi/2$ for 0 and $+\pi/2$ for 1, thus making the representations for 0 and 1 a total of π apart).
QAM	*See* Quadrature amplitude modulation.
QPSK	*See* Quaternary PSK.

(*Continued*)

TABLE 4A.1   (*Continued*)

Quadrature amplitude modulation (QAM)	To achieve higher-speed data communication, a combination of PSK and AM can be used, producing the QAM method. This makes use of 0-, $\pi/2$-, π-, $\frac{3}{2}\pi$-degree phase shifts together with ASK.
Quaternary PSK (QPSK)	*See also* (Multiple) phase-shift keying (M-PSK), Note 1. If we define four signals, each with a phase shift differing by 90°, we have quadrature phase-shift keying (QPSK).
	The input binary bit stream $\{dk\}$, $dk = 0,1,2,\ldots$, arrives at the modulator input at a rate of $1/T$ bps and is separated into two data streams $dI(t)$ and $dQ(t)$ containing odd and even bits, respectively:
	$$d_1(t) = d_0, d_2, d_4, \ldots$$ $$d_0(t) = d_1, d_3, d_5, \ldots$$

receiver because the phase shifts are from an absolute value (see Figure 4A.1). In differential PSK (DPSK) there is a phase shift relative to the previous logic bit transmitted. Binary 0 is a $\pi/2$-degree phase change from the previous logic bit and binary 1 is a $\frac{3}{2}\pi$-degree phase change from the previous logic bit; here, the receiver only needs to detect the phase change that took place from the preceding bit, rather than being compared to an absolute value that it, somehow, needs to know.

As noted in the preceding paragraph, it is often desirable to reduce the complexity of a receiver by removing some of the phase-tracking requirements on the demodulator. This can be done by differentially encoding the binary data prior to transmission; this process encodes the data not in the absolute phase of the transmitted symbol but in the phase difference between two consecutively transmitted symbols. In a PSK environment, this technique, known as *differentially encoded PSK* (DPSK), is typically applied to either BPSK or QPSK signal sets (see Figure 4A.2). There are two levels of simplification available in differentially encoded schemes.

1. *Coherently detected DPSK* uses a coherent signal to perform demodulation. The differential decoding simply allows for the removal of ambiguity at the receiver, due to symmetries in the transmitted signal set. The penalty taken

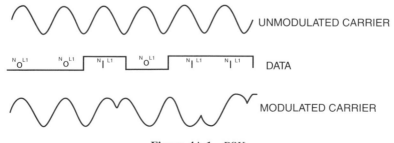

Figure 4A.1   PSK.

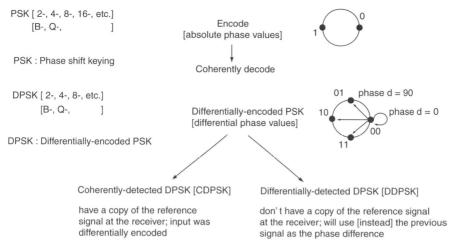

Figure 4A.2   Differentially encoded schemes.

for this simplification is typically about 1 to 2 dB of performance. Note that this performance loss is due to the fact that the effective bit error rate at the input of the decoder is approximately double that of nondifferentially encoded PSK. This is an inherent quality of the DPSK scheme, and cannot be avoided.

2. *Differentially detected DPSK* completely eliminates the coherent demodulation signal. This technique allows for further simplification at the receiver, but suffers slightly reduced performance relative to coherently detected DPSK. For both types of DPSK, the soft metric values depend on both the current and previous points received.

Encoding Capsule   As stated earlier, in the context of digital transmission and modulation, digital encoding is the topic of focus. The science of digital encoding was developed in the late 1940s, with the development of Shannon's information theory. In 1948, Claude Shannon derived theoretical machinery that allows one to define the highest rate at which information can be transmitted reliably over a channel, and to compare the performance of a physical encoder–decoder with the limits of the medium. It is the challenge of the implementation engineer to come up with an encoder–decoder that is as close to "best" as possible while maintaining a reasonable level of complexity and cost [4.28–4.33]. Communication-system designers deal routinely with trade-offs among data reliability, efficient use of available spectrum, data throughput, and cost.

Forward error correction (FEC) is one of the most powerful tools available to address channel performance trade-offs, and better FEC yields more design options. For example, with 3 dB of coding gain one could increase range by 40%, reduce antenna size by 30%, reduce transmitter power by a factor of 2, and reduce the required noise figure of the receiver by 3 dB. Alternatively, one can use a higher-order modulation scheme, which can reduce the required bandwidth by 50% or increase data throughput by a factor of 2 [4.38]. The most common FEC algorithms in use today are (1) the Reed–Solomon (RS) codes (developed the codes in 1960), (2) the convolutional codes, and (3) combinations of these two. Convolutional codes became popular with the introduction of the Viterbi decoding algorithm. The concatenation of these two codes, Reed–Solomon–Viterbi (RSV), has been for many years the best-in-breed algorithm.

Whereas the IEEE 802.11b/g standard does not include any FEC mechanisms, the 802.16a standard utilizes FEC to improve range and reduce bandwidth congestion associated with packet retransmissions. The default form of FEC is the well-known Reed–Solomon concatenated with Viterbi (RSV); as an option, the 802.16a standard also supports the higher-performing block product codes (BPCs), alternatively called block turbo codes (BTCs), turbo product codes (TPCs), or Tanner product codes [4.39]. As noted in the body of the chapter, the IEEE 802.16a standard offers significantly improved bit rates and distances.

Until 1990 it was a widely accepted that the performance point of a practical encoder–decoder operating on an additive white Gaussian noise (AWGN) channel could be no closer than about 3 dB from the theoretical performance limit, known as *channel capacity*. This point, for a fixed SNR, is known as the *practical capacity*, and until recently, it was regarded as a barrier beyond which practical systems could not perform (as we note below, progress has been made in this arena). Multipath channels represent an even greater technical challenge, as discussed above, since communicating over these channels is even more difficult.

The decade of the 1990s gave rise to changes in digital transmission practices that have been utilized since 1948. The concept of iterative decoding of concatenated codes, commonly referred to as *turbo coding*, has given the engineering community a way to rethink how one transmits information. In 1993, Claude Berrou, Alain Glavieux, and Punya Thitimajshima shattered the concept of practical capacity with an encoder–decoder that achieved a bit error rate of 10^{-5} within 1 dB of channel capacity [4.35,4.36]. Essentially overnight, engineers were offered about 2 dB of additional coding performance on the AWGN channel at the expense of increased decoding complexity. Furthermore, for the first time in history, a reasonably practical coding alternative was offered that performed so close to the theoretical "best" that significant, additional performance gains were essentially impossible. Any additional improvements would have to come, not in the form of gains in coding performance (at least not gains over 1 dB), but in the form of reduced system complexity. The complete algorithm employed by Berrou et al. consists of two concatenated, recursive convolutional encoders. Because the encoder consists of concatenated convolutional encoders, this class of turbo codes is known as *turbo convolutional codes* (TCCs). The decoding algorithm employs two soft-in, soft-out (SISO) decoding modules to form the confidence metric for each transmitted information bit. These SISOs operate in an iterative manner that requires, for each iteration, a complete forward and reverse traversal of the trellis. Furthermore, for each iteration, the confidence of every data bit must be calculated using a very complicated summation over the paths and states of the current trellis stage. The complexity of such an algorithm, although acceptable for many software applications, is often restrictive in a hardware implementation [4.35,4.36].

Parallel, concatenated, convolutional turbo codes are good error-correction-coding technology but have limitations. The first problem was that more extensive computer simulation of the codes exposed a weakness in these codes. The performance of the codes at low bit error rates (BERs) was within 1 dB of capacity; however, the performance tailed off, or met an "error floor," at high BERs, such that legacy codes such as the Reed–Solomon were still superior. The second problem was that the complexity of the required SISO decoder was such that a cost-efficient decoder was unavailable for most commercial applications. In 1998, a new approach to turbo codes solved both of these problems. Using efficient SISO-decoder algorithm hardware developers recently introduced the first commercially viable turbo decoder based on the iterative decoding of product codes rather than the convolutional codes [4.38]. In the general sense, turbo product, or block turbo, codes are composed of a multidimensional array of block codes, such as Hamming and BCH (Bose–Chaudhuri–Hocquenghem) codes.

REFERENCES

[4.1] T. S. Rappaport, "An Introduction to Indoor Radio Propagation," http://sss-mag.com/indoor.html, 2001.

[4.2] R. H. Katz, "CS 294-7: Radio Propagation," White Paper, University of California–Berkeley, 1996.

[4.3] T. S. Rappaport, *Wireless Communications: Principles and Practice*, IEEE Press, Piscataway, NJ, 1996.

[4.4] B. Peters, "Sensing Without wires: Wireless Sensing Solves Many Problems, But Introduces a Few of Its Own," *Machine Design*, Penton Media, Cleveland, OH, http://www.machinedesign.com/ASP/viewSelectedArticle.asp?strArticleId=57795&str-&strSite=MDSite&Screen = & CURRENTISSUE&CatID=3.

[4.5] T. H. Lin, W. J. Kaiser, G. J. Pottie, "Integrated Low-Power Communication System Design for Wireless Sensor Networks," *IEEE Communications*, Dec. 2004, pp. 142ff.

[4.6] IEEE 802.15 WPAN Task Group 1 (TG1), WPAN, June 20, 2005.

[4.7] Bluetooth SIG, www.bluetooth.com (more information at www.bluetooth.org).

[4.8] G. Fleishman, "Inside Bluetooth 2.0," *Macworld*, Feb. 9, 2005.

[4.9] C. Perkins, "IP Mobility Support for IPv4," RFC 3344, IETF, Sterling, VA, Aug. 2002.

[4.10] http://www.wave-report.com/tutorials/OFDM.htm.

[4.11] ZigBee Alliance, Bishop Ranch, CA, http://www.zigbee.org/.

[4.12] W. C. Craig, "ZigBee: Wireless Control That Simply Works," *ZMD America*, Dec. 8, 2004; "Open House," ZigBee Alliance, Bishop Ranch, CA, http://www.zigbee.org/.

[4.13] IEEE 802.15 WPAN Task Group 4 (TG4), http://grouper.ieee.org/groups/802/15/pub/TG4.html.

[4.14] E. Callaway, P. Gorday, L. Hester, J. A. Gutierrez, M. Neave, B. Heile, V. Bahl, "Home Networking with IEEE 802.15.4: A Developing Standard for Low-Rate Wireless

Personal Area Networks," *IEEE Communications*, Vol. 40, No. 8, pp. 70–77, Aug. 2002.

[4.15] D. Minoli, *Hotspot Networks: Wi-Fi for Public Access Locations*, McGraw-Hill, New York, 2002.

[4.16] WiMAX Forum, Beaverton, OR, http://www.wimaxforum.org/home.

[4.17] N. Mavrakis, "3G Wireless Systems: A Comprehensive Study," White Paper, Stevens Institute of Technology, Hoboken, NJ, Dec. 15, 2002.

[4.18] G. Patel, S. Dennett, "The 3GPP and 3GPP2 Movements Toward an All-IP Mobile Network," *IEEE Personal Communications*, Aug. 2000.

[4.19] 3GPP, European Telecommunications Standards Institute, http://www.3GPP.org.

[4.20] 3GPP2, European Telecommunications Standards Institute, http://www.3GPP2.org.

[4.21] ITU, "The IMT-2000 initiative," ITU-R Draft Record M; "Detailed Specifications of the Radio Interfaces of MT-2000," Document 8/126; International Telecommunications Union, Geneva, Switzerland, http://www.itu.int/imt.

[4.22] L. Bos, S. Leroy, "Toward an All-IP-Based UMTS System Architecture," *IEEE Network*, Jan. 2001.

[4.23] UMTS TDD Alliance, http://www.umtstdd.org.

[4.24] B. Turner, "The Impact of VoIP, Wi-Fi and 3G Data on Wireless Telecom: How Fixed-Mobile Convergence Will Reshape the Wireless Industry," International Wireless Telecom-Carriers Global Edition, World Media Online Limited, London, info@wirtel.co.uk.

[4.25] Staff, "Broadband Wireless Threatens 3G Voice Ambitions," *Wireless Watch*, Oct. 26, 2004.

[4.26] Online Magazine staff, "Wireless Voice over IP: Technical and Commercial Prospects," *3G Online Magazine*, Mar. 22, 2005, http://www.3g.co.uk/PR/March2005/1232.htm.

[4.27] Mobile Pipeline staff, "Verizon Looking at VoIP over 3G," *Mobile Pipeline*, Apr. 1, 2005, http://www.commweb.com/showArticle.jhtml?articleID=160401527.

[4.28] A. J. Viterbi, *Principles of Coherent Communication*, New York, McGraw-Hill, 1966.

[4.29] W. C. Lindsey, M. K. Simon, *Telecommunication Systems Engineering*, Prentice-Hall, Englewood Cliffs, NJ, 1973.

[4.30] W. C. Lindsey, "Phase-Shift-Keyed Signal Detection with Noisy Reference Signals," *IEEE Transactions on Aerospace and Electronic Systems*, Vol. 2, No. 4, July 1966, pp. 393–401.

[4.31] P. C. Jain, "Detection of a PSK Signal Transmitted Through a Hard-Limited Channel," *IEEE Transactions on Information Theory*, Vol. 19, No. 5, Sept. 1973, pp. 623–630.

[4.32] V. K. Prabhu, "PSK Performance with Imperfect Carrier Phase Recovery," *IEEE Transactions on Aerospace and Electronic Systems*, Vol. 12, No. 2, Mar. 1976, pp. 275–285.

[4.33] N. M. Blachman, "The Effect of Phase Error on DPSK Error Probability," *IEEE Transactions on Communications*, Vol. 29, No. 3, Mar. 1981, pp. 364–365.

[4.34] *IEEE Authoritative Dictionary of IEEE Standards Terms*, 7th ed., ANSI/IEEE Std. 100 (formerly called *IEEE Standard Dictionary of Electrical and Electronics Terms*), IEEE Press, Piscataway, NJ, 2001.

[4.35] C. Berrou, A. Glavieux, "Near Optimum Error-Correcting Coding and Decoding: Turbo Codes," *IEEE Transactions on Communications*, Vol. 44, Oct. 1996.

[4.36] C. Berrou, A. Glavieux, P. Thitimajshima, "Near Shannon Limit Error-Correcting Coding and Decoding: Turbo Codes," *Conference Record of the IEEE International Conference on Communications*, 1993.

[4.37] "Communications Blockset," MathWorks, Natick, MA, http://www.mathworks.de/access/helpdesk/help/toolbox/commblks/ref/mpskmodulatorbaseband.shtml#245366.

[4.38] D. Williams, "Turbo-Product Codes Advance ECC Technology," *Advanced Hardware Architectures*, edn, www.ednmag.com, Feb. 3, 2000, pp. 77ff.

[4.39] B. A. Banister, "Using Turbo Product Codes in Client Station Uplink for Reduced Power Consumption," White Paper, Comtech AHA Corporation, Moscow, ID, 2005.

[4.40] G. Karayannis, "Emerging Wireless Standards: Understanding the Role of IEEE 802.15.4 and ZigBee™ in AMR and Submetering—Mapping Your Future: From Data to Value," *AMRA 2003 International Symposium Conference Record*.

[4.41] M. Roberti, "Glossary of RFID Terms," RFID Journal, Inc., Hauppauge, NY, http://www.rfidjournal.com/article/articleview/208.

[4.42] J. Adams, "Designing with 802.15.4 and ZigBee," presented at the Industrial Wireless Applications Summit, San Diego, Ca, Mar. 9, 2004.

[4.43] C. Cordeiro, "Wireless Communication and Networking (WiCAN)," Wireless Communication and Networking (WiCAN) Department, Philips Research, Briarcliff Manor, NY, Carlos.Cordeiro@philips.com; www.ececs.uc.edu/~cordeicm.

[4.44] IEEE 802 LAN/MAN Standards Committee, "802.22 WG on WRANs (Wireless Regional Area Networks)," www.ieee802.org/22, Oct. 1, 2005.

[4.45] P. Mannion, "Sharing Spectrum the Smarter Way," *EE Times*, Apr. 5, 2004, http://www.commsdesign.com/news/tech_beat/www.eet.com/showArticle.jhtml?articleID=18700443.

5

MEDIUM ACCESS CONTROL PROTOCOLS FOR WIRELESS SENSOR NETWORKS

5.1 INTRODUCTION

WSNs are typically composed of a large number of low-cost, low-power, multifunctional wireless devices deployed over a geographical area in an ad hoc fashion and without careful planning. Individually, sensing devices are resource-constrained and therefore are only capable of a limited amount of processing and communication. It is the coordinated effort of these sensing devices, however, that bears promise for a significant impact on a wide range of applications in several fields, including science and engineering, military settings, critical infrastructure protection, and environmental monitoring [5.1–5.3].

Harnessing the potential benefits of WSNs requires a high-level of self-organization and coordination among the sensors to perform the tasks required to support the underlying application. At the heart of this collaborative effort to achieve communications is the need for the wireless sensor nodes to self-organize into a multihop wireless network. Consequently, the design of efficient communications and network protocols for WSNs becomes crucial for wireless sensor nodes to carry out successfully the mission for which they are deployed.

The establishment of a multihop wireless network infrastructure for data transfer requires the establishment of communication links between neighboring sensor nodes. Unlike communication over a guided medium in wired networks, however, communication in wireless networks is achieved in the form of electromagnetic signal transmission through the air. This common transmission medium must

Wireless Sensor Networks: Technology, Protocols, and Applications, by Kazem Sohraby, Daniel Minoli, and Taieb Znati

therefore be shared by all sensor network nodes in a fair manner. To achieve this goal, a medium access control protocol must be utilized. The choice of the medium access control protocol is the major determining factor in WSN performance. A number of access control protocols have been proposed for WSNs. The objective of this chapter is to discuss the fundamental design issues of medium access control for WSN methods and to provide an overview of these protocols. In Section 5.2, a description of the basic requirements of access control protocols is provided. In Section 5.3 we categorize the major media access control techniques used in shared medium access networks. In Section 5.4 we discuss specific requirements of access control methods for WSNs and describes several media access control (MAC) protocols for these networks.

5.2 BACKGROUND

Communication among wireless sensor nodes is usually achieved by means of a unique channel. It is the characteristic of this channel that only a single node can transmit a message at any given time. Therefore, shared access of the channel requires the establishment of a MAC protocol among the sensor nodes. The objective of the MAC protocol is to regulate access to the shared wireless medium such that the performance requirements of the underlying application are satisfied [5.4–5.7]. From the perspective of the Open Systems Interconnection (OSI) Reference Model (OSIRM), the MAC protocol functionalities are provided by the lower sublayer of the data link layer (DLL). The higher sublayer of the DLL is referred as the logical link control (LLC) layer. The subdivision of the data link layer into two sublayers is necessary to accommodate the logic required to manage access to a shared access communications medium. Furthermore, the presence of the LLC sublayer allows support for several MAC options, depending on the structure and topology of the network, the characteristics of the communication channel, and the quality of service requirements of the supported application.

Figure 5.1 depicts the OSI reference model and the logical architecture of the DLL for shared medium access in wireless networks. The physical layer (PHY) typically includes a specification of the transmission medium and the topology of the network. It defines the procedures and functions that must be performed by the physical device and the communications interface to achieve bit transmission and reception. It also coordinates the various functions necessary to transmit a stream of bits over the wireless communication medium. The major services provided by the physical layer typically include the encoding and decoding of signals, preamble generation and removal to achieve synchronization, and the transmission and reception of bits.

The MAC sublayer resides directly above the physical layer. It supports the following basic functions:

- The assembly of data into a frame for transmission by appending a header field containing addressing information and a trailer field for error detection

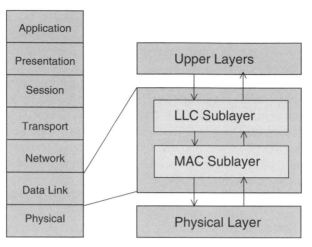

Figure 5.1   Open systems interconnection reference model and data link layer architecture.

- The disassembly of a received frame to extract addressing and error control information to perform address recognition and error detection and recovery
- The regulation of access to the shared transmission medium in a way commensurate with the performance requirements of the supported application

The LLC sublayer of the DDL provides a direct interface to the upper layer protocols. Its main purpose is to shield the upper layer protocols from the characteristics of the underlying physical network, thereby providing interoperability across different types of networks. The use of the LLC sublayer, however, has been very limited, as interoperability is typically achieved by other network layer protocols.

5.3   FUNDAMENTALS OF MAC PROTOCOLS

One major difficulty in designing effective MAC protocols for shared access media arises from the spatial distribution of the communicating nodes [5.8]. To reach agreement as to which node can access the communication channel at any given time, the nodes must exchange some amount of coordinating information. The exchange of this information, however, typically requires use of the communication channel itself. This recursive aspect of the multiaccess medium problem increases the complexity of the access control protocol and consequently, the overhead required to regulate access among the competing nodes. Furthermore, spatial distribution does not allow a given node on the network to know the instantaneous status of other nodes on the network. Any information explicitly or implicitly gathered by any node is at least as old as the time required for its propagation through the communication channel.

Two main factors, the intelligence of the decision made by the access protocol and the overhead involved, influence the aggregate behavior of a distributed multiple-access protocol. These two factors are unavoidably intertwined. An attempt to improve the quality of decisions does not necessarily reduce the overhead incurred. On the other hand, reducing the overhead is likely to lower the quality of the decision. Thus, a trade-off between these two factors must be made.

Determining the nature and extent of information used by a distributed multiple-access protocol is a difficult task, but potentially a valuable one. An understanding of exactly what information is needed could lead to an appreciation of its value. Most of the proposed distributed multiple-access protocols for WSNs operate somewhere along a spectrum of information ranging from a minimum amount of information to perfect information. Furthermore, the information can be predetermined, dynamic global, or local. *Predetermined information* is known to all communicating nodes. *Dynamic global information* is acquired by different nodes during protocol operation. *Local information* is known to individual nodes. Predetermined and dynamic global information may result in efficient, potentially perfect coordination among the nodes. However, there usually is a high price to pay in terms of wasted channel capacity. The use of local information has potential to reduce the overhead required to coordinate the competing nodes, but may result in poor overall performance of the protocol.

The trade-off between the efficiency of the MAC protocol and the overhead required to achieve it has been at the basis of most of the access techniques for shared-medium networks. In the remaining part of this section, the performance metrics for the MAC protocol are described and the major common techniques to regulate access to the medium are discussed.

5.3.1 Performance Requirements

In trying to determine the performance requirements of MAC protocols, the scope of research has been very broad [5.9]. Traditionally, issues such as delay, throughput, robustness, scalability, stability, and fairness have dominated the design of MAC protocols [5.10, 5.11]. Following is a brief discussion of these performance metrics.

Delay   *Delay* refers to the amount of time spent by a data packet in the MAC layer before it is transmitted successfully. Delay depends not only on the network traffic load, but also on the design choices of the MAC protocol. For time-critical applications, the MAC protocol is required to support delay-bound guarantees necessary for these applications to meet their QoS requirements [5.12]. The precise semantics of the QoS requirements are application-dependent. Guaranteed delay bounds are usually provided through careful message scheduling both locally within a communicating node and globally among all nodes in the network. Two types of delay guarantees can be identified, probabilistic and deterministic. *Probabilistic delay guarantees* are typically characterized by an expected value, a variance and a confidence interval. *Deterministic delay guarantees* ensure a predictable number of state transitions between message arrival and message transmission. Therefore, deterministic MAC schemes guarantee an upper bound for the access time. Determinism is a crucial requirement in a real-time environment, where the correctness of the application depends on the adherence of its underlying tasks to their specified execution deadline.

Throughput *Throughput* is typically defined as the rate at which messages are serviced by a communication system. It is usually measured either in messages per second or bits per second. In wireless environments it represents the fraction of the channel capacity used for data transmission. Throughput increases as the load on the communication system increases initially. After the load reaches a certain threshold, the throughput ceases to increase, and in some cases, it may start to decrease. An important objective of a MAC protocol is to maximize the channel throughput while minimizing message delay.

Robustness *Robustness*, defined as a combination of reliability, availability, and dependability requirements, reflects the degree of the protocol insensitivity to errors and misinformation. Robustness is a multidimensional activity that must simultaneously address issues such as error confinement, error detection and masking, reconfiguration, and restart. Achieving robustness in a time-varying network such as a WSN is difficult, as it depends strongly on the failure models of both the links and the communicating nodes.

Scalability *Scalability* refers to the ability of a communications system to meet its performance characteristics regardless of the size of the network or the number of competing nodes. In WSNs, the number of sensor nodes may be very large, exceeding thousands and in some cases millions of nodes. In these networks, scalability becomes a critical factor. Achieving scalability is challenging, especially in timevarying environments such as wireless networks. A common approach to achieve scalability is to avoid relying on globally consistent network states. Another approach is to localize interactions among the communicating nodes, through the development of hierarchical structures and information aggregation strategies. Grouping sensor nodes into clusters, for example, allows the design of shared medium access protocols which are highly scalable. Similarly, aggregating information from different sensors allows the development of traffic patterns which can be exploited efficiently to scale the MAC protocol to a large number of sensor nodes.

Stability *Stability* refers to the ability of a communications system to handle fluctuations of the traffic load over sustained periods of time. A stable MAC protocol, for example, must be able to handle instantaneous loads which exceed the maximum sustained load as long as the long-term load offered does not exceed the maximum capacity of the channel. Typically, the scalability of a MAC protocol is studied with respect to either delay or throughput. A MAC protocol is considered to be stable, with respect to delay, if the message waiting time is bounded. These systems can be characterized by a bounded backlog of messages in the transmission queue. With respect to throughput, a MAC protocol is stable if the throughput does not collapse as the load offered increases. Accommodating load fluctuations while maintaining system stability is difficult to achieve in time-varying large-scale WSNs. One possible approach is for the MAC protocol to adapt to high fluctuations in the traffic load through careful scheduling of bursty traffic.

Fairness   A MAC protocol is considered to be *fair* if it allocates channel capacity evenly among the competing communicating nodes without unduly reducing the network throughput. Achieving fairness among competing nodes is desirable to achieve equitable QoS and avoid situations where some nodes fare better than other nodes. As a result, no application is starved or penalized excessively. It is worth noting that the definition of fairness above assumes that the demands of all communicating nodes, expressed in terms of channel capacity, are equivalent. It could be the case, however, that the network must accommodate various traffic sources with different traffic generation patterns and a wide range of QoS requirements. To accommodate heterogeneous resource demands, communicating nodes are assigned different weights to reflect their relative resource share. Proportional fairness is then achieved based on the weights assigned. A MAC protocol is considered to be proportionally fair if it is not possible to increase the allocation of any competing node without reducing the service rate of another node below its proportional fair share.

Fair resource allocation in wireless networks is difficult to achieve, as global information may be required to coordinate access to the communication medium among all contending stations. The time-varying characteristics of the wireless links makes it difficult to compute the fair share of each contending node, even if a centralized resource allocation approach is used.

Energy Efficiency   A sensor node is equipped with one or more integrated sensors, embedded processors with limited capability, and short-range radio communication ability as discussed in Chapter 3. These sensor nodes are powered using batteries with small capacity. Unlike in standard wireless networks, wireless sensor nodes are often deployed in unattended environments, making it difficult to change their batteries. Furthermore, recharging sensor batteries by energy scavenging is complicated and volatile. These severe constraints have a direct impact on the lifetime of a sensor node. As a result, energy conservation becomes of paramount importance in WSNs to prolong the lifetime of sensor nodes. One possible approach to reducing energy consumption at a sensor node is to use low-power electronics. The integration of low-power chips in the design of sensor nodes is a necessary step toward achieving high levels of power efficiency. Energy gains resulting from energy-efficient chip design, however, can easily be squandered if the processing and communication capabilities of the sensor node are not operated efficiently. Achieving this goal requires the design of energy-aware communication protocols.

Energy efficiency is one of the most important issues in the design of MAC protocol for wireless sensor nodes. Several sources contribute to energy inefficiency in

MAC-layer protocols [5.44]. The first source of energy waste is *collision*, which occurs when two or more sensor nodes attempt to transmit simultaneously. The need to retransmit a packet that has been corrupted by a collision increases energy consumption. The second source of energy waste is *idle listening*. A sensor node enters this mode when it is listening for a traffic that is not sent. This energy expended monitoring a silent channel can be high in several sensor network applications. The third source of energy waste is *overhearing* which occurs when a sensor node receives packets that are destined to other nodes. Due to their low transmitter output, receivers in sensor nodes may dissipate a large amount of power. The fourth major source of energy waste is caused by *control packet overhead*. Control packets are required to regulate access to the transmission channel. A high number of control packets transmitted, relative to the number of data packets delivered indicates low energy efficiency. Finally, *frequent switching* between different operation modes may result in significant energy consumption. Limiting the number of transitions between sleep and active modes, for example, leads to considerable energy saving.

Energy-efficient link-layer protocols achieve energy savings by controlling the radio to eliminate, or at least reduce, energy waste caused by the sources noted above. Further energy gains can be achieved using comprehensive energy management schemes which focus not only on the sensor node radio, but equally important, on other sources of energy consumption.

5.3.2 Common Protocols

The choice of the MAC method is the major determining factor in the performance of a WSN. Several strategies have been proposed to solve the shared medium access problem. These strategies attempt, by various mechanisms, to strike a balance between achieving the highest-quality resource allocation decision and the overhead necessary to reach this decision. These strategies can be classified in three major categories: fixed assignment, demand assignment, and random assignment.

Fixed-Assignment Protocols   In fixed-assignment strategies, each node is allocated a predetermined fixed amount of the channel resources. Each node uses its allocated resources exclusively without competing with other nodes. Typical protocols that belong in this category include frequency-division multiple access (FDMA), time-division multiple access (TDMA), and code-division multiple access (CDMA) [5.13].

FDMA   The FDMA scheme is used by radio systems to share the radio spectrum. Based on this scheme, the available bandwidth is divided into subchannels. Multiple channel access is then achieved by allocating communicating nodes with different carrier frequencies of the radio spectrum. The bandwidth of each node's carrier is constrained within certain limits such that no interference, or overlap, occurs between different nodes. The scheme requires frequency synchronization among communicating nodes. Communication is achieved by having the receiver tune to the channel used by the transmitter.

TDMA TDMA is digital transmission technology that allows a number of communicating nodes to access a single radio-frequency channel without interference. This is achieved by dividing the radio frequency into time slots and then allocating unique time slots to each communicating node. Nodes take turns transmitting and receiving in a round-robin fashion. It is worth noting, however, that only one node is actually using the channel at any given time for the duration of a time slot.

CDMA CDMA is a spread spectrum (SS)–based scheme that allows multiple communicating nodes to transmit simultaneously. Spread spectrum is a radio-frequency modulation technique in which the radio energy is spread over a much wider bandwidth than that needed for the data rate. Systems based on spread-spectrum technology transmit an information signal by combining it with a noiselike signal of a much larger bandwidth to generate a wideband signal. Consequently, the signal transmitted occupies a larger bandwidth than that normally required to transmit the original information. Using wideband noiselike signals makes it hard to detect, intercept, or demodulate the original signal.

Most common spread spectrum–based systems use either frequency hopping (FH) or direct sequence (DS), although hybrid systems may use some combination of the two types. Frequency-hopping spectrum systems (FHSS) modulate the data signal with a narrowband carrier signal that hops over time from one frequency to another in a pseudorandom but predictable pattern selected from a wideband of frequencies [5.14]. For the signal to be decoded properly, the hopping patterns of the transmitting and receiving radios must be synchronized.

Direct-sequence spread-spectrum systems (DSSSs) divide the stream of information to be transmitted into small chunks, each of which is allocated to a frequency channel across the spectrum. A data signal at the sending node is combined with a higher-data-rate bit sequence, referred to as *chipping code*, which divides the user data according to a spreading ratio. The chipping code is a redundant bit pattern for each bit that is transmitted. This redundancy increases the resistance to interference of the signal transmitted and improves the likelihood of recovering the original data if one or more bits in the pattern are damaged during transmission.

Demand Assignment Protocols The main objective of demand assignment protocols is to improve channel utilization by allocating the capacity of the channel to contending nodes in an optimum or near-optimum fashion. Unlike fixed-assignment schemes, where channel capacity is assigned exclusively to the network nodes in a predetermined fashion regardless of their current communication needs, demand assignment protocols ignore idle nodes and consider only nodes that are ready to transmit. The channel is allocated to the node selected for a specified amount of time, which may vary from a fixed-time slot to the time it takes to transmit a data packet.

Demand assignment protocols typically require a network control mechanism to arbitrate access to the channel between contending nodes. Furthermore, a logical control channel, other than the data channel, may be required for contending stations to dynamically request access to the communication medium. Depending on the characteristics of the protocol, the need to request access to the channel may delay data transmission. Demand assignment protocols may be further classified as centralized or distributed. Polling schemes are representative of centralized control, whereas token- and reservation-based schemes use distributed control.

Polling A widely used demand assignment scheme is polling. In this scheme, a master control device queries, in some predetermined order, each slave node about whether it has data to transmit. If the polled node has data to transmit, it informs the controller of its intention to transmit. In response, the controller allocates the channel to the ready node, which uses the full data rate to transmit its traffic. If the node being polled has no data to transmit, it declines the controller's request. In response, the controller proceeds to query the next network node. The main advantage of polling is that all nodes can receive equal access to the channel. Preference can, however, be given to high-priority nodes by polling them more often. The major drawback of polling is the substantial overhead caused by the large number of messages generated by the controller to query the communicating nodes. Furthermore, the efficiency of the polling scheme depends on the reliability of the controller.

Reservation The basic idea in a reservation-based scheme is to set some time slots for carrying reservation messages. Since these messages are usually smaller than data packets, they are called *minislots*. When a station has data to send, it requests a data slot by sending a reservation message to the master in a reservation minislot. In some schemes, such as in fixed-priority-oriented demand assignment, each station is assigned its own minislot. In other schemes, such as in packet demand assignment multiple access, stations contend for access to a minislot using one of the distributed packet-based contention schemes, such as slotted ALOHA [5.15,5.16]. When the master receives the reservation request, it computes a transmission schedule and announces the schedule to the slaves.

In a reservation-based scheme, if each station has its own reservation minislot, collision can be avoided. Moreover, if reservation requests have a priority field, the master can schedule urgent data before delay-insensitive data. Packet collisions can happen only when stations contend for the minislot, which use only a small fraction of the total bandwidth. Thus, the largest part of the bandwidth assigned to data packets is used efficiently.

Random Assignment Protocols In fixed-assignment schemes, each communicating node is assigned a frequency band in FDMA systems or a time slot in TDMA systems. This assignment is static, however, regardless of whether or not the node has data to transmit. These schemes may therefore be inefficient if the traffic source is bursty. In the absence of data to be transmitted, the node remains idle, thereby resulting in the allocated bandwidth to be wasted. Random assignment strategies attempt to address this shortcoming by eliminating preallocation of bandwidth to communicating nodes.

Random assignment strategies do not exercise any control to determine which communicating node can access the medium next. Furthermore, these strategies do not assign any predictable or scheduled time for any node to transmit. All back-logged nodes must contend to access the transmission medium. Collision occurs when more than one node attempts to transmit simultaneously. To deal with collisions, the protocol must include a mechanism to detect collisions and a scheme to schedule colliding packets for subsequent retransmissions.

Random access protocols were first developed for long radio links and for satellite communications. The ALOHA protocol, also referred to as *pure ALOHA*, was one of the first such media access protocols. ALOHA simply allows nodes to transmit whenever they have data to transmit. Efforts to improve the performance of pure ALOHA lead to the development of several schemes, including carrier-sense multiple access (CSMA), carrier-sense multiple access with collision detection (CSMA/CD), and carrier-sense multiple access with collision avoidance (CSMA/CA).

ALOHA   ALOHA is a simple random assignment protocol developed to regulate access to a shared transmission medium among uncoordinated contending users. The protocol was originally developed for ground-based packet broadcasting networks and was used to connect remote users to mainframe computers [5.15,5.16]. Channel access in pure ALOHA is completely asynchronous and independent of the current activity on the transmission medium. A node is simply allowed to transmit data whenever it is ready to do so. Upon completing the data transmission, the communicating node listens for a period of time equal to the longest possible round-trip propagation time on the network. This is typically the time it takes for the signal to travel between the two most distant nodes in the network. If the node receives an acknowledgment for data transmitted before this period of time elapses, the transmission is considered successful. The acknowledgment is issued by the receiving station after it determines the correctness of the data received by examining the error check sum. In the absence of an acknowledgment, however, the communicating node assumes that the data are lost due to errors caused by noise on the communication channel or because of collision, and retransmits the data. If the number of transmission attempts exceeds a specified threshold, the node refrains from retransmitting the data and reports a fatal error.

ALOHA is simple protocol that requires no central control, thereby allowing nodes to be added and removed easily. Furthermore, under light-load conditions, nodes can gain access to the channel within short periods of time. The main drawback of the protocol, however, is that network performance degrades severely as the number of collisions rises rapidly with increased load. To improve the performance of pure ALOHA, *slotted ALOHA* was proposed. In this scheme, all communication nodes are synchronized and all packets have the same length. Furthermore, the communication channel is divided into uniform time slots whose duration is equal to the transmission time of a data packet. Contrary to pure ALOHA, transmission can occur only at a slot boundary. Consequently, collision can occur only in the beginning of a slot, and colliding packets overlap totally in time.

Limiting channel access to slot boundaries results in a significant decrease in the length of collision intervals, resulting in increased utilization of the underlying communication channel. Despite this performance improvement, however, ALOHA and pure ALOHA remain inefficient under moderate to heavy load conditions. Furthermore, in networks where the propagation delay is much shorter than the transmission time of a data packet, nodes can become aware almost immediately of an ongoing packet transmission. This observation led to the development of a new class of media access schemes, whereby before a transmission is attempted, a station that has a packet to transmit first "listens" to the channel to determine if it is busy. Carrier sensing forms the basis of the CSMA protocol.

CSMA   CSMA operates both in continuous time, unslotted CSMA, and in discrete time, slotted CSMA. Furthermore, the class of CSMA protocols can be divided into two categories, nonpersistent CSMA and persistent CSMA, depending on the strategy used to acquire a free channel and the strategy used to wait for a busy channel to become free. In *nonpersistent* CSMA protocol, when a node becomes ready to transmit a packet, it first senses the carrier to determine if another transmission is in progress. If the channel is idle, the node transmits its packet immediately and waits for an acknowledgment. In setting the acknowledgment timeout value, the node must take into account the round-trip propagation delay and the fact that the receiving node must also contend for the channel to transmit the acknowledgment. Estimating the average contention time required for a successful transmission is difficult, as it depends on the traffic load and the number of stations contending. In the absence of an acknowledgment, before a timeout occurs, the sending node assumes that the data packet is lost due to collision or noise interference. The station schedules the packet for retransmission. If the channel is busy, the transmitting node "backs off" for a random period of time after which it senses the channel again. Depending on the status of the channel, the station transmits its packet if the channel is idle, or enters the back-off mode if the channel is busy. This process is repeated until the data packet is transmitted successfully.

The nonpersistent CSMA protocol minimizes the interference between packet transmissions, as it requires stations that find the channel busy to reschedule their transmissions randomly. The major drawback of the nonpersistent CSMA scheme, however, results from the fact that a channel may become idle during the back-off time of a contending station. The unnecessary waste of channel capacity can reduce significantly the overall network throughput. The need to address the shortcomings of nonpersistent CSMA led to the development of a class of *p*-persistent CSMA schemes. These schemes differ in the algorithm they use to acquire a free channel. The 1-persistent scheme never allows the channel to remain idle if a node is ready to transmit. Based this scheme, a node ready to transmit a data packet first senses the channel. If the channel is free, the node transmits its message immediately. If the channel is busy, however, the node *persistently* continues to listen until the channel becomes idle. Transmission is attempted immediately after the channel is sensed idle.

The *p*-persistent algorithm represents a compromise between the nonpersistent and 1-persistent schemes. Based on this algorithm, a node that senses the channel idle transmits its packet with probability p. With probability $(1 - p)$, the station waits for a specific time period before attempting to transmit the packet again. The value of the waiting period is typically set to equal the maximum propagation delay between the most distant nodes in unslotted CSMA or to a slot time in slotted ALOHA. At the end of the waiting period, the node senses the channel again. If the channel is busy, the node continues to listen until the channel becomes idle. When the channel becomes idle, the node repeats the foregoing p-persistent channel acquisition algorithm. This process continues until the data packet is transmitted successfully.

The value of p plays a major role in the stability of the protocol. Under heavy traffic load, if the value of p is large, multiple nodes will attempt to transmit, thereby increasing the likelihood of collisions. Since the value of p is high, colliding nodes will probably attempt to retransmit almost immediately after the collision occurs. Worst yet, these retransmissions may have to compete with new transmissions from other nodes, almost guaranteeing more collisions. Eventually, as the number of contending stations increases, the network throughput decreases drastically. To avoid this situation, the value of p must be small. As the value of p is made small, the number of collisions decreases. Under a light traffic load, however, a small value of p may cause a contending station to wait unnecessarily for long delays before transmitting its data packets. Careful consideration of the offered traffic rate is therefore necessary to select a value of p effectively.

CSMA/CD   In networks where the propagation delay is small relative to the packet transmission time, the CSMA scheme and its variants can result in smaller average delays and higher throughput than with the ALOHA protocols. This performance improvement is due primarily to the fact that carrier sensing reduces the number of collisions and, more important, the length of the collision interval. The main drawback of CSMA-based schemes, however, is that contending stations continue transmitting their data packets even when collision occurs. For long data packets, the amount of wasted bandwidth is significant compared with the propagation time. Furthermore, nodes may suffer unnecessarily long delays waiting for the transmission of the entire packet to complete before attempting to transmit the packet again.

To overcome the shortcomings of CSMA-based schemes and further reduce the collision interval, networks using CSMA/CD extend the capabilities of a communicating node to listen while transmitting. This allows the node to monitor the signal on the channel and detect a collision when it occurs. More specifically, if a node has data to send, it first listens to determine if there is an ongoing transmission over the communication channel. In the absence of any activity on the channel, the node starts transmitting its data and continues to monitor the signal on the channel while transmitting. If an interfering signal is detected over the channel, the transmitting station immediately aborts its transmission. This reduces the amount of bandwidth wasted due to collision to the time it takes to detect a collision. When a collision occurs, each contending station involved in the collision waits for a time period of random length before attempting to retransmit the packet. The length of time that a colliding node waits before it schedules packet retransmission is determined by a probabilistic algorithm, referred to as the *truncated binary exponential back-off algorithm.* The algorithm derives the waiting time after collision from the slot time and the current number of attempts to retransmit.

The major drawback of CSMA/CD is the need to provision sensor nodes with collision detection capabilities. Sensor nodes have a very limited amount of storage, processing power, and energy resources. These limitations impose severe constraints on the design of the MAC layer. Support for collision detection in WSNs is not possible without additional circuitry. In particular, wireless transceivers are typically half-duplex. To detect collision, the sensor node must therefore be capable of "listening" while "talking." The complexity and cost of sensor nodes, however, are intended to be low and scalable to enable broad adaptations of the technology in cost-sensitive applications where deployment of large number of sensors is expected. Consequently, the design of physical layer must be optimized to keep the cost low.

Another important factor that works against using a CSMA/CD-based strategy to regulate access to a shared medium in a wireless environment is the difficulty of detecting collision in a wireless environment. In a wired medium, the low attenuation of the signal makes it such that the values of signal-to-noise ratio are nearly the same at the transmitter and the receiver. As a result, the detection of a collision at the transmitter can be used to infer unambiguously that a collision also occurred at the receiver. In wireless environments, the time-varying properties of the wireless channel, coupled with the rapid decrease of the signal power over distance, makes it difficult for the transmitting sensor node to infer unambiguously either the occurrence or the absence of a collision at the receiving node [5.17]. This drawback severely limits the applicability of collision detection–based schemes in WSNs.

CSMA/CA   Carrier sensing prior to transmission is an effective approach to increase the throughput efficiency in shared-medium access environments. Although applicable in wireless environments, the scheme is susceptible to two problems, commonly referred to as the *hidden-* and *exposed-node problems* [5.4,5.16,5.18]. The hidden- and exposed-node problems result indirectly from the time-varying properties of the wireless channel, which are caused by physical phenomena such as noise, fading, attenuation, and path loss. These interferences, combined with the rapid decrease in the power received with the distance between the sender and receiver, limit the maximum transmission range that can be achieved by a sending node. This limitation and the fact that CSMA is designed to avoid collision by sensing the signal in the vicinity of the transmitter give rise to the hidden- and exposed-node problems.

A *hidden node* is defined as a node that is within the range of the destination node but out of range of the transmitting node. To illustrate this example, consider Figure 5.2, where node B is within the transmission range of nodes A and C. Furthermore, assume that nodes A and C are outside their mutual transmission ranges. Consequently, any transmission from either of the two nodes will not reach the other node. Given this network configuration, assume that node A needs to transmit a data packet to node B. According to the CSMA protocol, node A senses the channel and determines that it is free. Node A then proceeds to transmit its

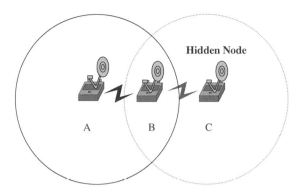

Figure 5.2 Hidden-node scenario in wireless sensor networks.

packet. Assume now that before node A completes its transmission to node B, node C decides to transmit a data packet to node B. Using the CSMA protocol, node C senses the channel and also determines that the channel is free, since node C, which is outside the transmission range of node A, cannot hear the signal transmitted by node A. As a result, both transmissions collide at node B, thereby causing the loss of both data packets. Notice that neither node A nor node C is aware of the collision, since it happens at the receiver. This feature is intrinsic to wireless networks and constitutes a fundamental difference in the way that collisions are dealt with in wired and wireless environments.

The exposed-node problem is also the result of the intrinsic property of the wireless channel. An *exposed node* is a node that is within the range of the sender but out of the range of the destination. To illustrate the exposed-node problem, consider the network depicted in Figure 5.3, where node B is within the transmission range of nodes A and C, nodes A and C are outside their mutual transmission ranges, and node D is within the transmission range of node C. Assume that node B wants to transmit a message to node A. Node B executes the CSMA protocol to sense the channel, determines that the channel is free, and proceeds to transmit the data packet

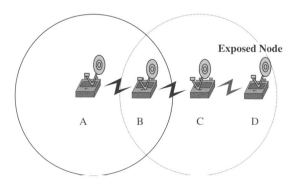

Figure 5.3 Exposed node scenario in wireless sensor networks.

to node A. Assume now that node C needs to send a packet to D. Node C follows the CSMA rule and first senses the channel. Due to the ongoing transmission between nodes B and A, node C determines that the channel is busy and delays the transmission of its packet to a later time. It is clear, however, that this delay is unnecessary, since the transmission from node C to node D would have been completed successfully, as node D is outside the range of node B.

Several approaches have been proposed to eliminate, or at least reduce, the impact of the hidden- and exposed-node problems on the network throughput. The first approach is based on the use of a *busy tone*. The basic idea of the busy-tone approach stems from the observation that collisions occur at the receiving node whereas CSMA is performed at the transmission node. To address the disparity between the design goals of CSMA as originally specified and application of the protocol to wireless environments, the busy-tone approach requires the use of two separate channels: a data channel and a control channel. The data channel is used to transmit data exclusively. The control channel is used by the receiver to signal to the remaining nodes in the network that it is in the process of receiving data.

Immediately after the node starts to receive a data packet, which carries its address in the destination address field, the node initiates the emission of an unmodulated wave on the control channel, indicating that its receiver is busy. The node continues to transmit the busy tone at the same time that it is receiving the data packet until the packet is fully received. Before transmitting a data packet, the sending node must first sense the control channel for the presence of a busy tone. The node proceeds to transmit the data packet only if the control channel is free. Otherwise, the sending node defers its transmission until the control channel is no longer busy.

The busy-tone approach solves both the hidden- and exposed-node problems, assuming that the busy-tone signal is emitted at a level such that it is not too weak not to be heard by a node within the range of a receiver and not too strong to force more nodes than necessary to suppress their transmissions. The major drawback of the approach, however, is a node's need to operate in duplex mode, to be able to transmit and receive simultaneously. This requirement increases the design complexity of a node significantly, thereby increasing its cost and power consumption.

The second approach to deal with the hidden-node problem is based on collision avoidance [5.19]. This is achieved using a procedure referred to as the ready-to-send (RTS), clear-to-send (CTS) handshake. Using this handshake procedure, the CSMA/CA scheme requires that nodes apply a standard mechanism to avoid collision of wireless messages. Since a node cannot detect if a collision has occurred, it attempts to avoid collisions by waiting for the wireless medium to be clear for the amount of time it takes for a packet to propagate through the entire medium: the time required to send a packet between the most distant nodes in the network. When a node intends to transmit a data packet, it first senses the carrier to determine if another node is already transmitting. If no other transmissions are sensed, the node sends a short RTS packet to the intended recipient of the data packet. If the recipient is, in fact, idle and senses that the medium is clear, it sends

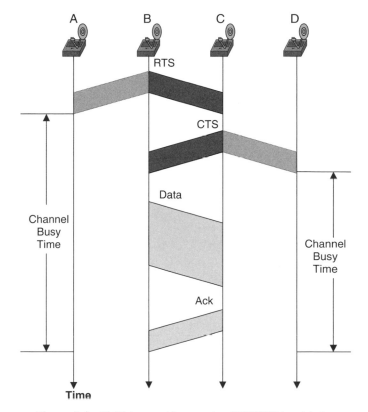

Figure 5.4   Collision avoidance using RTS/CTS handshake.

a short CTS packet in reply. Upon receiving the CTS packet, the transmitting node sends the actual data packet to its intended recipient. If after a predetermined period of time, the transmitting station does not receive a CTS packet in reply to its RTS packet, it waits a random period of time before repeating the RTS/CTS handshake procedure.

The use of the RTS/CTS handshake procedure in CSMA/CA schemes to avoid collisions is depicted in Figure 5.4. In this scenario, node B intends to transmit a data packet to node C. It senses the carrier to determine if any other node is already transmitting. After it determines that the channel is free, it transmits a RTS packet. In addition to the destination address, the packet also contains the duration field, which indicates the time necessary to complete the transmission of the packet and the receipt of the corresponding acknowledgment. In response, the intended recipient of the packet, node C in this case, transmits a CTS packet, which contains the remaining time until the completion of the transmission. Upon receiving the RTS packet, station A sets an internal timer to the remaining time until completion of the data packet transmission and avoids transmitting any packet until the timer expires. When node B receives the CTS packet, it proceeds to transmit its data packet. Upon receiving the CTS packet, node D sets an internal timer to the remaining time until completion of the data and defers the transmission of any packets until the timer expires.

In many environments, the RTS/CTS handshake procedure is sufficient to greatly reduce collisions and increase bandwidth utilization. This procedure, however, does not completely solve the hidden-node problem. To illustrate this limitation, consider the following scenarios, depicted in Figure 5.5. In the scenario depicted in Figure 5.5a, node A senses the channel to be free and sends an RTS packet to node B. In reply, node B sends a CTS packet. Node C, which is in the transmission range of node B, starts receiving the CTS packet. Before the reception of this packet is complete, however, node D, which is in the transmission range of node C, sends a RTS packet. The latter packet collides with the CTS packet sent by node B. Meanwhile, node A, which receives the CTS packet correctly, proceeds to transmit its data packet to node B. Node D later times-out and retransmits its RTS packet. Since node C never received node B's CTS packet, it assumes that the channel is free and replies with a CTS packet to node D. Since node B is within the transmission range of node C, the latter packet collides with the data packet being transmitted by node A.

The scenario depicted in Figure 5.5b shows another case where collision avoidance fails, using the RTS/CTS handshake. In this scenario, node A senses the channel to be free and sends an RTS packet to node B. In reply, node B sends a CTS packet to node A. The CTS packet is received correctly by node A, which allows it to transmit its packet. The CTS packet is also received by node C, which is within the transmission range of node B. Since node C has started transmitting an RTS packet to node D, nearly at the same time that node B is transmitting its CTS packet; node C does not receive correctly the CTS packet sent by B. Node D, however, receives correctly the RTS packet sent by node C. In response, it sends a CTS packet to node C, thereby allowing it to start transmitting its data packet. Since node A did not complete transmission of its data packet to node B, node C's data transmission causes a collision at node B. Despite its failure to solve the hidden-node problem completely, the RTS/CTS handshake is used widely in wireless networks to avoid packet collisions and increase network throughput.

5.4   MAC PROTOCOLS FOR WSNs

The need to conserve energy is the most critical issue in the design of scalable and stable MAC layer protocols for WSNs. Several factors contribute to energy waste, including excessive overhead, idle listening, packet collisions, and overhearing. Regulating access to the media requires the exchange of control and synchronization information among the competing nodes. The explicit exchange of a large number of these control and synchronization packets may result in significant energy consumption. Long periods of idle listening may also increase energy consumption and decrease network throughput. In some cases, energy wasted by idle listening accounts for over one-half of the total energy consumed by a sensor during

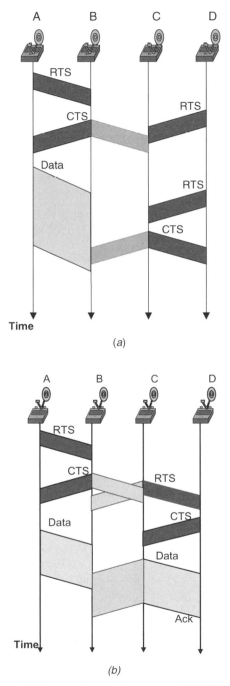

Figure 5.5 Collision avoidance failure using RTS/CTS handshake.

its lifetime. The retransmission of colliding packets is yet another source of significant energy waste. A high number of these collisions may lead to severe performance degradation of the MAC-layer protocol. Similarly, excessive overhearing, which causes a node to receive and decode packets intended for other sensor nodes, unnecessarily increases energy consumption and can severely degrade the network throughput. These packets are eventually dropped after the node realizes that the destination address is different from its own address.

The main objective of most MAC-layer protocols is to reduce energy waste caused by collisions, idle listening, overhearing, and excessive overhead. These protocols can be categorized into two main groups: schedule- and contention-based MAC-layer protocols. *Schedule-based protocols* are a class of deterministic MAC-layer protocols in which access to the channel is based on a schedule. Channel access is limited to one sensor node at a time. This is achieved based on preallocation of resources to individual sensor nodes. *Contention-based MAC-layer protocols* avoid preallocation of resources to individual sensors. Instead, a single radio channel is shared by all nodes and allocated ondemand. Simultaneous attempts to access the communications medium, however, results in collision. The main objective of contention-based MAC layer protocols is to minimize, rather than completely avoid, the occurrence of collisions. To reduce energy consumption, these protocols differ in the mechanisms used to reduce the likelihood of a collision while minimizing overhearing and control traffic overhead.

Resolving collisions is typically achieved using distributed, randomized algorithms to reschedule channel access among competing sensor nodes. The basic approach used to reduce overhearing is to force nodes into a sleep state when they become inactive. Un-coordinating sleeping, however, can make communications among neighboring nodes difficult. To address this shortcoming, a variety of less restrictive schedules have been proposed by different MAC-layer protocols to coordinate the activity of the network sensors.

In the following section we first discuss schedule-based MAC-layer protocols for WSNs. We then briefly review a variety of contention-based MAC-layer protocols. We conclude the section with two cases studies. The first study focuses on S-MAC, a low-duty-cycle contention-based MAC-layer protocol specifically designed for WSNs. S-MAC strives to retain the flexibility of contention-based MAC-layer protocols, such as IEEE 802.11, while reducing energy waste caused by idle listening, collisions, overhearing, and excessive control overhear. S-MAC uses the concepts of low-duty-cycle coordinated sleep and wakeup time periods to reduce power consumption while achieving high throughput.

The second case study focuses on the IEEE MAC-layer protocol specification for a low-data-rate wireless personal area network standard: IEEE 802.15.4, Wireless Medium Access Control (MAC) and Physical Layer (PHY) Specifications for Low Rate Wireless Personal Area Networks (LR-WPANs). The IEEE 802.15.4 specification supports three traffic types: periodic, intermittent, and repetitive. Furthermore, the protocol specification supports fixed, portable, and moving devices operating at data rates ranging from 20 to 250 kbps. When lines of communication exceed 30ft, the standard allows for the creation of self-configuring multihop network topologies [5.20]. It also provides features that allow devices operating under the standard to coexist with other wireless devices, such as those that comply with IEEE 802.11 and 802.15.1.

5.4.1 Schedule-Based Protocols

Schedule-based MAC protocols for WSNs assume the existence of a schedule that regulates access to resources to avoid contention between nodes. Typical resources include time, a frequency band, or a CDMA code. The main objective of schedule-based MAC protocols is to achieve a high level of energy efficiency in order to prolong the network lifetime. Other attributes of interest include scalability, adaptability to changes in traffic load, and network topology [5.21,5.22]. Most of the scheduled-based protocols for WSNs use a variant of a TDMA scheme whereby the channel is divided into time slots, as depicted in Figure 5.6. A set of N contiguous slots, where N is a system parameter, form a logical frame. This logical frame repeats cyclically over time. In each logical frame, each sensor node is assigned a set of specific time slots. This set constitutes the schedule according to which the sensor node operates in each logical frame. The schedule can be either fixed, constructed on demand on a per-frame basis by the base station to reflect the current requirements of sensor nodes and traffic pattern, or hybrid, in which case the structure varies over different time scales and sensor behavior.

Based on its assigned schedule, a sensor alternates between two modes of operation: active mode and sleep mode. In the active mode, the sensor uses its assigned slots within a logical frame to transmit and receive data frames. Outside their assigned slots, sensor nodes move into sleep mode. In this mode the sensor nodes switch their radio transceivers off to conserve energy.

Many variations on the basic TDMA protocol have been proposed for media access control in WSNs [5.23,5.24]. Next, we provide a brief review of some of these protocols.

Self-Organizing Medium Access Control for Sensornets (SMACS)   SMACS is a medium access control protocol to enable the formation of random network topologies without the need to establish global synchronization among the network

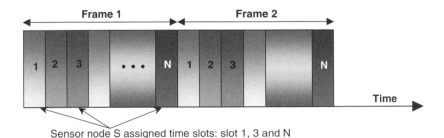

Sensor node S assigned time slots: slot 1, 3 and N

Figure 5.6   TDMA-based MAC protocols for wireless sensor networks.

nodes [5.11,5.25]. A key feature of SMACS is its use of a hybrid TDMA/FH method referred to as *nonsynchronous scheduled communication* to enable links to be formed and scheduled concurrently throughout the network without the need for costly exchange of global connectivity information or time synchronization. Each node in the network maintains a TDMA-like frame, referred to as a *superframe*, for communication with known neighbors. The length of a superframe is fixed. Furthermore, the superframe is divided into smaller frames. The size of each frame is not fixed and may vary in time for a single node and also from node to node. SMACS requires that each node regularly execute a neighborhood discovery procedure to detect neighboring nodes. Each node establishes a link to each neighbor discovered by assigning a time slot to this link. The selection of time slots is such that the node talks only to neighbors at each time slot. However, since a node and its neighbors are not required to transmit at different slot times, the link establishment procedure must ensure that no interference occurs between adjacent links. This is achieved by randomly assigning a channel, selected from a large number of channels (FDMA), or spreading code (CDMA) to each link. Using the superframe structure, each node maintains its own time slot schedules with all its neighbors, and nodes are required to tune their radios to the proper frequency channel or CDMA code to achieve communication.

Bluetooth   Bluetooth is an emerging technology whose primary media access control is a centralized TDMA-based protocol [5.26]. Bluetooth is designed to replace cables and infrared links used to connect disparate electronic devices such as cell phones, headsets, PDAs, digital cameras, notebook computers, and their peripherals with one universal short-range radio link [5.27]. Bluetooth operates in the 2.45-GHz ISM frequency band. Its physical layer is based on a pseudorandom frequency-hopping scheme with a hopping frequency of 1.6 kHz and a scheme for hopping sequence allocation. A set of 79 hop carriers are defined with 1-MHz spacing. Each hop sequence defines a Bluetooth channel, which can support 1 Mbps.

A group of devices sharing a common channel is called a *piconet* [5.28]. Each piconet has a master unit which controls access to the channel, and at most seven slave devices as group participants. Each channel is divided into 625-ms slots. Each piconet is assigned a unique frequency-hopping pattern determined by the master's Bluetooth device address (48 bits) and clock. All slave devices follow their piconet-assigned hopping sequence. Different piconets use different hopping sequences, thereby guaranteeing their coexistence. Piconets can be interconnected, via *bridge* nodes, to form larger ad hoc networks, referred to as *scatternets*. Within a piconet, the master assigns each slave device a unique internal address of 3 bits. Access to the channel is regulated using a slotted time-division duplex (TDD) protocol in which the master uses a polling protocol to allocate time slots to slave nodes. A *Bluetooth frame*, representing one polling epoch, consists of two slots during which a packet can be exchanged between the piconet master and the slave being polled. The master polls the slave devices continuously for communication. A slave can communicate in a slot only if the master has addressed it in a previous slot. Packets can be one, three, or five time slots long and are transmitted in consecutive slots. A packet can be more than one slot long if the communication is asynchronous.

To reduce energy consumption, Bluetooth specifies four operational modes: active, sniff, hold, and park. In the *active mode*, the slave listens for packet transmission from the master. On receiving a packet, it checks the address and packet length field of the packet header. If the packet does not contain its own address, the slave sleeps for the duration of the remaining packet transmission. The intended slave, however, remains active and receives the packet payload in the following reserved slot. The *sniff mode* is intended to reduce the duty cycle of a slave's listen activity. In this mode, the master transmits to the slave only in specified periodic time slots within a predefined sniff time interval. A slave in sniff mode listens for the master transmissions only during the specified time slots for any possible transmission to it. In the *hold mode*, a slave goes into the sleep mode for a specified amount of time, referred to as the *hold time*. When the hold time expires, the slave returns to the active mode. In the *park mode*, the slave stays in the sleep state for an unspecified amount of time. The master has to awake the slave explicitly and bring it into the active mode at a future time.

Bluetooth specifies four types of communication between nodes within and across piconets: *intra piconet unicast*, for slave-to-slave communication within a piconet; *intra piconet broadcast*, to support broadcasting by a slave to all participants within its piconet; *inter piconet unicast*, for piconet-to-piconet communications; and *inter piconet broadcast*, for piconet-to-all scatternet node communications.

For intra piconet unicast communication, the source slave writes its own MAC address in the corresponding field of the data packet and sets the forward field to 1 and the destination address of the packet to the targeted destination node. Upon receiving the message, the master checks the forward field. If it is set, the master replaces the MAC address field with its MAC address and sends the message to the intended slave device indicated by the destination address of the original packet.

For intra piconet broadcast communication, the source slave writes its own MAC address and sets the forward field to 1 and the destination address to 000. Upon receiving the message, the master notices that the forward field is set. In response, the master replaces the MAC address with its own address and sends the message to all nodes in its piconet.

For inter piconet unicast communication, the source device sends the data packet with its own MAC address and sets the forward field to 1, the broadcast field to 1 and the destination address to the relay of the next piconet. Furthermore, the source device sets the routing vector field (RVF) of the packet to contain the logical path to the targeted destination device in the intended piconet. The RVF is a sequence of tuples of the form (LocId, Mac_Addr), where LocId represents the identity of the local master and Mac_Addr its corresponding piconet MAC address. Upon receiving the message, the master forwards it to the relay node. The relay extracts from the RVF the next pair, containing the local identity and the MAC address of the master, and sends the message to this master. This process is repeated until the RVF becomes empty, signaling that the destination device has been reached.

For inter piconet broadcast communication, the source device creates a packet containing its own MAC address and sets the forward and broadcast fields of the packet to 1 and the destination address to 000. The packet is then sent to the master. When the master notices that the broadcast field is set to 1, it sends the packet to all the slaves within its piconet, including relay nodes. When a relay node receives the broadcast packet, it forwards it to all masters to which it is connected, except the one from which it came.

Low-Energy Adaptive Clustering Hierarchy (LEACH) LEACH takes a hierarchical approach and organizes nodes into clusters. Within each cluster, nodes take turns to assume the role of a cluster head. LEACH uses TDMA to achieve communication between nodes and their cluster head [5.29–5.31]. The cluster head forwards to the base station messages received from its cluster nodes.

The cluster head node sets up a TDMA schedule and transmits this schedule to all nodes in its cluster. The schedule prevents collisions among data messages. Furthermore, the schedule can be used by the nodes to determine the time slots during which they must be active. This allows each cluster node, except for the head cluster, to turn off their radio components until its allocated time slots. LEACH assumes that cluster nodes start the cluster setup phase at the same time and remain synchronized thereafter. One possible mechanism to achieve synchronization is to have the base station send out synchronization pulses to the all the nodes [5.32,5.33].

To reduce intercluster interference, LEACH uses a *transmitter-based code assignment* scheme. Communications between a node and its cluster head are achieved using direct-sequence spread spectrum (DSSS), whereby each cluster is assigned a unique spreading code, which is used by all nodes in the cluster to transmit their data to the cluster head. Spreading codes are assigned to cluster heads on a first-in, first-served basis, starting with the first cluster head to announce its position, followed by subsequent cluster heads. Nodes are also required to adjust their transmit powers to reduce interference with nearby clusters.

Upon receiving data packets from its cluster nodes, the cluster head aggregates the data before sending them to the base station. The communication between a cluster head and a base station is achieved using fixed spreading code and CSMA. Before transmitting data to the base station, the cluster head must sense the channel to ensure that no other cluster head is currently transmitting data using the base station spreading code. If the channel is sensed busy, the cluster head delays the data transmission until the channel becomes idle. When this event occurs, the cluster head sends the data using the base station spreading code.

In general, schedule-based protocols are contention-free, and as such, they eliminate energy waste caused by collisions. Furthermore, sensor nodes need only turn their radios on during those slots where data are to be transmitted or received. In all other slots, the sensor node can turn off its radio, thereby avoiding overhearing. This results in low-duty-cycle node operations, which may extend the network lifetime significantly. Schedule-based MAC protocols have several disadvantages, however, which limit their use in WSNs. The use of TDMA requires the organization

of nodes into clusters. This hierarchical structure often restricts nodes to communicate only with their cluster head. Consequently, peer-to-peer communication cannot
be supported directly, unless nodes are required to listen during all time slots. Most
of the schedule-based schemes depend on distributed, fine-grained time synchronization to align slot boundaries. Achieving time synchronization among distributed
sensor nodes is difficult and costly, especially in energy-constrained wireless networks. Schedule-based schemes also require additional mechanisms such as FDMA
or CDMA to overcome intercluster communications and interference. Finally,
TDMA-based MAC-layer protocols have limited scalability and are not easily adaptable to node mobility and changes in network traffic and topology. As nodes join or
leave a cluster, the frame length as well as the slot assignment must be adjusted.
Frequent changes may be expensive or slow to take effect.

5.4.2　Random Access-Based Protocols

Traditional random access MAC-layer protocols, also known as contention-based
protocols, require no coordination among the nodes accessing the channel. Colliding nodes back off for a random duration of time before again attempting to access
the channel. These protocols, however, are not well suited for WSN environments.
The enhancement of these protocols with collision avoidance and request-to-send
(RTS) and clear-to-send (CTS) mechanisms improves their performance and makes
them more robust to the hidden terminal problem [5.34]. The energy efficiency of
contention-based MAC-layer protocols, however, remains low due to collisions,
idle listening, overhearing, and excessive control overhead. To address this shortcoming, efforts in the design of random access MAC-layer protocols focused on
reducing energy waste in order to extend the network lifetime.

The power aware multiaccess protocol with signaling (PAMAS) avoids overhearing among neighboring nodes by using a separate signaling channel
[5.4,5.35]. The protocol combines the use of a busy tone with RTS and CTS packets
to allow nodes currently not actively transmitting or receiving packets to turn off
their radio transceivers. The protocol does not, however, provide mechanisms to
reduce energy waste caused by idle listening.

The sparse topology and energy management (STEM) protocol trades latency
for energy efficiency [5.36]. This is achieved using two radio channels: a data radio
channel and a wake-up radio channel. A variant of STEM uses a busy tone instead
of encoded data for the wake-up signal. STEM is known as a pseudoasynchronous
scheduled scheme. Based on this scheme, a node turns off its data radio channel
until communication with another node is desired. When a node has data to transmit, it begins transmitting on the wake-up radio channel. The wake-up signal channel acts like a paging signal. The transmission of this signal lasts long enough to
ensure that all neighboring nodes are paged. When a node is awakened from its
sleeping mode, it may remain awake long enough to receive a "session" of packets.
A node can also be awakened to receive all of its pending packets before going into
the sleep mode again. The STEM protocol is general and can be used in conjunction
with other MAC-layer scheduling protocols. The scheme is, however, effective only

in network environments where events do not happen very frequently. If events occur frequently, the energy wasted by continuously transmitting wake-up signals may offset, or may exceed, the energy gained in sleeping modes.

A variety of IEEE 802.11-inspired contention-based protocols prevent overhearing by using RTS and CTS packets [5.37–5.40]. A common feature of these protocols is to use the overhearing of RTS and CTS packet exchange between two other contending nodes to force a contending node to go into sleep mode. These protocols also rely on synchronized schedules between neighboring nodes to avoid idle listening. These protocols differ in the way they maintain low duty cycles and the way they achieve energy efficiency, especially when the size of the data packets is of the same order of magnitude as the size of the RTS and CTS packets. They also differ in the mechanisms used to reduce packet latency, as a sending node may have to wait a significant period of time before the receiver wakes up. Finally, the protocols also differ in the level and the way in which they achieve fairness among nodes.

The timeout-MAC (T-MAC) is a contention-based MAC-layer protocol designed for applications characterized by low message rate and low sensitivity to latency [5.5]. To avoid collision and ensure reliable transmission, T-MAC nodes use RTS, CTS, and acknowledgment packets to communicate with each other. Furthermore, the protocol uses an adaptive duty cycle to reduce energy consumption and adapt to traffic load variations. The basic idea of the T-MAC protocol is to reduce idle listening by transmitting all messages in bursts of variable length. Nodes are allowed to sleep between bursts. Furthermore, the protocol dynamically determines the optimal length of the active time, based on current load. Since messages between active times must be buffered, the buffer capacity determines an upper bound on the maximum frame time.

Based on the T-MAC protocol, nodes alternate between sleep and wake-up modes. Each node wakes up periodically to communicate with its neighbors. A node keeps listening and potentially transmitting as long as it is in the active period. An active period ends when no active event occurs for a predetermined time interval. Active events include the hearing of a periodic frame timer, the reception of data over the radio, the sensing of an activity such as collision on the channel, the end of transmission of a node's own data packet or acknowledgment, and the end of a neighboring node's data exchange, determined through overhearing of prior RTS and CTS packets. At the end of the active period, the node goes into sleep mode.

The Berkeley media access control (B-MAC) is a lower-power carrier-sense media access protocol for WSNs [5.41–5.43]. In contrast to traditional IEEE 802.11-inspired MAC-layer protocols, which include mechanisms for network organization and clustering, the B-MAC protocol embodies a small core of media access functionality. B-MAC uses clear channel assessment (CCA) and packet back-offs for channel arbitration, link-layer acknowledgments for reliability, and listening for low-power communication. B-MAC does not provide direct support for multipacket mechanisms to address the hidden terminal problem, handle message fragmentation, or enforce a particular low-power policy. However, in addition to the standard message interface, provides, B-MAC, a set of interfaces that allow services to tune its operation. By exposing a set of configurable mechanisms, protocols built on B-MAC make local policy decisions to optimize power consumption, latency, throughput, fairness, or reliability.

To achieve low-power operation, B-MAC employs an adaptive preamble sampling scheme to reduce duty cycle and minimize idle listening. Each time the node wakes up, it turns on the radio and checks for activity. If it detects activity, the node powers its radio transceiver up and stays awake for the time required to receive the incoming packet. After reception, the node returns to sleep. If no packet is received within the specified timeout, the node goes to sleep. B-MAC supports on-the-fly reconfiguration and provides bidirectional interfaces for system services to optimize performance, whether it is for throughput, latency, or power conservation.

In the remaining sections of this chapter, we first discuss in detail a common IEEE 802.11-inspired protocol MAC-layer protocol, referred to as S-MAC. We then describe the basic architecture and protocols of the IEEE 802.15.4 wireless MAC- and physical-layer specifications for low-data-rate wireless personal area networks.

5.5 SENSOR-MAC CASE STUDY

The sensor-MAC (S-MAC) protocol is designed explicitly to reduce energy waste caused by collision, idle listening, control overhead, and overhearing [5.44–5.46]. The goal is to increase energy efficiency while achieving a high level of stability and scalability. In exchange, the protocol incurs some performance reduction in per-hop fairness, and latency S-MAC uses multiple techniques to reduce energy consumption, control overhead, and latency, in order to improve application-level performance. In the following we provide an overview of the S-MAC-layer protocol and discuss the techniques it proposes to achieve energy efficiency while keeping latency low.

5.5.1 Protocol Overview

The protocol design assumes a large number of sensor nodes, with limited storage, communication, and processing capabilities. The nodes are configured in an ad hoc, self-organized, and self-managed wireless network. Data generated by sensors are processed and communicated in a store-and-forward manner. The applications supported by the network are assumed to alternate between long idle periods, during which no events occur, and bursty active periods, during which data flow toward the base station through message exchange among peer sensor nodes. Furthermore, the applications are assumed to tolerate increased latency for an extended network lifetime. Typical applications that fall into this category include surveillance and monitoring of natural habitats and protection of critical infrastructure. In these applications the sensors must be vigilant over long periods of time, during which they remain inactive until some event occurs. The frequency at which these events occur is typically orders of magnitude slower than the time it takes to transmit a message across the network toward the base station.

S-MAC exploits the bursty profile of sensor applications to establish low-duty-cycle operation on nodes in a multihop network and to achieve significant energy savings. During the long periods of time during which no sensing occurs, S-MAC nodes alternate periodically between listening and sleep modes. Each node sets a wakeup time and sleeps for a certain period of time, during which its radio is turned off. At the expiration of the timer, the node becomes active again. To further reduce control overhead while keeping message latency low, the protocol uses coordinated sleeping among neighboring nodes. Periodic sleeping reduces energy consumption at the expense of increased latency. The importance of message latency strongly depends on the requirements of the sensing application. S-MAC focuses on applications that can tolerate latency on the order of seconds. However, when nodes follow their schedule strictly, latency can increase significantly. To address this shortcoming and keep message delay within the targeted-second-level latency, S-MAC uses adaptive listening.

As stated previously, S-MAC design is focused on cooperating applications, such as monitoring and surveillance applications. The applications cooperate to achieve a common single task, such as protecting a critical infrastructure. The nature of these applications is such that at any particular point in time, one sensor node may have a large amount of information to communicate to its neighbors. To accommodate this requirement while further reducing overhead, S-MAC sacrifices channel access fairness and uses the concept of message passing, whereby a node is allowed to send a long message in burst. Message passing reduces control overheard and avoids overhearing.

5.5.2 Periodic Listen and Sleep Operations

One of the S-MAC design objectives is to reduce energy consumption by avoiding idle listening. This is achieved by establishing low-duty-cycle operations for sensor nodes. Periodically, nodes move into a sleep state during which their radios are turned off completely. Nodes become active when there is traffic in the network. The basic periodic listen and sleep scheme is depicted in Figure 5.7. Based on this scheme, each node sets a wake-up timer and goes to sleep for the specified period of time. At the expiration of the timer, the node wakes up and listens to determine if it needs to communicate with other nodes. The complete listen- and-sleep cycle is referred to as a *frame*. Each frame is characterized by its *duty cycle*, defined as the listening interval-to-frame length ratio. Although the length of the listening interval can be selected independently by sensor nodes, for simplicity the protocol assumes the value to be the same for all nodes.

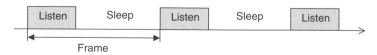

Figure 5.7 S-MAC period listen and sleep modes of operations [5.44].

Nodes are free to schedule their own sleep and listen intervals. It is preferable, however, that the schedules of neighboring nodes be coordinated in order to reduce the control overhead necessary to achieve communications between these nodes. Contrary to other protocols in which coordination is achieved through a master node such as a cluster head, S-MAC nodes form virtual clusters around schedules but communicate directly with their peers to exchange and synchronize their sleep and listen schedules.

5.5.3   Schedule Selection and Coordination

The neighboring nodes coordinate their listen and sleep schedules such that they all listen at the same time and all sleep at the same time. To coordinate their sleeping and listening, each node selects a schedule and exchanges it with it neighbors during the synchronization period. Each node maintains a schedule table that contains the schedule of all its known neighbors.

To select a schedule, a node first listens to the channel for a fixed amount of time, at least equal to the synchronization period. At the expiration of this waiting period, if the node does not hear a schedule from another node, it immediately chooses its own schedule. The node announces the schedule selected by broadcasting a SYNC packet to all its neighbors. It is worth noting that the node must first perform physical carrier sensing before broadcasting the SYNC packet. This reduces the likelihood of SYNC packet collisions among competing nodes. If during the synchronization period the node receives a schedule from a neighbor before choosing and announcing its own schedule, the node sets its schedule to be the same as the schedule received. The node waits until the next synchronization period to announce the schedule to its neighboring nodes.

It is worth noting that a node may receive a different schedule after it chooses and announces its own schedule. This may occur if the SYNC packet is corrupted by either collision or channel interference. If the node has no neighbor with whom it shares a schedule, the node simply discards its own schedule and adopts the new one. On the other hand, if the node is aware of other neighboring nodes that have already adopted its schedule, the node adopts both schedules. The node is then required to wake up at the listen intervals of the two schedules adopted. This is illustrated in Figure 5.8. The advantage of carrying multiple schedules is that border nodes are required to broadcast only one SYNC packet. The disadvantage of this approach is that border nodes consume more energy, as they spend less time in the sleep mode.

It is to be noted that neighboring nodes may still fail to discover each other, due to the delay or loss of a SYNC packet. To address this shortcoming, S-MAC nodes are required to perform frequent neighbor discovery, whereby a node listens periodically to the entire synchronization period. Nodes that currently do not have any neighbors are expected to perform neighbor discovery more frequently.

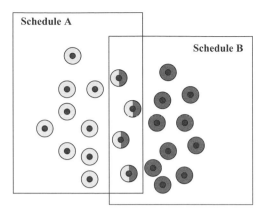

Figure 5.8 Border node schedule selection and synchronization.

5.5.4 Schedule Synchronization

Neighboring nodes need to synchronize their schedules periodically to prevent long-term clock drift. Schedule updating is accomplished by sending a SYNC packet. For a node to receive both SYNC packets and data packets, the listen interval is divided into two subintervals as depicted in Figure 5.9. This figure illustrates three cases. In the first case the sender sends only a SYNC packet; in the second the sender sends only a data packet; and in the third the sender sends a SYNC packet in addition to the data packet.

Access to the channel by contending nodes during these subintervals is regulated using a multislotted contention window. The first subinterval is dedicated to the transmission of SYNC packets; the second subinterval is used for the transmission of data packets. In either of these subintervals, a contending station randomly selects a time slot, performs carrier sensing, and starts sending its packet if it detects that the channel is idle. Transmission of data packets uses the RTS/CTS handshake

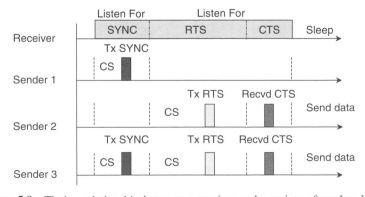

Figure 5.9 Timing relationship between a receiver and a variety of senders [5.44].

to secure exclusive access to the channel during transmission of the data. This access procedure guarantees that the neighboring nodes receive both the synchronization and data packets.

5.5.5 Adaptive Listening

A closer look at the periodic listen and sleep scheme reveals that a message may incur increased latency as it is stored and forwarded between adjacent network nodes. If a sensor is to follow its sleep schedule strictly, data packets may be delayed at each hop. To address this shortcoming and improve latency performance, the protocol uses an aggressive technique referred to as *adaptive listening*. Based on this technique, a node that overhears, during its listen period, the exchange of a CTS or RTS packet between a neighboring node and another node assumes that it may be the next hop along the routing path of the overheard RTS/CTS packet, ignores its own wake-up schedule, and schedules an extra listening period around the time the transmission of the packet terminates. The overhearing node determines the time necessary to complete the transmission of the packet from the duration field of the overheard CTS or RTS packet. Immediately upon receiving the data packet, the node issues an RTS packet to initiate an RTS/CTS handshake with the overhearing node. Ideally, the latter node is awake, in which case the packet forwarding process proceeds immediately between the two nodes. If the overhearing node does not receive an RTS packet during adaptive listening, it reenters its sleep state until the next scheduled listen interval.

5.5.6 Access Control and Data Exchange

To regulate access to the communication channel among contending sensor nodes, S-MAC uses a CSMA/CA-based procedure, including physical and virtual carrier sensing and the use of RTS/CTS handshake to reduce the impact of the hidden and exposed terminal problems. Virtual carrier sensing is achieved through use of the network allocation vector (NAV), a variable whose value contains the remaining time until the end of the current packet transmission. Initially, the NAV value is set to the value carried in the duration field of the packet transmitted. The value is decremented as time passes and eventually reaches zero. A node cannot initiate its own transmission until the NAV value reaches zero. Physical carrier sensing is performed by listening to the channel to detect ongoing transmission. Carrier sensing is randomized within a contention window to avoid collisions and starvation. A node is allowed to transmit if both virtual and physical carrier sensing indicate that the channel is free.

To perform virtual carrier sensing effectively, nodes may be required to listen to all transmissions from their neighbors. As a result, nodes may be required to listen to packets that are destined for other nodes. Packet overhearing may lead to significant energy waste. To avoid overhearing, S-MAC allows nodes to move into sleep mode after they hear the exchange of an RTS or a CTS packet between two other nodes. The node initializes its NAV with the value contained in the duration field of

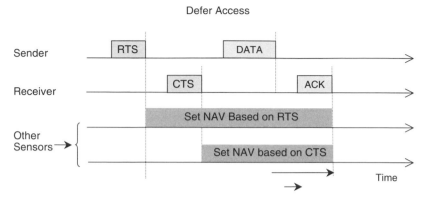

Figure 5.10   S-MAC collision avoidance scheme [5.46].

the RTS or CTS packets and enters the sleep state until the NAV value reaches zero. Since data packets are typically larger than control packets, the overhearing avoidance process may lead to significant energy savings. The scheme used by S-MAC to avoid collisions is illustrated in Figure 5.10.

A node attempting to transmit a message must first sense the channel. If the channel is busy, the node goes to sleep and wakes up when the channel becomes free again. If the channel is idle, a node, sending a data packet, first issues an RTS packet and waits for a CTS packet from the receiver. When it receives the CTS packet, the node sends its data packet. The transaction is completed when the node receives an acknowledgment from the receiver. It is worth noting that after successful exchange of the RTS and CTS packets, the communicating nodes use their normal sleep time to exchange data packets. The nodes do not resume their regular sleep schedule until the data transmission is completed. Furthermore, the transmission of a broadcast packet, such as a SYNC packet, does not require the exchange of the RTS and CTS packet.

5.5.7   Message Passing

To improve application-level performance, S-MAC introduces the concept of *message passing*, where a message is a meaningful unit of data that a node can process. Messages are divided into small fragments. These fragments are then transmitted in a single burst. The fragments of a message are transmitted using only one RTS/CTS exchange between the sending and receiving nodes. At the completion of this exchange, the medium is reserved for the time necessary to complete the transfer of the entire message successfully. Furthermore, each fragment carries in its duration field the time needed to transmit all the subsequent fragments and their corresponding acknowledgments. This procedure is depicted in Figure 5.11.

Upon transmitting a fragment, the sender waits for an acknowledgment from the receiver. If it receives the acknowledgment, the sender proceeds with transmission

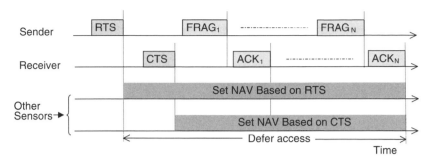

Figure 5.11   S-MAC message passing [5.46].

of the next fragment. If it fails to receive the acknowledgment, however, the sender extends the time required to complete transmission of the segment to include the time to transmit one more fragment and its corresponding acknowledgment and retransmits the unacknowledged frame immediately. It is worth noting that sleeping nodes can hear about this extension only if they hear extended fragments or their corresponding acknowledgments. Nodes that only heard the initial RTS and CTS packet exchange remain unaware of the transmission extension. The S-MAC has the potential to achieve significant energy savings. It is well suited for applications where fairness is not a critical design goal and increased latency is tolerable.

5.6   IEEE 802.15.4 LR-WPANs STANDARD CASE STUDY

The IEEE 802.15.4 specification complements the IEEE 802 set of wireless standards [5.47]. The main design objective of the IEEE 802.15.4 open standard is to support the wireless connectivity of a vast number of industrial, home, and medical applications, including automotive monitoring and control, home automation, ubiquitous and pervasive health care, gaming, and sensor-rich environments. Such applications require a small, low-cost, highly reliable technology that offers long battery life, measured in months or even years, and automatic or semiautomatic installation. The IEEE 802.15.4 standard supports these requirements by trading off higher speed and performance for architectures that benefit from low power consumption and low cost, as noted in Chapter 4.

The IEEE 802.15.4 standard has been adopted by the ZigBee Alliance for wireless personal area network technology [5.48]. The alliance is an association of hundreds of members from around the world, working together to enable the reliable and cost-effective networking of wireless devices for monitoring and control, based on an open global standard. The reference model, depicted in Figure 5.12(*a*), shows the various layers of the ZigBee wireless technology architecture the relationship of the IEEE 802.15.4 standard to the ZigBee alliance MAC layer protocol model. These layers facilitate the features that make ZigBee very attractive: low cost, very low power consumption, reliable data transfer, and easy implementation. Using the IEEE 802.15.4 specifications, the alliance focuses on the design issues related to the network, security, and applications layers. It also provides specification for interoperability and testing. Figure 5.12(*b*) shows the ZigBee stack reference model and lists the basic functionalities at each layer.

The physical layer (PHY) of the reference model specifies the network interface components, their parameters, and their operation. Furthermore, to support operation of the MAC layer, the PHY layer includes a variety of features, such as receiver energy detection (RED), link quality indicator (LQI), and clear channel assessment

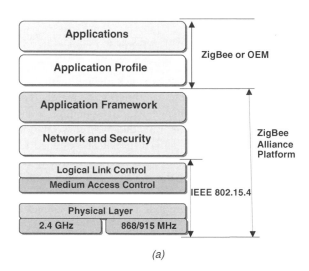

(*a*)

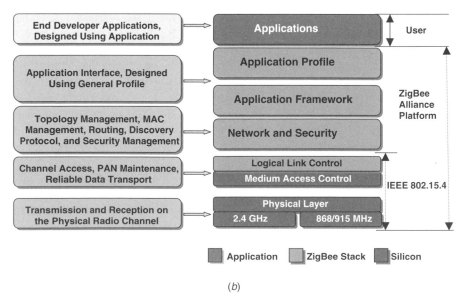

(*b*)

Figure 5.12 (*a*) IEEE 802.15.4 and ZigBee reference model; (*b*) ZigBee stack reference model.

TABLE 5.1   IEEE 80215.4 PHY Layer Main Parameters

Parameter	2.4-GHz PHY	868/915-MHz PHY
Sensitivity @ 1% PER	−85 dBm	−92 dBm
Receiver maximum input level	−20 dBm	
Adjacent channel rejection	0 dB	
Alternate channel rejection	30 dB	
Output power, lowest maximum	−3 dBm	
Transmission modulation accuracy	EMV < 35% for 1000 chips	
Number of channels	16	1/10
Channel spacing	5 MHz	NA^a/2 MHz
Transmission rates		
Date rate	250 kbps	20/40 kbps
Symbol rate	62.5 kilosymbols/sec	20/40 kilosymbols/sec
Chip rate	2 megachips/sec	300/60 kilochips/sec
Chip modulation	O-QPSK with half-sine pulse shaping (MKS)	BPSK with raised cosine pulse shaping
RX-TX and TX-RX turnaround time	12 symbols	

^{a}Single channel.

(CCA). The PHY layer is also specified with a wide range of operational low-power features, including low-duty-cycle operations, strict power management, and low transmission overhead. These parameters are listed in Table 5.1.

The MAC layer handles network association and disassociation. It also regulates access to the medium. This is achieved through two modes of operation: beaconing and nonbeaconing. The beaconing mode is specified for environments where control and data forwarding is achieved by an always-active device. The nonbeaconing mode specifies the use of unslotted, nonpersistent CSMA-based MAC protocol.

The network layer provides the functionality required to support network configuration and device discovery, association and disassociation, topology management, MAC-layer management, routing, and security management. Three network topologies—star, mesh, and cluster tree—are supported.

The security layer leverages the basic security services specified by the IEEE 802.15.4 security model to provide support for infrastructure security and application data security. The first security service of the IEEE 802.15.4 security model provides support for access control. This basic security service prevents unauthorized parties from participating in the network. It allows a legitimate device to maintain a list of trusted devices in the network. A legitimate device uses this list to detect and reject messages from unauthorized devices. The second security service supports message integrity protection to prevent an adversary from tampering with a data message from an authorized sender, while the message is in transit. The third security service provides data confidentiality to keep the information carried by a data message secret from unauthorized parties. This is achieved using the advanced encryption standard (AES) as its core cryptographic algorithm. The encryption scheme uses a 128-bit key to encrypt messages. The fourth security service deals with sequential data freshness to prevent replay attacks. An adversary engages in a replay attack by eavesdropping on legitimate messages sent between authorized users and replaying them at a later time.

Using the basic security services, the MAC layer describes a variety of security suites. Each suite offers a different set of security properties and guarantees. By default, security is not enabled. The application must therefore explicitly set the appropriate control parameters into the radio stack to enable the desired level of security.

The application layer consists of the application support sublayer (APS), the ZigBee device object (ZDO), and the manufacturer-defined application objects. The responsibilities of the APS sublayer include maintaining tables for binding devices together, based on their services and their needs, and forwarding messages between bound devices. The ZDO can be thought of as a special application object that is resident on all nodes. It has its own profile, referred to as the ZigBee device profile (ZDP), which user application endpoints and other ZigBee nodes can access. The ZDO is responsible for overall device management and security keys and policies, including defining the role of the device within the network, initiating and responding to binding requests, and establishing a secure relationship between network devices. The manufacturer-defined application objects implement the actual applications according to the ZigBee-defined application descriptions.

In the following section we focus on physical- and MAC-layer design issues, mechanisms, and protocols. First, the overall characteristics of the PHY layer are highlighted. Following, a description of the different components and operations of the MAC layer is provided.

5.6.1 PHY Layer

The design of the PHY layer is driven by the need for a low-cost power-effective physical layer for cost-sensitive low-data-rate monitoring and control applications. Under IEEE 802.15.4, wireless links can operate in three unlicensed frequency bands: 858 MHz, 902 to 928 MHz, and 2.4 GHz. Based on these frequency bands, the IEEE 802.15.4 standard defines three physical media:

1. Direct-sequence spread spectrum using BPSK operating in the 868-MHz band at a data rate of 20 kbps
2. Direct-sequence spread spectrum using BPSK operating in the 915-MHz band at a data rate of 40 kbps
3. Direct-sequence spread spectrum using O-QPSK operating in the 2.4-GHz band at a data rate of 140 kbps

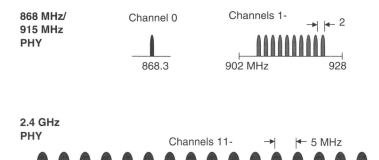

Figure 5.13 IEEE 802.15.4 PHY-layer operating frequency bands.

These operating frequency bands are depicted in Figure 5.13. The spreading code of the 868- and 915-MHz PHY layers is a 15-chip m-sequence. Both specifications use BPSK with a differential encoding data modulation scheme. The data rate of 868-MHz layer is 20 kbps while the data rat of the 915 MHz specification is 40 kbps. The chip modulation used by both specifications is BPSK with raised cosine shaping ($a = 1.0$). The resulting chip rate is 300 kilochips/sec for the 868-MHz PHY layer and 600 kilochips/sec for the 915-MHz PHY layer.

The data modulation of the 2.4-GHz PHY layer is a 16-ary orthogonal modulation. Consequently, 16 symbols are an orthogonal set of 32-chip PN codes. The resulting data rate is 250 kbps (4 bits/symbol, 62.5 kilosymbols/sec). The specification uses O-QPSK with half-sine pulse shaping, which is equivalent to minimum shift keying. The resulting chip rate is 2.0 megachips/sec. The packet structure of the IEEE 802.15.4 PHY layer is depicted in Figure 5.14. The first field of this structure contains a 32-bit preamble. This field is used for symbol synchronization. The next field represents the start of a packet delimiter. This field of 8 bits is used for frame synchronization. The 8-bit PHY header field specifies the length of the PHY service data unit (PSDU). The PSDU field can carry up to 127 bytes of data.

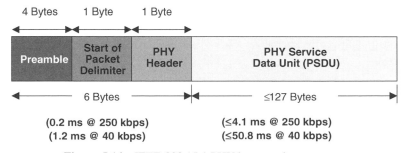

Figure 5.14 IEEE 802.15.4 PHY-layer packet structure.

5.6.2 MAC Layer

The IEEE 802.15.4 MAC-layer specification is designed to support a vast number of industrial and home applications for control and monitoring. These applications typically require low to medium data rates and moderate average delay requirements with flexible delay guarantees. Furthermore, the complexity and implementation cost of the IEEE 802.15.4 standard compliant devices must be low to minimize energy consumption and enable the deployment of these devices on a large scale.

To address the needs of its intended applications while enabling the deployment of a large number of monitoring and control devices at a reduced implementation cost, the IEEE 802.15.4 MAC-layer specification embeds in its design several unique features for flexible network configurations and low-power operations. These features include:

- Support for various network topologies and network devices
- The availability of an optional superframe structure to control the network devices' duty cycle
- Support for direct and indirect data transmissions
- Contention- and schedule-based media access control methods
- Beaconed and nonbeaconed modes of operation (In the beacon mode, the protocol uses a superframe structure to coordinate access to the medium— both contention-based access and guaranteed time slots allocation are supported; in the nonbeaconed mode, the protocol uses an unslotted CSMA/CA-based access scheme.)
- Efficient energy management schemes for an extended battery life, including adaptive sleep for extended period of time over multiple beacons
- Flexible addressing scheme to support the deployment of large-scale networks, theoretically over 65,000 nodes per network

In the following, the classes of network devices supported by the IEEE 802.15.4 MAC standard and the network topologies that can be achieved using these devices are discussed. The optional superframe structure is then described and the two modes of operations, beaconed and nonbeaconed modes, are discussed. Depending on the mode of operations used, two MAC layer protocols are specified. The basic operations of these two MAC layer protocols, including the procedures that govern the data transfer between the network devices in each mode of operations, are described.

Device Types and Network Topologies To accommodate the MAC protocol, the IEEE 802.15.4 standard distinguishes devices based on their hardware complexity and capability. Accordingly, the standard defines two classes of physical devices: a full-function device (FFD) and a reduced-function device (RFD). These device types differ in their use and how much of the standard they implement. An FFD

TABLE 5.2   Device Types in ZigBee Networks

Physical Device Type	Logical Device Type		
	Coordinator	Router	End Device
Full-function device	Yes	Yes	Yes
Reduced-function device	No	No	Yes

is equipped with the adequate resources and memory capacity to handle all the functionalities and features specified by the standard. It can therefore assume multiple network responsibilities. It can also communicate with any other network device. An RFD is a simple device that carries a reduced set of functionalities, for lower cost and complexity. It typically contains a physical interface to the wireless modem and executes the specified IEEE 802.15.4 MAC layer protocol. Furthermore, it can only associate and communicate with an FFD.

Based on these physical device types, ZigBee defines a variety of logical device types. These logical devices are distinguished based on their physical capabilities and the role they play in the network deployed. Table 5.2 illustrates the possible combinations of device types that can be deployed in a ZigBee-enabled network.

There are three categories of logical devices:

1. *Network coordinator*: an FFD device responsible for network establishment and control. The coordinator is responsible for choosing key parameters of the network configuration and for starting the network. It also stores information about the network and acts as the repository for security keys.

2. *Router*: an FFD device that supports the data routing functionality, including acting as an intermediate device to link different components of the network and forwarding message between remote devices across multihop paths. A router can communicate with other routers and end devices.

3. *End devices*: an RFD device that contains just enough functionality to communicate with its parent node: the network coordinator or a router. An end device does not have the capability to relay data messages to other end devices.

Based on these logical devices types, a ZigBee wireless personal area network (PAN) can be organized in one of three possible topologies: a star, a mesh (peer-to-peer), or a cluster tree. The three network configurations are illustrated in Figure 5.15. The star network topology supports a single coordinator, with up to 65,536 devices. In this topology configuration, one of the FFD-type devices assumes the role of network coordinator. All other devices act as end devices. The coordinator selected is responsible for initiating and maintaining the end devices on the network. Upon initiation, the end devices can only communicate with the coordinator. The mesh configuration allows path formation from any source device to any destination device, using tree and table-driven routing algorithms. The table-driven routing algorithm employs a simplified version of the on-demand distance vector routing (AODV) and Internet Engineering Task Force

● PAN Coordinator
● Full Function Device (FFD)
○ Reduced Function Device (RFD)

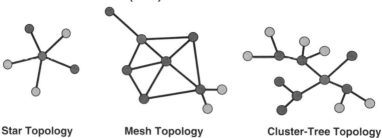

  Star Topology   Mesh Topology   Cluster-Tree Topology

Figure 5.15   Network topologies.

(IETF) proposal for mobile ad hoc networking (MANET). In the mesh topology, the radio receivers of the coordinator and the routers must always be on.

Cluster tree networks enable a peer-to-peer network to be formed with a minimum of routing overhead, using multihop routing. The topology is suitable for latency-tolerant applications. A cluster tree network is self-organized and supports network redundancy to achieve a high degree of fault resistance and self-repair. The cluster can be significantly large, comprising up to 255 clusters of up to 254 nodes each, for a total of 64,770 nodes. It may also span large physical areas. Any FFD can be a coordinator. Only one coordinator is selected for the PAN. The PAN coordinator forms the first cluster and assigns to it a cluster identity (CID) of value zero. Subsequent clusters are then formed with a designated cluster head for each cluster.

Each PAN is uniquely identified by a 16-bit identifier. A PAN coordinator is the designated principal controller of the WPAN. Every network has exactly one PAN coordinator, selected from within all the coordinators of the network. A coordinator is a network device configured to support network functionalities and additional responsibilities, including:

- Managing a list of all associated network devices
- Exchanging data frames with network devices and a peer coordinator
- Allocating 16-bit short addresses to network devices (The short addresses, assigned on demand, are used by the associated devices in lieu of the 64-bit addresses for subsequent communications with the coordinator.)
- Generating beacon frames on a periodic basis (These frames are used to announce the PAN identifier, the list of outstanding frames, and other network and device parameters.)

Superframe Structure   The IEEE 802.15.4 MAC standard defines an optional superframe structure. It is initiated by the PAN coordinator. Furthermore, its format is decided by the coordinator. As Figure 5.16(*a*) shows, the superframe is bounded by network beacons and divided into 16 equally sized slots. The first time slot of each superframe is used to transmit the beacon. The main purpose of the beacon is to synchronize the attached devices, identify the PAN, and describe the superframe structure. The remaining time slots are used by competing devices for communications during the contention access period (CAP). The devices use a slotted CSMA-CA-based protocol to gain access to compete for the time slots. All communications between devices must complete by the end of the current CAP and the beginning of the next network beacon.

To satisfy the latency and bandwidth requirements of the supported applications, the PAN coordinator may dedicate groups of contiguous time slots of the active superframe to these applications. These slots are labeled as guaranteed time slots (GTSs). The number of GTSs cannot exceed seven. A single GTS allocation, however, may occupy more then one time slot. Together the GTSs form the contention-free period (CFP). As shown in Figure 5.16(b), the CFP always appears at the end of the active superframe and starts at a slot boundary immediately following the CAP. The CAP time slots remain for contention-based access between networked devices and new devices wishing to join the network. All communication transactions using contention-based access and GTS-based access must complete before the end their associated CAP and CFP, respectively.

Network devices, which need GTS allocation, can send requests during the CAP period to reserve a desired number of contiguous time slots. The requested slots can be of either the "receive" or the "transmit" type. The receive slots are used by the device to fetch data from the coordinator, while the transmit slots are used to send data to the coordinator. Devices that have no data to exchange with the coordinator can switch off their power and go into a sleep mode. Devices are expected to remain active, however, during their allocated GTSs. Devices are allowed to go into a sleep mode during the rest of the GTSs.

To reduce energy consumption, the coordinator may also issue a superframe containing both an active period and an idle period, as shown in Figure 5.16(c). The active period, composed of 16 time slots, contains the frame beacon, the CAP time slots, and if applicable, the CAP slots. The inactive period defines a time period during which all network nodes, including the coordinator, can go into a sleep mode. In this mode, the network devices switch off their power and set a timer to wake up immediately before the announcement of the next beacon frame.

It is worth noting that to accommodate a wide range of application requirements and network deployment, the length of the active and inactive periods, the time slot duration, and the number and usage of the slots designated as GTSs are configurable network parameters. Consequently, depending on the network activity, the types of devices connected to the network, and the nature of the application supported by the network, the length of the inactive period varies and may be set to zero.

Frame Types   The general MAC frame format of the IEEE 802.15.4 MAC-layer standard is depicted in Figure 5.17(a). It is composed of three basic components: the MAC header, the MAC payload, and the MAC footer. The MAC header

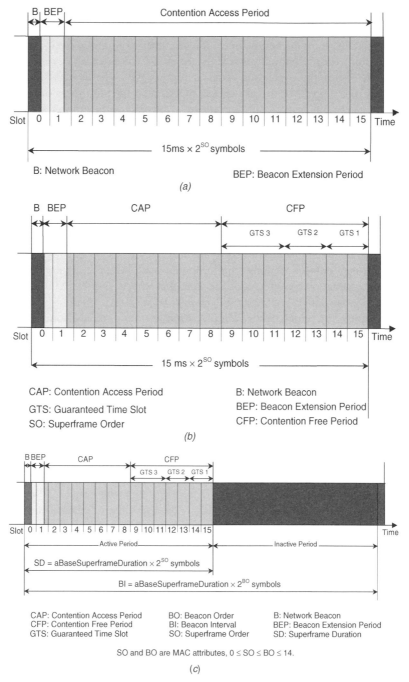

Figure 5.16 (*a*) Superframe structure; (*b*) QoS frame structure; (*c*) Superframe structure with energy saving.

Octets:	1	0/	0/2/	0/	0/2/	Variabl	2
Frame Control	Sequence number	Destinatio n PAN	Destinatio n	Source PAN Identifier	Source Address	Frame Payload	Frame Check Sequence
			Addressing				

		MAC				MAC Payload	MAC Footer

Bits: 0-	3	4	5	6	7-	10-	12-	14-
Frame Type	Security Enabled	Frame Pending	Ack Request	Intra PAN	Reserved	Destination Addressing Mode	Reserve	Source Addressing Mode

Frame Type Value $b_0 b_1 b_2$	Description
0 0 0	Beacon
0 0 1	Data
0 1 0	Acknowledgement
0 1 1	MAC Command
1 0 0 - 1 1 1	Reserved

(a)

Octets:2	1	4 or 10	2	Variable	Variable	Variable	2
Frame Control	Beacon Sequence Number	Source Address	Superframe Spec	GTS Fields	Pending Address Fields	Beacon Payload	Frame Check Sequence
	MAC Header				MAC Payload		MAC Footer

Superframe Specification

Bits: 0-3	4-7	8-11	12	13	14	15
Beacon Order	Superframe Order	Final CAP Slot	Battery Life Extension	Reserved	PAN Coordinator	Association Permit

Extension

GTS Fields

Octets: 1	0/1	Variable
GTS Spec	GTS Directions	GTS List

Pending Address Fields

Octets: 1	Variable
Pending Address Spec	Address List

GTS Specification

Bits: 0-2	3-6	7
GTS Descriptor Count	Reserved	GTS Permit

GTS Directions

Bits: 0-6	7
GTS Directions Mask	Reserved

GTS List

Bits: 0-15	16-19	20-23
Device Short Address	GTS Starting Slot	GTS Length

(b)

Figure 5.17 (*a*) General MAC frame format; (*b*) Beacon frame format; (*c*) Data and acknowledgment frame format; (*d*) MAC command frame format. (*Continued*)

Data Frame Format

Octets:2	1	4 to 20	variable	2
Frame Control	Data Sequence Number	Address Information	Data	Frame Check Sequence
MAC			MAC Payload	MAC Foote

Acknowledgement Frame Format

Octets:	1	2
Frame Control	Data Sequence Number	Frame Check Sequence
MAC		MAC Foote

(c)

Octets:	1	4 to	1	variabl	2
Frame Control	Data Sequence	Address Information	Command	Command	Frame Check Sequence
MAC			MAC		MAC foote

Command Frame Types

Command Frame Identifier	Command Name	RFD	
		Tx	Rx
0 x 01	Association Request	X	
0 x 02	Association Response		X
0 x 03	Dis-association Notification	X	X
0 x 04	Data Request	X	
0 x 05	PAN ID Conflict Notification	X	
0 x 06	Orphan Notification	X	
0 x 07	Beacon Request		
0 x 08	Coordinator Realignment		X
0 x 09	GTS Request		
0 x 0a – 0	Reserved		

(d)

Figure 5.17 *(continued)*

contains a frame control field and the addressing field. The control field carries the frame type and other information necessary for network control and operation. The addressing specifies the source PAN identifier, the source node address, the destination PAN identifier, and the destination address. The MAC payload contains the data frame to be exchanged between the communicating devices. The MAC footer contains the frame check sequence field. This field is used to detect frame errors.

The IEEE 802.15.4 defines four basic frame types: the beacon frame, the data frame, the acknowledgment frame, and the MAC command frame. The *beacon frame* is transmitted periodically by the coordinator. The frame serves multiple purposes, including identifying the network and its structure, waking up devices from the sleep mode to the listening mode, and synchronizing network operations. The beacon frames are particularly important in mesh and cluster-tree network topologies. They keep all the network nodes synchronized without requiring these nodes to remain awake for long period of times, thereby reducing considerably energy consumption and extending battery lifetime. The beacon frame and its fields are described in Figure 5.17(*b*).

The *data frame* carries a payload of up to 104 octets. Each data frame carries a sequence number that identifies the frame uniquely. The sequence number ensures that all frames are accounted for and are received in order. The FCS field is used to detect frames in error.

The *acknowledgment frame* is used by a receiver to acknowledge the receipt of a data frame. The receipt of the acknowledgment by the sender constitutes a confirmation that the corresponding data frame was received without error and in order. A data frame and its corresponding acknowledgment frame are matched by their respective sequence numbers. The data and acknowledgment frame formats are depicted in Figure 5.17(*c*).

The *MAC command frame* is used by the MAC entities in different devices for negotiation and communication. The frame provides the mechanism for a centralized network manager to control and configure devices remotely, irrespective of the network size and topology. Typical commands include device association and disassociation request and notification, data request, PAN ID conflict notification, orphan notification, beacon request, GTS request, and coordinator realignment. The MAC command frame format is shown in Figure 5.17(*d*). Upon receiving a frame, the MAC-layer entity must process the received frame to determine the actions required to handle the frame properly. To provide enough time for the MAC-layer entity to process the frame, the IEEE 802.15.4 MAC-layer standard requires that an interframe spacing (IFS) be inserted between two consecutive frames. The IFS duration depends on whether the transmission transaction is acknowledged or unacknowledged.

If the transmission is acknowledged, the IFS follows the acknowledgment frame. Furthermore, if the frame length does not exceed the threshold, aMaxSIFSFrameSize, the acknowledgment must be followed by a short IFS (SIFS) period, the duration of which should be at least aMinSIFSPeriod. If the frame length exceeds aMaxSIFSFrameSize, the acknowledgment must be followed by a long IFS (LIFS). The duration

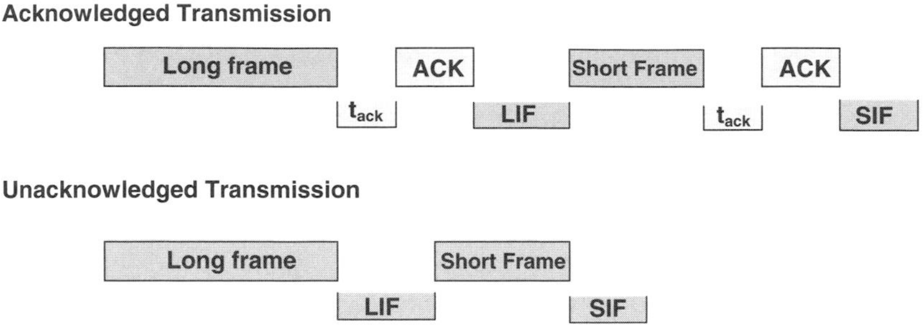

Acknowledged Transmission

Unacknowledged Transmission

Figure 5.18 Interframe spacing.

of the LIFS must be at least aMaxLIFSPeriod. The a MinSIFSPeriod is typically 12 symbols long, while the aMaxLIFSPeriod is 40 symbols long.

If the transmission is unacknowledged, the IFS immediately follows the data frame. Depending on whether the size of the frame exceeds or does not exceed the aMaxSIFSFrameSize threshold, a LIFS or a SIFS is used. The IFS procedure is depicted in Figure 5.18. As shown in this figure, an acknowledgment of the transmitted frame is expected to be received within a time interval, t_{ack}, such that aTurnaroundTime $\leq t_{ack} \leq$ (aTurnaroundTime + aUnitBackoff Period); aTurnaroundTime is typically 12 symbols long, while the aUnitBackoff Period is 20 symbols long.

Modes of Operation   The IEEE 802.15.4 MAC protocol is designed to meet the requirements of multiple types of traffic. Each traffic type is characterized by its unique characteristics in terms of the data profile and latency requirement. Based on this characterization, the IEEE 802.15.4 standard identifies three types of traffic: periodic data, intermittent data, and repetitive low-latency data. *Periodic data* characterize wireless sensor applications, whereby the sensor alternates between two modes of operations, active and idle. In the active mode, the sensor wakes up and exchanges data with a coordinator or another network device. In the idle mode, the sensor goes to sleep. *Intermittent data* are defined by an external stimulus or by an application such as a wireless light switch controlling a lamp. In this example, the lamp, typically mains powered, can monitor the channel in a continuous manner. On the other hand, the switch remains idle until it is toggled, in which case it transmits the information to the lamp. *Repetitive low-latency data* are defined by critical applications such as security monitoring systems. This type of application requires time slot allocations to guarantee access to the channel within its latency tolerance. To accommodate the three types of traffic, the IEEE 802.15.4 MAC-layer standard specifies a beaconed and a beaconless mode of operation. In the following we discuss the basic operations of these two modes.

Beacon Mode Operation   The beacon mode allows devices within an extended network, such as mesh or cluster tree, to synchronize their actions and coordinate communications with each other. Devices wake up only when a beacon is broadcast. A device registers with the network coordinator and looks for any messages addressed to it. If no messages are pending, the device returns to sleep. The coordinator may also go to sleep when the communications with the end devices are completed.

To regulate access to the channel, the network coordinator uses a superframe structure. As discussed above, the superframe is divided into 16 equally sized slots, the first of which is dedicated to the transmission of the beacon frame. Network devices can compete to access the channel during the contention access period (CAP), using a slotted CSMA-CA mechanism. Applications requiring low latency or a specific data rate may issue a request to the PAN coordinator for the allocation of GTSs during the contention-free period (CFP). The allocation of these GTSs are such that all contention-based transactions complete before the start of the CAP, and all GTS-based transactions finish within their allocated time slots and before the end of the CFP. In the following we first discuss the GTS allocation. We then describe the CSMA-CA access mechanism used by network devices to compete for the channel during the CAP.

GTS Allocation   To reserve GTS for data exchange, a device sends an explicit request to the network coordinator. The request specifies the type, transmit or receive, and the number of the contiguous slots desired. A transmit slot is used by the device to send data to the coordinator. A request of a receive slot, on the other hand, expresses the readiness of the device to receive data from the coordinator. Immediately upon receiving a GTS request, the coordinator acknowledges the reservation frame but the acknowledgment constitutes neither a confirmation nor a denial of the reservation request. Upon receiving the acknowledgment, the device sets a timer to a specific value referred to as a GTSDescPersistenceTime and monitors transmission of the subsequent coordinator's beacon.

Depending on slot availability, three scenarios are possible. In the first scenario, the coordinator responds favorably to the request within the GTSDescPersistence-Time interval and inserts a GTS descriptor into a subsequent beacon. The GTS descriptor contains the short address of the requesting device, the number of the allocated GTSs, and their position within the GTS interval. These slots remain allocated to the device and are used every time they are announced in the GTS descriptor until either voluntarily relinquished by the device or explicitly revoked by the coordinator.

The device can request deallocation of the GTSs by sending an explicit control frame. The coordinator can also revoke the GTSs reserved if it observes that the slots have not been used within a specified number of superframes. The coordinator informs the device of its decision to deallocate the GTSs reserved by generating a GTS descriptor with a start slot of value zero. The second scenario occurs when not enough slots are available to honor the device's requests. In this case, the coordinator generates a GTS descriptor with an invalid time slot value of zero, but specifies the amount of available slots in the descriptor length field. Depending on the type of data to be exchanged and the nature of the transaction with the coordinator, the device may renegotiate its request. The third scenario occurs when the interval GTSDescPersistenceTime elapses and the coordinator does not insert an appropriate GTS descriptor into one of the beacon frames. The device concludes that the resource request failed.

If the GTS allocation request is confirmed by the coordinator, the device uses these slots for communication. Depending on the type of the request, the GTSs allocated can be either transmit or receive. In the case of a transmit GTS, the device wakes up before the start of the allocated GTSs and uses the reserved contiguous slots to transmit its data to the coordinator. The latter acknowledges the receipt of the data immediately. Similarly, if the allocated GTS is of the receive type, the device wakes up at the beginning of the reserved time slots and receives the data transmitted by the coordinator at the start of the same slot. Upon receiving the data, the device completes the data transfer transaction by acknowledging receipt of the data. It is worth noting, however, that the receive transaction can be carried out successfully only if the device has reserved enough time slots to cover the transmission of the data packet and its corresponding acknowledgment, along with the required interframe spacing.

If the number of allocated time slots for a transmit transaction is not sufficient to cover the entire cycle of the transaction, the device must send its data during the CAP of the active period. This is also the case when the device does not have an allocated slot. In the case of a receive transaction, if the coordinator cannot use a receive GTS allocation, it announces the buffered packet to the intended recipient by including its address in the pending address field of the beacon frame. In response, the receiving device sends a special data packet request during the CAP of the active period and sets its transceiver on in preparation of the incoming packet. The coordinator acknowledges the request and proceeds to transmit the packet immediately. Notice that the coordinator continues to include the device's address in the pending address field, as long as the packet is still buffered or until its associated timer expires. If the data request fails to trigger an acknowledgment from the coordinator, the device may reiterate its attempt in one of the subsequent superframes. It can also switch off its transceiver until the next beacon transmission.

Contention-Based Channel Access The access to the medium during the CAP of an active period is regulated using a slotted, nonpersistent CSMA protocol hardware. The protocol, however, does not address the hidden terminal problem. As such, it does not use any mechanisms, such as a RTS/CTS handshake, to alleviate this problem. To reduce the likelihood of collisions, the protocol uses random delays. The basic steps of the contention-based MAC protocol are described in Figure 5.19.

Contrary to the traditional protocols, the CAP access protocol does not use the superframe slots for its back-off procedure. Instead, it uses the back-off period to accelerate the contention resolution process. A device attempting to transmit a date packet during the CAP of an active period first synchronizes its transceiver with the coordinator's beacon. It then initializes and maintains three main variables: NB,

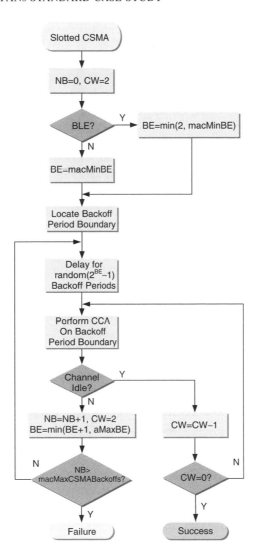

BLE: Battery Life Extension

Figure 5.19 Slotted CSMA algorithm.

BE, and CW. The variable NB, initially set to zero, counts the number of back-offs. The variable BE denotes the current back-off exponent. The initial value of this variable is set to macBinBE, a protocol-specific parameter. The last variable, CW, represents the current congestion window.

Prior to transmitting the new packet, the device first aligns itself with the boundary of the next back-off period. It then draws an integer number, i, between $[0, 2^{BE}-1]$ and waits for i back-off periods. At the expiration of this waiting interval, the contending device performs a clear channel assessment (CCA) operation. If the channel is idle, the contending device decrements CW by 1. If the value of CW is positive, the device performs a second CCA operation to determine the current state of the channel. If the channel is clear, the device decrements CW, which then becomes zero, declares the channel idle, and proceeds with the transmission of the packet.

If the second CCA operation indicates a busy channel, however, the device increases the number of back-offs, BE, by 1 and sets CW to 2. It then computes a new BE as min*(BE+1, aMaxBE)*, where *aMaxBE* is a protocol-specific parameter. If the number of back-offs, NB, exceeds the maximum back-off threshold, MaxCS MABack-offs, the device drops the frame and declares failure. Otherwise, the device waits for a number of back-off periods, randomly drawn from $[0, 2^{BE}-1]$, before it makes a new attempt to transmit the frame.

It is worth noting that a contending device must assert that the channel is clear for two consecutive back-off periods to account for the hardware nonzero receive-to-transmit turnaround time. Failure to do so may cause the contending device to senses the medium as clear during the turnaround time of an exchange between two devices and wrongly declares the channel idle. The attempt to transmit the frame results in a collision.

As stated previously, the CAP access control protocol is designed to greatly reduce, when adequate, the device duty cycle, especially in low-activity networks. For applications running on low-activity networks, even the minimum duration of a CAP interval exceeds the length of the time interval necessary to perform the low activity of the network. To address this situation, the CAP access control protocol offers a battery life extension (BLE) mode of operation. The BLE mode allows devices to go into sleep mode in the presence of low activity. To trigger a BLE mode of operation, the coordinator sets a BLE flag in its beacon frame. It then limits its monitoring of the CAP to six back-off periods. If no activity is heard within this period, the coordinator returns to sleep. Upon detecting the BLE flag, a device attempting to communicate with the coordinator sets the initial value of the back-off exponent to 2 or less, thereby reducing considerably the channel sensing period. Although such an action may increase the likelihood of a collision, it has been shown that using BLE with the superframe order (SO) set to 14 greatly reduces the total system duty cycle.

Beaconless Mode of Operation A beaconless mode is better suited for applications where remote units such as intrusion sensors and motion detectors wake up on a regular, yet random basis to report on events as they occur. The network coordinator, mains powered, is continuously awake waiting to hear from each of these units. In the beaconless mode, the network coordinator does not send a beacon frame on a regular basis. Furthermore, this mode of operation does not support allocation of guaranteed slots for low-latency applications and applications requiring a specific data rate. Instead, devices compete for channel access using an unslotted, nonpersistent CSMA/CA protocol. When a device wishes to transmit data, it waits for a random number of back-off periods before sensing the channel. If the channel

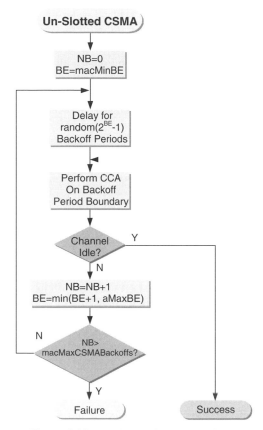

Figure 5.20 Unslotted CSMA algorithm.

is busy, the device increases the number of attempts by one and checks if the maximum number of attempts has been reached. If the limit is exceeded, the device generates a channel access error and reports this event to upper layers. If the number of attempts is below the limit, the device reiterates this procedure until it either captures the channel successfully or the number of attempts exceeds the limit. The basic steps of this protocol are described in Figure 5.20. It is worth noting that the absence of a synchronization mechanism, such as the beacon, coupled with the unslotted nature of the access protocol, devices are no longer required to locate the back-off period boundaries, as was the case in the slotted CSMA access protocol. Furthermore, the devices are no longer required to execute only a single CCA operation. In summary, the IEEE 802.15.4 MAC-layer standard is designed for secure, low-power operations. Due to the low data throughput of the applications envisioned to use the IEEE 802.15.4 standard, the MAC-layer protocol includes several mechanisms to keep the average power consumption of the network devices low. This is achieved largely by allowing nodes to operate at low duty cycles while maintaining network-level connectivity.

By adopting a loosely synchronized sleep and wake-up cycle, network nodes can operate for extended periods of time with minimum energy consumption. To accommodate different data types, the standard defines different modes of operations. Support for these traffic types is achieved by the use of an optional superframe for the beacon-enabled operation mode, with the possibility of guaranteed slot allocation to accommodate low-latency applications. The standard also offers a beaconless mode of operations to support star-based topologies for monitoring and security applications.

5.7 CONCLUSION

Sensor networking is an emerging technology that has a wide range of potential applications, including critical infrastructure protection, environmental monitoring, smart spaces, ubiquitous and pervasive health care, and robotic exploration. A WSN normally consists of a large number of distributed, battery-operated nodes equipped with one or more sensors, embedded processors, and low-power radios. These nodes cooperate to organize themselves into a multihop wireless network. The design of efficient MAC-layer protocols for WSNs is crucial for the wireless sensor nodes to carry out successfully the mission for which they are deployed.

The choice of the medium access control protocol is the major determining factor in WSN performance. Several attributes must be considered in the design of an efficient MAC layer protocol for WSN. Sensor network are likely to be battery powered, and it is often difficult, if not impossible, to change or recharge the batteries on these nodes. An efficient design of a MAC-layer protocol for a WSN must therefore be energy efficient to extend the lifetime of the network [5.10]. The MAC-layer protocol must also be scalable to accommodate changes in he network size, node density, and network topology. Finally, access fairness, reduced latency, high throughput, and bandwidth utilization are also important attributes in the design of MAC layer protocols for WSNs.

A number of access control protocols have been proposed for WSNs. In this chapter we discuss the fundamental design issues of medium access control for WSN methods and provide an overview of these protocols. In the first part of the chapter, a description of the basic requirements of access control protocols is provided. The following section categorizes the major media access control techniques used in shared medium access networks. In the last part of the chapter we discuss specific requirements of access control methods for WSNs and describe several MAC layer protocols for these networks. Two cases studies focused on a detailed description of two IEEE 802.11 inspired protocols, S-MAC and IEEE 802.15.4. S-MAC is designed explicitly for WSNs. The MAC-layer protocol uses periodic and coordinated sleeping to reduce energy consumption. It also uses message passing to reduce contention latency for delay-sensitive sensor applications.

The IEEE 802.15.4 standard is designed to provide support for low-data-rate connectivity among relatively inexpensive, minimally powered devices, typically residing within 30 ft or less of each other. The type of these devices ranges from sensors and switches for industrial and residential use to interactive toys, inventory tracking, smart tags, and badges. The IEEE 802.15.4 wireless MAC- and PHY-layer specifications allow a collection of portable and moving devices, which operate at a data rate of 10 to 250 kpbs, to form ad hoc personal area networks within which these devices and interact directly.

The development of a MAC-layer protocol for sensor networks is likely to continue to be a topic of interest as new WSNs continue to emerge. Furthermore, recent developments in cognitive radio are likely to bring a new perspective to the design of MAC-layer protocol for WSNs. The ability of a cognitive radio to interact directly with its environment will enhance the ability of a wireless network device equipped with such a radio to better adapt to and interact with its environment while carefully managing its energy consumption.

REFERENCES

[5.1] J. Hill, R. Szewczyk, A. Woo, S. Hollar, D. Culler, K. Pister, "System Architecture Directions for Networked Sensors," *Proceedings of the 9th International Conference on Architectural Support for Programming Languages and Operating Systems*, Cambridge, MA, Nov. 2000, pp. 93–104.

[5.2] Y.-C. Tseng, C.-S. Hsu, T.-Y. Hsieh, "Power-Saving Protocols for IEEE 802.11-Based Multi-hop Ad Hoc Networks," *Proceedings of the 21st Annual Joint Conference of the IEEE Computer and Communications Societies* (InfoCom'02), New York, June 2002, pp. 200–209.

[5.3] J. Zhao, R. Govindan, "Understanding Packet Delivery Performance in Dense Wireless Sensor Networks," *Proceedings of the 1st ACM Conference on Embedded Networked Sensor Systems* (SenSys'03), Los Angeles, Nov. 2003, pp. 1–13.

[5.4] V. Bharghavan, A. Demers, S. Shenker, L. Zhang, "MACAW: A Media Access Protocol for Wireless LANs," *Proceedings of the ACM SIGCOMM*, Portland, Oregon, 1994, pp. 212–225.

[5.5] T. V. Dam, K. Langendoen, "An Adaptive Energy-Efficient MAC Protocol for Wireless Sensor Networks," *Proceedings fo the 1st ACM Conference on Embedded Networked Sensor Systems* (SenSys'03), Los Angeles, Nov. 2003.

[5.6] J. Ding, K. Sivalingam, R. Kashyapa, L. J. Chuan, "A Multi-layered Architecture and Protocols for Large-Scale Wireless Sensor Networks," *Proceedings of the IEEE 58th Vehicular Technology Conference* (VTC'03), Oct. 2003, Vol. 3, pp. 1443–1447.

[5.7] P. Karn, "MACA: A New Channel Access Method for Packet Radio," *Proceedings of the ARRL/CRRL Amateur Radio 9th Computer Networking Conference*, Sept. 1990, pp. 134–140.

[5.8] "Wireless LAN Medium Access Control (MAC) and Physical Layer (PHY) Specifications," ANSI/IEEE Std. 802.11-1999, 1999.

[5.9] L. Bao, J. J. Garcia-Luna-Aceves, "A New Approach to Channel Access Scheduling for Ad Hoc Networks," 2001, *Proceedings of the 7th ACM International Conference on Mobile Computing and Networking* (MobiCom'01), Rome, Italy, July 2001. pp. 210–221.

[5.10] J. Monks, V. Bharghavan, W.-M. Hwu, "A Power Controlled Multiple Access Protocol for Wireless Packet Networks," *Proceedings of the 20th Annual Joint Conference of the IEEE Computer and Communications Societies* (InfoCom'01), Anchorage, AK, Apr. 2001.

[5.11] K. Sohrabi, G. J. Pottie, "Performance of a Novel Self-Organization Protocol for Wireless Ad Hoc Sensor Networks," *Proceedings of the IEEE 50th Vehicular Technology Conference* (VTC'99), 1999, pp. 1222–1226.

[5.12] G. D. Bacco et al., "A MAC Protocol for Delay-Bounded Applications in Wireless Sensor Networks," *Proceedings of the 3rd Annual Mediterranean Ad Hoc Networking Workshop*, June 2004.

[5.13] A. El-Hoiydi, "Spatial TDMA and CSMA with Preamble Sampling for Low Power Ad Hoc Wireless Sensor Networks," *Proceedings of the 7th IEEE International Symposium on Computers and Communications* (ISCC'02), July 2002, pp. 685–692.

[5.14] E. McCune, "DSSS vs. FHSS Narrowband Interference Performance Issues," *RF Signal Processing*, Sept. 2000.

[5.15] L. Kleinrock, Fouad Tobagi, "Packet Switching in Radio Channels, Part I: Carrier Sense Multiple Access Modes and Their Throughput Delay Characteristics," *IEEE Transactions on Communications*, Vol. 23, No. 12, Dec. 1975, pp. 1400–1416.

[5.16] F. A. Tobagi, L. Kleinrock, "Packet Switching in Radio Channels: Part II: The Hidden Terminal Problem in Carrier Sense Multiple Access and the Busy Tone Solution," *IEEE Transactions on Communications*, Vol. 23, Dec. 1975, pp. 1417–1433.

[5.17] T. S. Rappaport, *Wireless Communications: Principles and Practice*, 1996, Prentice Hall, Upper Saddle River, NJ.

[5.18] S. Xu, T. Saadawi, "Does the IEEE 802.11 MAC Protocol Work Well in Multihop Wireless Ad Hoc Networks?" *IEEE Communications*, June 2001, pp. 130–137.

[5.19] Y. C. Tay, K. Jamieson, H. Balakrishnan, "Collision-Minimizing CSMA and Its Applications to Wireless Sensor Networks," *IEEE Journal on Selected Areas in Communications*, Vol. 22, No. 6, Aug. 2004, pp. 1048–1057.

[5.20] D. J. Baker, J. Wieselthier, "A Distributed Algorithm for Scheduling the Activation of Links in a Self-Organizing, Mobile, Radio Network," *Proceedings of the International Conference on Communications*, 1982, pp. 2F.6.1–2F.6.5.

[5.21] P. Lin, C. Qiao, X. Wang, "Medium Access Control with a Dynamic Duty Cycle for Sensor Networks", *IEEE Wireless Communications and Networking Conference* (WCNC'04), Mar. 2004, Vol. 3, pp. 1534–1539.

[5.22] G. Lu, B. Krishnamachari, C. Raghavendra, "An Adaptive Energy-Efficient and Low-Latency MAC for Data Gathering in Sensor Networks," presented at the Workshop on Energy-Efficient Wireless Communications and Networks (EWCN'04), held in conjunction with the IEEE International Performance Computing and Communications Conference (IPCCC), Apr. 2004.

[5.23] S. S. Kulkarni, "TDMA Services for Sensor Networks," *Proceedings of the 24th International Conference on Distributed Computing Systems Workshops*, Mar. 2004, pp. 604–609.

[5.24] K. Sohrabi, G. J. Pottie, "Performance of a Novel Self-Organization Protocol for Wireless Ad Hoc Sensor Networks," *Proceedings of the IEEE 50th Vehicular Technology Conference* (VTC'99), 1999, pp. 1222–1226.

[5.25] K. Sohrabi, J. Gao, V. Ailawadhi, G. J. Pottie, "Protocols for Self-Organization of a Wireless Sensor Network," *IEEE Personal Communications*, Vol. 7, No. 5, Oct. 2000, pp. 16–27.

[5.26] J. C. Haartsen, "The Bluetooth Radio System," *IEEE Personal Communications*, Feb. 2000, pp. 28–36.

[5.27] "Specification of the Bluetooth System: Core," http://www.bluetooth.org/, 2001.

[5.28] F. Bennett, D. Clarke, J. B. Evans, A. Hopper, A. Jones, D. Leask, "Piconet: Embedded Mobile Networking," *IEEE Personal Communications*, Vol. 4, Oct. 1997, pp. 8–15.

[5.29] W. R. Heinzelman, A. Chandrakasan, H. Balakrishnan, "Energy Efficient Communication Protocols for Wireless Microsensor Networks," *Proceedings of the 33rd Hawaii International Conference Systems Sciences* (HICSS'00), Maui, HI, Jan. 2000, pp. 3005–3014.

[5.30] W. Heinzelman, A. Sinha, A. Wang, A. Chandrakasan, "Energy-Scalable Algorithms and Protocols for Wireless Microsensor Networks," *Proceedings of the International Conference on Acoustics, Speech, and Signal Processing (ICASSP '00)*, June 2000.

[5.31] W. Heinzelman, "Application-Specific Protocol Architectures for Wireless Networks," Ph.D. dissertation, Massachusetts Institute of Technology, June 2000.

[5.32] J. Elson, D. Estrin, "Time Synchronization for Wireless Sensor Networks," *Proceedings of the 15th International Symposium on Parallel and Distributed Processing*, San Francisco, CA, Apr. 2001.

[5.33] S. Ganeriwal, R. Kumar, M. B. Srivastava, "Timing-Sync Protocol for Sensor Networks," *Proceedings of the 1st ACM Conference on Embedded Networked Sensor Systems* (SenSys'03), Los Angeles, Nov. 2003.

[5.34] C. L. Fullmer, J. J. Garcia-Luna-Aveces, "Solutions to Hidden Terminal Problems in Wireless Networks," *Proceedings of the ACM SIGCOMM Conference*, 1997, pp. 39–49.

[5.35] S. Singh, C. S. Raghavendra, "PAMAS: Power Aware Multi-access Protocol with Signalling for Ad Hoc Networks," *ACM Computers in Communications Review*, Vol. 28, No. 3, July 1998, pp. 5–26.

[5.36] C. Schurgers, V. Tsiatsis, S. Ganeriwal, M. Srivastaval., "Optimizing Sensor Networks in the Energy-Latency-Density Design Space," *IEEE Transactions on Mobile Computing*, Vol. 1, No. 1, Jan.–Mar. 2002.

[5.37] A. El-Hoiydi, J.-D. Decotignie, "Wisemac: An Ultra Low Power Mac Protocol for Multi-hop Wireless Sensor Networks," *Algorithmic Aspects of Wireless Sensor Networks: 1st International Workshop* AlgoSensors'04, Turku, Finland, July 2004.

[5.38] C. C. Enz, A. El-Hoiydi, J.-D. Decotignie, V. Peiris, "WiseNET: An Ultralow-Power Wireless Sensor Network Solution," *IEEE Computer*, Vol. 37, No. 8, Aug. 2004.

[5.39] K. Jamieson, H. Balakrishnan, Y. C. Tay, "Sift: A MAC Protocol for Event-Driven Wireless Sensor Networks," Technical Report 894, MIT Laboratory for Computer Science, Cambridge, MA, http://www.lcs.mit.edu/publications/pubs/pdf/MIT-LCS-TR-894.pdf, May 2003.

[5.40] V. Rajendran, K. Obraczka, J. J. Garcia-Luna-Aceves, "Energy-Efficient, Collision-Free Medium Access Control for Wireless Sensor Networks," *Proceedings of the 1st ACM Conference on Embedded Networked Sensor Systems* SenSys'03, Los Angeles, Nov. 2003, pp. 181–192.

[5.41] J. Polastre, J. Hill, D. Culler, "Versatile Low Power Media Access for Wireless Sensor Networks," *Proceedings of the 2nd ACM Conference on Embedded Networked Sensor Systems* (SenSys'04), Baltimore, MD, Nov. 2004.

[5.42] A. Woo, D. Culler, "A Transmission Control Scheme for Media Access in Sensor Networks," *Proceedings of the 7th ACM/IEEE International Conference on Mobile Computing and Networking* (MobiCom'01), Rome, Italy, July 2001, pp. 221–235.

[5.43] A. Woo, T. Tong, D. Culler, "Taming the Underlying Challenges of Reliable Multihop Routing in Sensor Networks," *Proceedings of the 1st ACM Conference on Embedded Networked Sensor Systems* (SenSys'03), Nov. 2003, Los Angeles, pp. 14–27.

[5.44] W. Ye, J. Heidemann, D. Estrin, "A Flexible and Reliable Radio Communication Stack on Motes," Technical Report ISI-TR-565, Information Sciences Institute, University of Southern California, Los Angeles, Sept. 2002.

[5.45] W. Ye, J. Heidemann, D. Estrin, "Medium Access Control with Coordinated Adaptive Sleeping for Wireless Sensor Networks," *IEEE/ACM Transactions on Networking*, Vol. 12, No. 3, June 2004, pp. 493–506.

[5.46] W. Ye, J. Heidemann, D. Estrin, "An Energy-Efficient MAC Protocol for Wireless Sensor Networks," *Proceedings of the 21st Annual Joint Conference of the IEEE Computer and Communications Societies* (InfoCom'02), New York, June 2002, pp. 1567–1576.

[5.47] IEEE 802.15.4 Standard-2003, "Part 15.4: Wireless Medium Access Control (MAC) and Physical Layer (PHY) Specifications for Low-Rate Wireless Personal Area Networks (LRWPANs)," IEEE-SA Standards Board, 2003, http://grouper.ieee.org/groups/802/15/pub/TG4b.html.

[5.48] "ZigBee Specification," Document 053474r05, Version 1.0, ZigBee Alliance, Bishop Ranch, CA, Dec. 14, 2004, available (since June 2005) at http://www.zigbee.org/.

6

ROUTING PROTOCOLS FOR WIRELESS SENSOR NETWORKS

6.1 INTRODUCTION

WSNs are extremely versatile and can be deployed to support a wide variety of applications in many different situations, whether they are composed of stationary or mobile sensor nodes. The way these sensors are deployed depends on the nature of the application. In environmental monitoring and surveillance applications, for example, sensor nodes are typically deployed in an ad hoc fashion so as to cover the specific area to be monitored (e.g., C1WSNs). In health care–related applications, smart wearable wireless devices and biologically compatible sensors can be attached to or implanted strategically within the human body to monitor vital signs of the patient under surveillance. Once deployed, sensor nodes self-organize into an autonomous wireless ad hoc network, which requires very little or no maintenance. Sensor nodes then collaborate to carry out the tasks of the application for which they are deployed.

Despite the disparity in the objectives of sensor applications, the main task of wireless sensor nodes is to sense and collect data from a target domain, process the data, and transmit the information back to specific sites where the underlying application resides. Achieving this task efficiently requires the development of an energy-efficient routing protocol to set up paths between sensor nodes and the data sink. The path selection must be such that the lifetime of the network is maximized. The characteristics of the environment within which sensor nodes typically operate, coupled with severe resource and energy limitation, make the routing problem very challenging.

Wireless Sensor Networks: Technology, Protocols, and Applications, by Kazem Sohraby, Daniel Minoli, and Taieb Znati
Copyright © 2007 John Wiley & Sons, Inc.

Our objective in this chapter is to discuss issues central to routing in WSNs and describe different strategies used to develop routing protocols for these networks. To this end, we first discuss a representative class of sensor applications. The goal is to highlight the unique and distinctive features of the nature of the traffic typically generated in WSNs. In the second part of the chapter we provide a brief taxonomy of the basic routing strategies used to strike a balance between responsiveness and energy efficiency. Achieving this balance brings about new challenges that span the network layers in a manner that differs from infrastructured as well as ad hoc wireless networks. In the third part of the chapter we review a number of protocols that address the problem of routing in today's WSNs. Although the field is in its infancy and routing in WSNs remains largely relegated to research, multiple strategies have emerged as workable solutions to the routing problem. As the application of WSNs to different fields becomes more apparent, advances in network hardware and battery technology will pave the way to practical cost-effective implementations of these routing protocols.

6.2 BACKGROUND

WSNs have created new opportunities across the spectrum of human endeavors, including engineering design and manufacturing, monitoring and control of environmental systems, forest fire tracking, health care, battlefield surveillance, disaster management, and critical infrastructure protection. Several applications involving sensed data collection and dissemination are depicted in Figure 6.1, along with

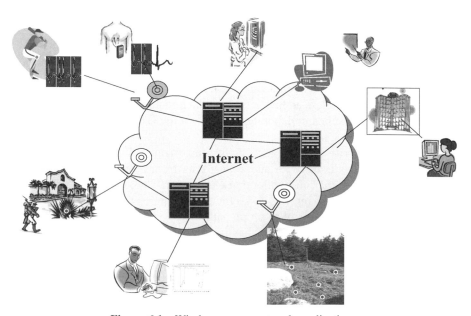

Figure 6.1   Wireless sensor network applications.

the way that data flow from its source to the anticipated sink [6.1–6.7]. Many applications (see Chapter 3) exhibit strong similarity in the way that sensors are used to collect and disseminate data to carry out the objectives for which they are deployed. In these applications, sensors are designed primarily to sense the environment and to record and possibly process the sensor readings before forwarding the information collected, through the base station toward the data sink and eventually, to where the application resides. These readings may be light levels, temperature, vital signs, or levels of environmental disturbance. The process of data collection and forwarding is either triggered by the occurrence of specific events in the environment where the sensors are deployed or is initiated in response to a query issued by the application supported. It is worth noting that in many cases it is useful to aggregate data collected by various sensors before forwarding the data to the base station. Data aggregation reduces the number of messages transmitted, leading to a significant decrease in energy consumption due to communication.

6.3 DATA DISSEMINATION AND GATHERING

The way that data and queries are forwarded between the base station and the location where the target phenomena are observed is an important aspect and a basic feature of WSNs. A simple approach to accomplishing this task is for each sensor node to exchange data directly with the base station. A single-hop-based approach, however, is costly, as nodes that are farther away from the base station may deplete their energy reserves quickly, thereby severely limiting the lifetime of the network. This is the case particularly where the wireless sensors are deployed to cover a large geographical region or where the wireless sensors are mobile and may move away from the base station.

To address the shortcomings of the single-hop approach, data exchange between the sensors and the base stations is usually carried out using multihop packet transmission over short communication radius. Such an approach leads to significant energy savings and reduces considerably communication interference between sensor nodes competing to access the channel, particularly in highly dense WSNs. Data forwarding between the sensors where data are collected and the sinks where data are made available is illustrated in Figure 6.2. In response to queries issued by the sinks or when specific events occur within the area monitored, data collected by the sensors are transmitted to the base station using multihop paths. It is worth noting that depending on the nature of the application, sensor nodes can aggregate data correlated on their way to the base station.

In a multihop WSN, intermediate nodes must participate in forwarding data packets between the source and the destination. Determining which set of intermediate nodes is to be selected to form a data-forwarding path between the source and the destination is the principal task of the routing algorithm. In general, routing in large-scale networks is inherently a difficult problem whose solution must address multiple challenging design requirements, including correctness, stability, and optimality with respect to various performance metrics. The intrinsic properties

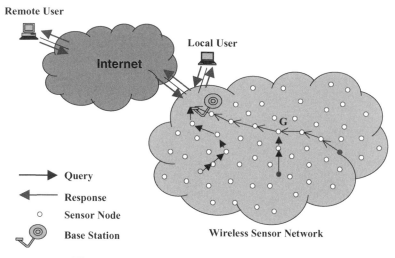

Figure 6.2   Multihop data and query forwarding.

of WSNs, combined with severe energy and bandwidth constraints, bring about additional challenges that must be addressed to satisfy the traffic requirements of the application supported, while extending the lifetime of the network.

In the following sections, we first discuss the primary routing challenges and design goals of routing in WSNs. We then discuss various strategies, approaches, and techniques that have been proposed to design efficient routing protocols for WSNs. Following that, we survey the state of the art in WSN routing protocols.

6.4   ROUTING CHALLENGES AND DESIGN ISSUES IN WIRELESS SENSOR NETWORKS

Although WSNs share many commonalities with wired and ad hoc networks, they also exhibit a number of unique characteristics which set them apart from existing networks. These unique characteristics bring to sharp focus new routing design requirements that go beyond those typically encountered in wired and wireless ad hoc networks. Meeting these design requirements presents a distinctive and unique set of challenges. These challenges can be attributed to multiple factors, including severe energy constraints, limited computing and communication capabilities, the dynamically changing environment within which sensors are deployed, and unique data traffic models and application-level quality of service requirements.

6.4.1   Network Scale and Time-Varying Characteristics

Sensor nodes operate with limited computing, storage, and communication capabilities under severe energy constraints, as discussed in Chapter 4. Due to the large number of conceivable sensor-based applications, the densities of the WSNs may vary widely, ranging from very sparse to very dense. Furthermore, in many applications, the sensor nodes, in some cases numbering in the hundreds if not thousands, are deployed in an ad hoc and often unsupervised manner over wide coverage areas. In these networks, the behavior of sensor nodes is dynamic and highly adaptive, as the need to self-organize and conserve energy forces sensor nodes to adjust their behavior constantly in response to their current level of activity or the lack thereof. Furthermore, sensor nodes may be required to adjust their behavior in response to the erratic and unpredictable behavior of wireless connections caused by high noise levels and radio-frequency interference, to prevent severe performance degradation of the application supported.

6.4.2 Resource Constraints

Sensor nodes are designed with minimal complexity for large-scale deployment at a reduced cost. Energy is a key concern in WSNs, which must achieve a long lifetime while operating on limited battery reserves. Multihop packet transmission over wireless networks is a major source of power consumption. Reducing energy consumption can be achieved by dynamically controlling the duty cycle of the wireless sensors. The energy management problem, however, becomes especially challenging in many mission-critical sensor applications. The requirements of these applications are such that a predetermined level of sensing and communication performance constraints must be maintained simultaneously. Therefore, a question arises as to how to design scalable routing algorithms that can operate efficiently for a wide range of performance constraints and design requirements. The development of these protocols is fundamental to the future of WSNs.

6.4.3 Sensor Applications Data Models

The data model describes the flow of information between the sensor nodes and the data sink. These models are highly dependent on the nature of the application in terms of how data are requested and used. Several data models have been proposed to address the data-gathering needs and interaction requirements of a variety of sensor applications [6.8,6.9]. A class of sensor applications requires data collection models that are based on periodic sampling or are driven by the occurrence of specific events. In other applications, data can be captured and stored, possibly processed and aggregated by a sensor node, before they are forwarded to the data sink. Yet a third class of sensor applications requires bidirectional data models in which two-way interaction between sensors and data sinks is required [6.10,6.11].

The need to support a variety of data models increases the complexity of the routing design problem. Optimizing the routing protocol for an application's specific data requirements while supporting a variety of data models and delivering the highest performance in scalability, reliability, responsiveness, and power efficiency becomes a design and engineering problem of enormous magnitude.

6.5 ROUTING STRATEGIES IN WIRELESS SENSOR NETWORKS

The WSN routing problem presents a very difficult challenge that can be posed as a classic trade-off between responsiveness and efficiency. This trade-off must balance the need to accommodate the limited processing and communication capabilities of sensor nodes against the overhead required to adapt to these. In a WSN, overhead is measured primarily in terms of bandwidth utilization, power consumption, and the processing requirements on the mobile nodes. Finding a strategy to balance these competing needs efficiently forms the basis of the routing challenge. Furthermore, the intrinsic characteristics of wireless networks gives rise to the important question of whether or not existing routing protocols designed for ad hoc networks are sufficient to meet this challenge [6.12].

Routing algorithms for ad hoc networks can be classified according to the manner in which information is acquired and maintained and the manner in which this information is used to compute paths based on the acquired information. Three different strategies can be identified: proactive, reactive, and hybrid [6.13,6.14]. The *proactive strategy*, also referred to as *table driven*, relies on periodic dissemination of routing information to maintain consistent and accurate routing tables across all nodes of the network. The structure of the network can be either flat or hierarchical. Flat proactive routing strategies have the potential to compute optimal paths. The overhead required to compute these paths may be prohibitive in a dynamically changing environment. Hierarchical routing is better suited to meet the routing demands of large ad hoc networks.

Reactive routing strategies establish routes to a limited set of destinations on demand. These strategies do not typically maintain global information across all nodes of the network. They must therefore, rely on a dynamic route search to establish paths between a source and a destination. This typically involves flooding a route discovery query, with the replies traveling back along the reverse path. The reactive routing strategies vary in the way they control the flooding process to reduce communication overhead and the way in which routes are computed and reestablished when failure occurs.

Hybrid strategies rely on the existence of network structure to achieve stability and scalability in large networks. In these strategies the network is organized into mutually adjacent clusters, which are maintained dynamically as nodes join and leave their assigned clusters. Clustering provides a structure that can be leveraged to limit the scope of the routing algorithm reaction to changes in the network environment. A hybrid routing strategy can be adopted whereby proactive routing is used within a cluster and reactive routing is used across clusters. The main challenge is to reduce the overhead required to maintain the clusters.

In summary, traditional routing algorithms for ad hoc networks tend to exhibit their least desirable behavior under highly dynamic conditions. Routing protocol overhead typically increases dramatically with increased network size and dynamics. A large overhead can easily overwhelm network resources. Furthermore, traditional routing protocols operating in large networks require substantial internodal coordination, and in some cases global flooding, to maintain consistent and accurate information, which is necessary to achieve loop-free routing. The use of these techniques increases routing protocol overhead and convergence times. Consequently, although they are well adapted to operate in environments where the computation and communications capabilities of the network nodes are relatively high compared to sensor nodes, the efficiency of these techniques conflict with routing requirements in WSNs. New routing strategies are therefore required for sensor networks that are capable of effectively managing the trade-off between optimality and efficiency.

6.5.1   WSN Routing Techniques

The design of routing protocols for WSNs must consider the power and resource limitations of the network nodes, the time-varying quality of the wireless channel, and the possibility for packet loss and delay. To address these design requirements, several routing strategies for WSNs have been proposed. One class of routing protocols adopts a flat network architecture in which all nodes are considered peers. A flat network architecture has several advantages, including minimal overhead to maintain the infrastructure and the potential for the discovery of multiple routes between communicating nodes for fault tolerance.

A second class of routing protocols imposes a structure on the network to achieve energy efficiency, stability, and scalability. In this class of protocols, network nodes are organized in clusters in which a node with higher residual energy, for example, assumes the role of a cluster head. The cluster head is responsible for coordinating activities within the cluster and forwarding information between clusters. Clustering has potential to reduce energy consumption and extend the lifetime of the network.

A third class of routing protocols uses a data-centric approach to disseminate interest within the network. The approach uses attribute-based naming, whereby a source node queries an attribute for the phenomenon rather than an individual sensor node. The interest dissemination is achieved by assigning tasks to sensor nodes and expressing queries to relative to specific attributes. Different strategies can be used to communicate interests to the sensor nodes, including broadcasting, attribute-based multicasting, geo-casting, and anycasting.

A fourth class of routing protocols uses location to address a sensor node. Location-based routing is useful in applications where the position of the node within the geographical coverage of the network is relevant to the query issued by the source node. Such a query may specify a specific area where a phenomenon of interest may occur or the vicinity to a specific point in the network environment.

In the following sections, several routing algorithms that have been proposed for data dissemination in WSNs are described. The design trade-offs and performance of these algorithms are also discussed.

6.5.2   Flooding and Its Variants

Flooding is a common technique frequently used for path discovery and information dissemination in wired and wireless ad hoc networks. The routing strategy is

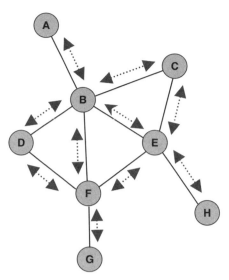

Figure 6.3   Flooding in data communications networks.

simple and does not rely on costly network topology maintenance and complex route discovery algorithms. Flooding uses a reactive approach whereby each node receiving a data or control packet sends the packet to all its neighbors. After transmission, a packet follows all possible paths. Unless the network is disconnected, the packet will eventually reach its destination. Furthermore, as the network topology changes, the packet transmitted follows the new routes. Figure 6.3 illustrates the concept of flooding in data communications network. As shown in the figure, flooding in its simplest form may cause packets to be replicated indefinitely by network nodes.

To prevent a packet from circulating indefinitely in the network, a hop count field is usually included in the packet. Initially, the hop count is set to approximately the diameter of the network. As the packet travels across the network, the hop count is decremented by one for each hop that it traverses. When the hop count reaches zero, the packet is simply discarded. A similar effect can be achieved using a time-to-live field, which records the number of time units that a packet is allowed to live within the network. At the expiration of this time, the packet is no longer forwarded. Flooding can be further enhanced by identifying data packets uniquely, forcing each network node to drop all the packets that it has already forwarded. Such a strategy requires maintaining at least a recent history of the traffic, to keep track of which data packets have already been forwarded.

Despite the simplicity of its forwarding rule and the relatively low-cost maintenance that it requires, flooding suffers several deficiencies when used in WSNs. The first drawback of flooding is its susceptibility to *traffic implosion,* as shown in Figure 6.4. This undesirable effect is caused by duplicate control or data packets being sent repeatedly to the same node. The second drawback of flooding is the

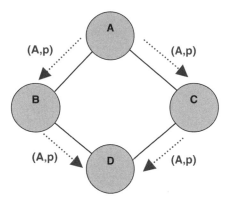

Figure 6.4 Flooding traffic implosion problem.

overlap problem to which it gives rise, as depicted in Figure 6.5. Overlapping occurs when two nodes covering the same region send packets containing similar information to the same node. The third and most severe drawback of flooding is *resource blindness*. The simple forwarding rule that flooding uses to route packets does not take into consideration the energy constraints of the sensor nodes. As such, the node's energy may deplete rapidly, reducing considerably the lifetime of the network.

To address the shortcomings of flooding, a derivative approach, referred to as *gossiping*, has been proposed [6.15,6.16]. Similar to flooding, gossiping uses a simple forwarding rule and does not require costly topology maintenance or complex route discovery algorithms. Contrary to flooding, where a data packet is broadcast to all neighbors, gossiping requires that each node sends the incoming packet to a randomly selected neighbor. Upon receiving the packet, the neighbor selected randomly chooses one of its own neighbors and forwards the packet to the neighbor chosen. This process continues iteratively until the packet reaches its intended

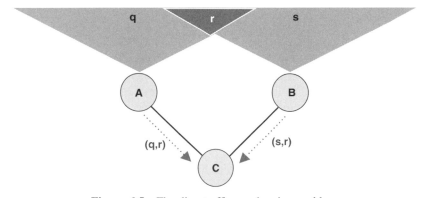

Figure 6.5 Flooding traffic overlapping problem.

destination or the maximum hop count is exceeded. Gossiping avoids the implosion problem by limiting the number of packets that each node sends to its neighbor to one copy. The latency that a packet suffers on its way to the destination may be excessive, particularly in a large network. This is caused primarily by the random nature of the protocol, which, in essence, explores one path at a time.

6.5.3   Sensor Protocols for Information via Negotiation

Sensor protocols for information via negotiation (SPIN) is a data-centric negotiation-based family of information dissemination protocols for WSNs [6.17]. The main objective of these protocols is to efficiently disseminate observations gathered by individual sensor nodes to all the sensor nodes in the network. Simple protocols such as flooding and gossiping are commonly proposed to achieve information dissemination in WSNs. Flooding requires that each node sends a copy of the data packet to all its neighbors until the information reaches all nodes in the network. Gossiping, on the other hand, uses randomization to reduce the number of duplicate packets and requires only that a node receiving a data packet forward it to a randomly selected neighbor.

The simplicity of flooding and gossiping is appealing, as both protocols use simple forwarding rules and do not require topology maintenance. The performance of these algorithms in terms of packet delay and resource utilization, however, quickly deteriorates with the size of the network and the traffic load. This performance drawback is typically caused by traffic implosion and geographical overlapping. Traffic implosion results in multiple copies of the same data being delivered to the same sensor node. Geographical overlapping, on the other hand, causes nodes covering the same geographical area to disseminate, unnecessarily, similar data information items to the network sensor nodes. Simple protocols such as flooding and gossiping do not alter their behavior to adapt communication and computation to the current state of their energy resource. This lack of resource awareness and adaptation may reduce the lifetime of the network considerably, as highly active nodes may rapidly deplete their energy resources.

The main objective of SPIN and its related family members is to address the shortcomings of conventional information dissemination protocols and overcome their performance deficiencies. The basic tenets of this family of protocols are data negotiation and resource adaptation. Semantic-based *data negotiation* requires that nodes running SPIN "learn" about the content of the data before any data are transmitted between network nodes. SPIN exploits data naming, whereby nodes associate *metadata* with data they produce and use these descriptive data to perform negotiations before transmitting the actual data. A receiver that expresses interest in the data content can send a request to obtain the data advertised. This form of negotiation assures that data are sent only to interested nodes, thereby eliminating traffic implosion and reducing significantly the transmission of redundant data throughout the network. Furthermore, the use of meta data descriptors eliminates the possibility of overlap, as nodes can limit their requests to name only the data that they are interested in obtaining.

Resource adaptation allows sensor nodes running SPIN to tailor their activities to the current state of their energy resources. Each node in the network can probe its associated resource manager to keep track of its resource consumption before transmitting or processing data. When the current level of energy becomes low, the node may reduce or completely eliminate certain activities, such as forwarding third-party metadata and data packets. The resource adaptation feature of SPIN allows nodes to extend their longevity and consequently, the lifetime of the network.

To carry out negotiation and data transmission, nodes running SPIN use three types of messages. The first message type, ADV, is used to advertise new data among nodes. A network node that has data to share with the remaining nodes of the network can advertise its data by first transmitting an ADV message containing the metadata describing the data. The second message type, REQ, is used to request an advertised data of interest. Upon receiving an ADV containing metadata, a network node interested in receiving specific data sends a REQ message the metadata advertising node, which then delivers the data requested. The third message type, DATA, contains the actual data collected by a sensor, along with a metadata header. The data message is typically larger than the ADV and REQ messages. The latter messages only contain metadata that are often significantly smaller than the corresponding data message. Limiting the redundant transmission of data messages using semantic-based negotiation can result in significant reduction of energy consumption.

The basic behavior of SPIN is illustrated in Figure 6.6, in which the data source, sensor node A, advertises its data to its immediate neighbor, sensor node B, by sending an ADV message containing the metadata describing its data. Node B

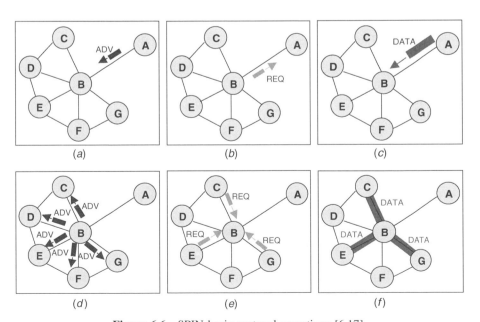

Figure 6.6 SPIN basic protocol operations [6.17].

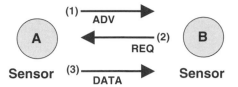

Figure 6.7   SPIN-PP three-way handshake protocol.

expresses interest in the data advertised and sends a REQ message to obtain the data. Upon receiving the data, node B sends an ADV message to advertise the newly received data to its immediate neighbors. Only three of these neighbors, nodes C, E, and G, express interest in the data. These nodes issue a REQ message to node B, which eventually delivers the data to each of the requesting nodes.

The simplest version of SPIN, referred to as SPIN-PP, is designed for a point-to-point communications network. The three-step handshake protocol used by SPIN-PP is depicted in Figure 6.7. In step 1, the node holding the data, node A, issues an advertisement packet (ADV). In step 2, node B expresses interest in receiving the data by issuing a data request (REQ). In step 3, node A responds to the request and sends a data packet to node B. This completes the three-step handshake procedure. SPIN-PP uses negotiation to overcome the implosion and overlap problems of the traditional flooding and gossiping protocols. A simulation-based performance study of SPIN-1 shows that the protocol reduces energy consumption by a factor of 3.5 compared to flooding. The protocol also achieves high data dissemination rates, nearing the theoretical optimum.

An extension of this basic protocol, SPIN-EC, additionally incorporates a threshold-based resource-awareness mechanism to complete data negotiation. When its energy level approaches the low threshold, a node running SPIN-EC reduces its participation in the protocol operations. In particular, a node engages in protocol operations only if it concludes that it can complete all the stages of the protocol operations without causing its energy level to decrease below the threshold. Consequently, if a node receives an advertisement, it does not send out an REQ message if it determines that its energy resource is not high enough to transmit an REQ message and receive the corresponding DATA message. The simulation results of this protocol show that SPIN-EC disseminates 60% more data per unit energy than flooding. Furthermore, the data show that SPIN-EC comes very close to the ideal amount of data that can be disseminated per unit energy.

Both SPIN-PP and SPIN-EC are designed for point-to-point communication. A third member of the SPIN family, SPIN-BC, is designed for broadcast networks. In these networks, nodes share a single channel for communications. In this class of networks, when a node sends out a data packet on the broadcast channel, the packet transmitted is received by all the other nodes within a certain range of the sending node. The SPIN-BC protocol takes advantage of the broadcasting capability of the channel and requires that a node which has received an ADV message does not respond immediately with an REQ message. Instead, the node waits for a certain amount of time, during which it monitors the communications channel. If the node

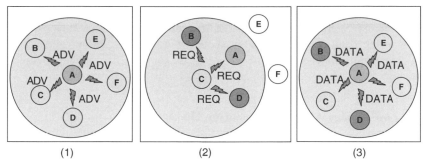

Figure 6.8 SPIN-BC protocol basic operations [6.17].

hears an REQ message issued by another node which is interested in receiving the data, it cancels its own request, thereby eliminating any redundant requests for the same message. Furthermore, upon receiving an REQ message, the advertising node sends the data message only once, even when it receives multiple requests for the same message.

The basic operations of the SPIN-BC protocol are depicted in Figure 6.8. In this configuration, the node holding the data, node A, sends a ADV packet to advertise the data to its neighbors. All nodes hear the advertisement, but node C is first to issue a REQ packet to request the data from node A. Nodes B and D hear the broadcast request and refrain from issuing their own REQ packets. Nodes E and F either have no interest in the data advertised or intentionally delay their requests. Upon hearing node C's request, node A replies by sending the data packet. All nodes within the transmission range of A receive the data packet, including nodes E and F. In broadcast environments, SPIN-BC has the potential to reduce energy consumption by eliminating redundant exchange of data requests and replies.

The last protocol of the SPIN family, SPIN-RL, extends the capabilities of SPIN-BC to enhance its reliability and overcome message transmission errors caused by a lossy channel. Enhanced reliability is achieved by periodic broadcasting of ADV and REQ messages. Each node in SPIN-BC keeps track of the advertisements it hears and the nodes where these advertisements originate. If a node requesting specific data of interest does not receive the data requested within a certain period of time, it sends the request again. Furthermore, improved reliability can be provided by readvertising metadata periodically. Finally, SPIN-RL nodes limit the frequency with which they resend the data messages. After sending out a data message, a node waits for a certain time period before it responds to other requests for the same data message.

The SPIN protocol family addresses the major drawbacks of flooding and gossiping. Simulation results show that SPIN is more energy efficient than flooding or gossiping. Furthermore, the results also show that the rate at which SPIN disseminates data is greater than or equal to the rate of either of these protocols. SPIN achieves these gains by localizing topology changes and eliminating dissemination of redundant information through semantic negotiation. It is worth noting, however, that localized negotiation may not be sufficient to cover the entire network and ensure that all interested nodes receive the data advertisement and eventually, the data of interest. Such a situation may occur if intermediate nodes may not express interest in the data and drop the corresponding ADV message upon receiving it. This shortcoming may prevent the use of SPIN for specific applications such as monitoring for intrusion detection and critical infrastructure protection.

6.5.4 Low-Energy Adaptive Clustering Hierarchy

Low-energy adaptive clustering hierarchy (LEACH) is a routing algorithm designed to collect and deliver data to the data sink, typically a base station [6.19]. The main objectives of LEACH are:

- Extension of the network lifetime
- Reduced energy consumption by each network sensor node
- Use of data aggregation to reduce the number of communication messages

To achieve these objectives, LEACH adopts a hierarchical approach to organize the network into a set of clusters. Each cluster is managed by a selected cluster head. The cluster head assumes the responsibility to carry out multiple tasks. The first task consists of periodic collection of data from the members of the cluster. Upon gathering the data, the cluster head aggregates it in an effort to remove redundancy among correlated values [6.19,6.20]. The second main task of a cluster head is to transmit the aggregated data directly to the base station. The transmission of the aggregated data is achieved over a single hop. The network model used by LEACH is depicted in Figure 6.9. The third main task of the cluster head is to create a TDMA-based schedule whereby each node of the cluster is assigned a time slot

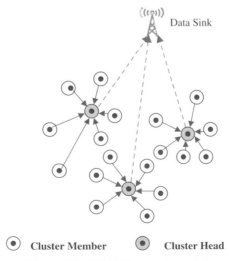

Figure 6.9 LEACH network model.

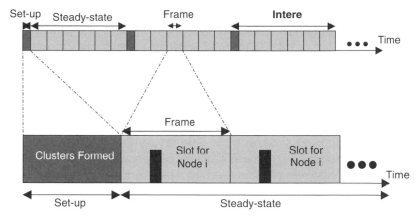

Figure 6.10 LEACH phases [6.18].

that it can use for transmission. The cluster head advertises the schedule to its cluster members through broadcasting. To reduce the likelihood of collisions among sensors within and outside the cluster, LEACH nodes use a code-division multiple access–based scheme for communication.

The basic operations of LEACH are organized in two distinct phases. These phases are illustrated in Figure 6.10. The first phase, the setup phase, consists of two steps, cluster-head selection and cluster formation. The second phase, the steady-state phase, focuses on data collection, aggregation, and delivery to the base station. The duration of the setup is assumed to be relatively shorter than the steady-state phase to minimize the protocol overhead.

At the beginning of the setup phase, a round of cluster-head selection starts. The cluster-head selection process ensures that this role rotates among sensor nodes, thereby distributing energy consumption evenly across all network nodes. To determine if it is its turn to become a cluster head, a node, n, generates a random number, v, between 0 and 1 and compares it to the cluster-head selection threshold, $T(n)$. The node becomes a cluster head if its generated value, v, is less than $T(n)$. The cluster-head selection threshold is designed to ensure with high probability that a predetermined fraction of nodes, P, is elected cluster heads at each round. Further, the threshold ensures that nodes which served in the last $1/P$ rounds are not selected in the current round.

To meet these requirements, the threshold $T(n)$ of a competing node n can be expressed as follows:

$$T(n) = \begin{cases} 0 & \text{if } n \notin G \\ \dfrac{P}{1 - P(r \bmod(1/P))} & \forall n \in G \end{cases}$$

The variable G represents the set of nodes that have not been selected to become cluster heads in the last $1/P$ rounds, and r denotes the current round. The predefined parameter, P, represents the cluster-head probability. It is clear that if a node has served as a cluster head in the last $1/P$ rounds, it will not be elected in this round.

At the completion of the cluster-head selection process, every node that was selected to become a cluster head advertises its new role to the rest of the network. Upon receiving the cluster-head advertisements, each remaining node selects a cluster to join. The selection criteria may be based on the received signal strength, among other factors. The nodes then inform their selected cluster head of their desire to become a member of the cluster.

Upon cluster formation, each cluster head creates and distributes the TDMA schedule, which specifies the time slots allocated for each member of the cluster. Each cluster head also selects a CDMA code, which is then distributed to all members of its cluster. The code is selected carefully so as to reduce intercluster interference. The completion of the setup phase signals the beginning of the steady-state phase. During this phase, nodes collect information and use their allocated slots to transmit to the cluster head the data collected. This data collection is performed periodically.

Simulation results show that LEACH achieves significant energy savings. These savings depend primarily on the data aggregation ratio achieved by the cluster heads. Despite these benefits, however, LEACH suffers several shortcomings. The assumption that all nodes can reach the base station in one hop may not be realistic, as capabilities and energy reserves of the nodes may vary over time from one node to another. Furthermore, the length of the steady-state period is critical to achieving the energy reduction necessary to offset the overhead caused by the cluster selection process. A short steady-state period increases the protocol's overhead, whereas a long period may lead to cluster head energy depletion. Several algorithms have been proposed to address these shortcomings. The extended LEACH (XLEACH) protocol takes into consideration the node's energy level in the cluster-head selection process [6.20]. The resulting threshold cluster-head selection, $T(n)$, used by n to determine if it will be a cluster head in the current round is defined as

$$T(n) = \frac{P}{1 - P(r \bmod (1/P))} \left[\frac{E_{n,\mathrm{current}}}{E_{n,\max}} + \left(r_{n,s} \mathrm{div} \frac{1}{P} \right) \left(1 - \frac{E_{n,\mathrm{current}}}{E_{n,\max}} \right) \right]$$

In this equation, $E_{n,\mathrm{current}}$ is the current energy, and $E_{n,\max}$ is the initial energy of the sensor node. The variable $r_{n,s}$ is the number of consecutive rounds in which a node has not been a cluster head. When the value of $r_{n,s}$ approaches $1/P$, the threshold $T(n)$ is reset to the value it had before the inclusion of the remaining energy onto the threshold equation. Additionally, $r_{n,s}$ is set to 0 when a node becomes a cluster head.

LEACH exhibits several properties which enable the protocol to reduce energy consumption. Energy requirement in LEACH is distributed across all sensor nodes, as they assume the cluster head role in a round-robin fashion based on their residual energy. LEACH is a completely distributed algorithm, requiring no control information from the base station. The cluster management is achieved locally, which obliterates the need for global network knowledge. Furthermore, data aggregation by the cluster also contributes greatly to energy saving, as nodes are no longer required to send their information directly to the sink. It has been shown using simulation that LEACH outperforms conventional routing protocols, including direct transmission and multihop routing, minimum-transmission-energy routing, and static clustering–based routing algorithms.

6.5.5 Power-Efficient Gathering in Sensor Information Systems

Power-efficient gathering in sensor information systems (PEGASIS) and its extension, hierarchical PEGASIS, are a family of routing and information-gathering protocols for WSNs [6.21]. The main objectives of PEGASIS are twofold. First, the protocol aims at extending the lifetime of a network by achieving a high level of energy efficiency and uniform energy consumption across all network nodes. Second, the protocol strives to reduce the delay that data incur on their way to the sink.

The network model considered by PEGASIS assumes a homogeneous set of nodes deployed across a geographical area. Nodes are assumed to have global knowledge about other sensors' positions. Furthermore, they have the ability to control their power to cover arbitrary ranges. The nodes may also be equipped with CDMA-capable radio transceivers. The nodes' responsibility is to gather and deliver data to a sink, typically a wireless base station. The goal is to develop a routing structure and an aggregation scheme to reduce energy consumption and deliver the aggregated data to the base station with minimal delay while balancing energy consumption among the sensor nodes. Contrary to other protocols, which rely on a tree structure or a cluster-based hierarchical organization of the network for data gathering and dissemination, PEGASIS uses a chain structure.

Based on this structure, nodes communicate with their closest neighbors. The construction of the chain starts with the farthest node from the sink. Network nodes are added to the chain progressively, starting from the closest neighbor to the end node. Nodes that are currently outside the chain are added to the chain in a greedy fashion, the closest neighbor to the top node in the current chain first, until all nodes are included. To determine the closest neighbor, a node uses the signal strength to measure the distance to all its neighboring nodes. Using this information, the node adjusts the signal strength so that only the closest node can be heard.

A node within the chain is selected to be the chain leader. Its responsibility is to transmit the aggregated data to the base station. The chain leader role shifts in positioning the chain after each round. Rounds can be managed by the data sink, and the transition from one round to the next can be tripped by a high-powered beacon issued by the data sink. Rotation of the leadership role among nodes of the chain ensures on average a balanced consumption of energy among all the network nodes. It is worth noting, however, that nodes assuming the role of chain leadership may be arbitrarily far away from the data sink. Such a node may be required to transmit with high power in order to reach the base station.

Data aggregation in PEGASIS is achieved along the chain. In its simplest form, the aggregation process can be performed sequentially as follows. First, the chain leader issues a token to the last node in the right end of the chain. Upon receiving the token, the end node transmits its data to its downstream neighbor in the chain toward the leader. The neighboring node aggregates the data and transmits them to its downstream neighbor. This process continues until the aggregated data reach the leader. Upon receiving the data from the right side of the chain, the leader issues a token to the left end of the chain, and the same aggregation process is carried out until the data reach the leader. Upon receiving the data from both sides of the chain, the leader aggregates the data and transmits them to the data sink. Although simple, the sequential aggregation scheme may result in long delays before the aggregated data are delivered to the base station. Such a sequential scheme, however, may be necessary if arbitrarily close simultaneous transmission cannot be carried out without signal interference.

A potential approach to reduce the delay required to deliver aggregated data to the sink is to use parallel data aggregation along the chain. A high degree of parallelism can be achieved if the sensor nodes are equipped with CDMA-capable transceivers. The added ability to carry out arbitrarily close transmissions without interference can be used to "overlay" a hierarchical structure onto the chain and use the embedded structure to perform data aggregation. At each round, nodes at a given level of the hierarchy transmit to a close neighbor in the upper level of the hierarchy. This process continues until the aggregated data reach the leader at the top level of the hierarchy. The latter transmits the final data aggregate to the base station.

To illustrate the chain-based approach, consider the example depicted in Figure 6.11. In this example it is assumed that all nodes have global knowledge

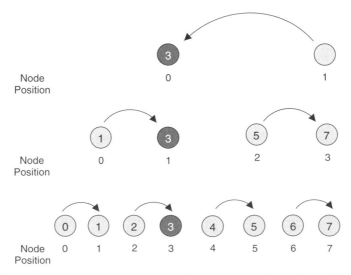

Figure 6.11   Chain-based data gathering and aggregation scheme [6.21].

of the network and employ a greedy algorithm to construct the chain. Furthermore, it is assumed that nodes take turns in transmitting to the base station such that node $i \bmod N$, where N represents the total number of nodes, is responsible for transmitting the aggregate data to the base station in round i. Based on this assignment, node 3, in position 3 in the chain, is the leader in round 3. All nodes in an even position must send their data to their neighbor to the right. At the next level, node 3 remains in an odd position. Consequently, all nodes in an even position aggregate their data and transmit them to their right neighbors. At the third level, node 3 is no longer in an odd position. Node 7, the only node beside node 3 to rise to this level, aggregates its data and sends them to node 3. Node 3, in turn, aggregates the data received with its own data and sends them to the base station.

The chain-based binary approach leads to significant energy reduction, as nodes operate in a highly parallel manner. Furthermore, since the hierarchical, treelike structure is balanced, the scheme guarantees that after $\log_2 N$ steps, the aggregated data arrive at the leader. The chain-based binary aggregation scheme has been used in PEGASIS as an alternative to achieving a high degree of parallelism. With CDMA-capable sensor nodes, it has been shown that the scheme performs best with respect to the energy-delay product needed per round of data gathering, a metric that balances the energy and delay cost.

The sequential scheme and the CDMA-based fully parallel scheme constitute two endpoints of the design spectrum. A third scheme, which does not require the node transceivers to be equipped with CDMA capabilities, strikes a balance between the two extreme schemes and achieves some level of parallelism. The basic idea of the scheme is to restrict simultaneous transmission to nodes that are spatially separated. Based on this restriction, hierarchical PEGASIS creates a three-level hierarchy in which the total number of network nodes is divided into three groups. Data are aggregated simultaneously within each group and exchanged between groups. The data aggregated eventually reach the leader, which delivers them to the data sink. It is worth noting that simultaneous transmission must be carefully scheduled to avoid interference. Furthermore, the three-level hierarchy must be restructured properly to allow leadership rotation among group nodes.

The simulation results of the hierarchical extension of PEGASIS show considerable improvement over schemes such as LEACH. Further, the hierarchical scheme has been shown to outperform the original PEGASIS scheme by a factor of 60.

6.5.6 Directed Diffusion

Directed diffusion is a data-centric routing protocol for information gathering and dissemination in WSNs [6.22]. The main objective of the protocol is to achieve substantial energy savings in order to extend the lifetime of the network. To achieve this objective, directed diffusion keeps interactions between nodes, in terms of message exchanges, localized within a limited network vicinity. Using localized interaction, direct diffusion can still realize robust multipath delivery and adapt to a minimal subset of network paths. This unique feature of the protocol, combined

TABLE 6.1 Interest Description Using Value and Attribute Pairs

Attribute–Value Pair	Description
Type = Hummingbirds	Detect hummingbird location
Interval = 20 ms	Report events every 20 ms
Duration = 10 s	Report for the next 10 s
Field = $[(x_1, y_1), (x_2, y_2)]$	Report from sensors in this area

with the ability of the nodes to aggregate response to queries, results into significant energy savings.

The main elements of direct diffusion include interests, data messages, gradients, and reinforcements. Directed diffusion uses a publish-and-subscribe information model in which an inquirer expresses an interest using attribute–value pairs. An *interest* can be viewed as a query or an interrogation that specifies what the inquirer wants. Table 6.1 shows an example that illustrates how an interest in hummingbirds can be expressed using a set of attribute–value pairs. Sensor nodes, which can service the interest, reply with the corresponding data.

For each active sensing task, the data sink periodically broadcasts an interest message to each neighbor. The message propagates throughout the sensor network as an *interest* for named data. The main purpose of this exploratory interest message is to determine if there exist sensor nodes that can service the sought-after interest. All sensor nodes maintain an *interest cache*. *Each* entry of the interest cache corresponds to a different interest. The cache entry contains several fields, including a timestamp field, multiple gradient fields for each neighbor, and a duration field. The *timestamp field* contains the timestamp of the last matching interest received. Each *gradient field* specifies both the data rate and the direction in which data are to be sent. The value of the data rate is derived from the interval attribute of the interest. The *duration field* indicates the approximate lifetime of the interest. The value of the duration is derived from the timestamp of the attribute. Figure 6.12 illustrates interest propagation in a WSN.

A gradient can be thought of as a reply link pointing toward the neighboring node from which the interest is received. The diffusion of interests across the entire network, coupled with the establishment of gradients at the network nodes, allows the discovery and establishment of paths between the data sinks that are interested in the named data and the nodes that can serve the data. A sensor node that detects an event searches its interest cache for an entry matching the interest. If a match is identified, the node first computes the highest event rate requested among all its outgoing gradients. It then sets its sensing subsystem to sample the events at this highest rate. The node then sends out an event description to each neighbor for which it has a gradient. A neighboring node that receives a data searches for a matching interest entry in its cache. If no match is found, the node drops the data message with no further action. If such a match exists, and the data message received does not have a matching data cache entry, the node adds the message to the data cache and sends the data message to the neighboring nodes.

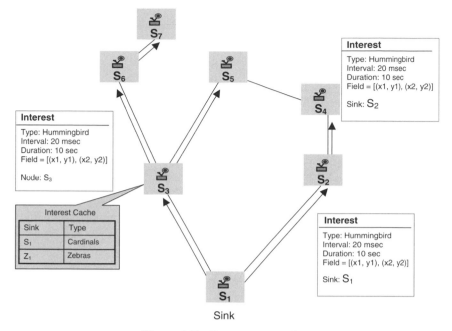

Figure 6.12 Interest propagation.

Upon receiving an interest, a node checks its interest cache to determine if an entry exists in its cache for this interest. If such an entry does not exist, the receiving node creates a new cache entry. The node then uses the information contained in the interest to instantiate the parameters of the newly created interest field. Furthermore, the entry is set to contain a single gradient field, with the event rate specified, pointing toward the neighboring node from which the interest is received. If a match exists between the interest received and a cache entry, the node updates the timestamp and duration fields of the matching entry. If the entry contains no gradient for the sender of the interest, the node adds a gradient with the value specified in the interest message. If the matching interest entry contains a gradient for the interest sender, the node simply updates the timestamp and duration fields. A gradient is removed from its interest entry when it expires. Figure 6.13 shows the initial gradient setup.

During the gradient setup phase, a sink establishes multiple paths. The sink can use these paths to higher-quality events by increasing its data rate. This is achieved through a path reinforcement process. The sink may choose to reinforce one or several particular neighbors. To achieve this, the sink resends the original interest message, at a higher data rate, across the paths selected, thereby reinforcing the source nodes on the paths to send data more frequently. The path performing most often can then be retained while negatively reinforcing the remaining paths. Negative reinforcement can be achieved by timing out all high-data-rate gradients in the network, except for those that are explicitly reinforced. Figure 6.14 shows data delivery along a reinforced path.

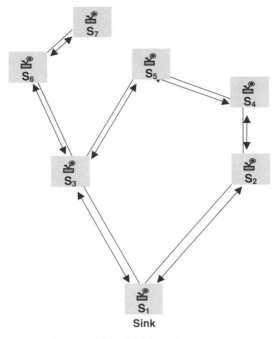

Figure 6.13 Initial gradient setup.

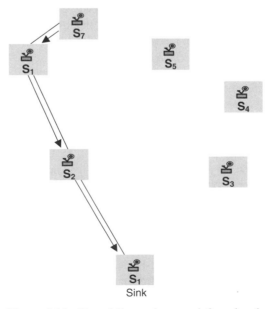

Figure 6.14 Data delivery along a reinforced path.

Link failures caused by environmental factors affecting the communications channel, as well as node failures or performance degradation caused by node energy dissipation or complete depletion, can be repaired in directed diffusion. These failures are typically detected by reduced rate or data loss. When a path between a sensing node and the data sink fails, an alternative path, which is sending at lower rates, can be identified and reinforced. Lossy links can also be negatively reinforced by either sending interests with the exploratory data rate or simply by letting the neighbor's cache expire over time.

Directed diffusion has the potential for significant energy savings. Its localized interactions allow it to achieve relatively high performance over unoptimized paths. Furthermore, the resulting diffusion mechanisms arc stable under a range of network dynamics. Its data-centric approach obliterates the need for node addressing. The directed diffusion paradigm, however, is tightly coupled into a semantically driven query-on-demand data model. This may limit its use to applications that fit such a data model, where the interest-matching process can be achieved efficiently and unambiguously.

6.5.7 Geographical Routing

The main objective of geographical routing is to use location information to formulate an efficient route search toward the destination. Geographical routing is very suitable to sensor networks, where data aggregation is a useful technique to minimize the number of transmissions toward the base station by eliminating redundancy among packets from different sources. The need for data aggregation to reduce energy consumption shifts the computation and communications model in sensor networks from a traditional address-centric paradigm, where the interaction is between two addressable endpoints of communications, to a data-centric paradigm, where the content of the data is more important than the identity of the node that gathers the data. In this new paradigm, an application may issue a query to inquire about a phenomenon within a specific physical area or near the vicinity of a landmark. For example, scientists analyzing traffic flow patterns may be interested in determining the average number, size, and speed of vehicles that travel on a specific section of a highway. The identity of the sensors that collect and disseminate information about traffic flow on a specific section of the highway is not as important as the data content. Furthermore, multiple nodes that happen to be located in the targeted section of the highway may participate in collecting and aggregating the data in order to answer the query. Traditional routing approaches, which are typically designed to discover a path between two addressable endpoints, are not well suited to handling geographically specific multidimensional queries. Geographical routing, on the other hand, leverages location information to reach a destination, with each node's location used as its address.

In addition to its compatibility with data-centric applications, geographical routing requires low computation and communication overhead. In traditional routing approaches such as the one used in distributed shortest-path routing protocols for wired networks, knowledge of the entire network topology, or a summary thereof, may be required for a router to compute the shortest path to each destination [6.28]. Furthermore, to maintain correct paths to all destinations, routers are called upon to update the state describing the current topology in a periodic fashion and when link failure occurs. The need to update the topology state constantly may lead to substantial overhead, proportional to the product of the number of routers and the rate of topological changes in the network.

Geographical routing, on the other hand, does not require maintaining a "heavy" state at the routers to keep track of the current state of the topology. It requires only the propagation of single-hop topology information, such as the position of the "best" neighbor to make correct forwarding decisions. The self-describing nature of geographical routing, combined with its localized approach to decision, obliterates the need for maintaining internal data structures such as routing tables. Consequently, the control overhead is reduced substantially, thereby enhancing its scalability in large networks. These attributes make geographical routing a feasible solution for routing in resource-constrained sensor networks.

Routing Strategies   The objective of geographical routing is to use location information to formulate a more efficient routing strategy that does not require flooding request packets throughout a network. To achieve this goal, a data packet is sent to nodes located within a designated forwarding region. In this scheme, also referred to as *geocasting*, only nodes that lie within the designated forwarding zone are allowed to forward the data packet [6.23,6.24]. The forwarding region can be statically defined by the source node, or constructed dynamically by intermediate nodes to exclude nodes that may cause a detour when forwarding the data packet. If a node does not have information regarding the destination, the route search can begin as a fully directed broadcast. Intermediate nodes, with better knowledge of the destination, may limit the forwarding zone in order to direct traffic toward the destination. The idea of limiting the scope of packet propagation to a designated region is commensurate with the data-centric property of sensor networks, in which the interest in the data content, rather than the sensor, provides the data. The efficacy of the strategy depends largely on the way the designated forwarding is defined and updated as data travel toward the destination. It also depends on the connectivity of the nodes within a designated zone.

A second strategy used in geographical routing, referred to as *position-based routing*, requires a node to know only the location information of its direct neighbors [6.25,6.26]. A greedy forwarding mechanism is then used whereby each node forwards a packet to the neighboring node that is "closest" to the destination. Several metrics have been proposed to define the concept of closeness, including the Euclidean distance to the destination, the projected distance to the destination on the straight line joining the current node and the destination, and the deviation from the straight direction toward the destination.

Position-based routing protocols have the potential to reduce control overhead and reduce energy, as flooding for node discovery and state propagation are localized to within a single hop. The efficiency of the scheme, however, depends on the network density, the accurate localization of nodes, and more important, on the forwarding rule used to move data traffic toward the destination. In the following section, various forwarding rules commonly used in position-based routing are described. Basic techniques used to overcome the lack of position information and obstacles are described.

Forwarding Approaches An important aspect of geographical routing is the rule used to forward traffic toward its final destination. In position-based routing, each node decides on the next hop based on its own position, the position of its neighbors, and the destination node. The quality of the decision clearly depends on the extent of the node's knowledge of the global topology [6.27]. Local knowledge of the topology may lead to suboptimal paths, as depicted in Figure 6.15, where the node currently holding the packet makes a forwarding decision based solely on local topology knowledge. Finding the optimal path requires global knowledge of the topology. The overhead that global knowledge of the topology entails, however, is prohibitive in resource-constrained WSNs. To overcome this problem, various forwarding strategies have been proposed [6.28–6.32].

The greedy routing scheme selects among its neighbors the one that is closest to the destination. In Figure 6.16, the node currently holding the message, node MH, selects node GRS as the next hop to forward the message. It is worth noting that the selection process used in this scheme considers only the set of nodes that are closer to the destination than the current message holder. If such a set is empty, the scheme fails to progress forward.

In the most-forward-within-R strategy (MFR), where R represents the transmission range, a node transmits its packet to the most forward among its neighbors toward the destination. Based on this approach, the next hop selected by MH to

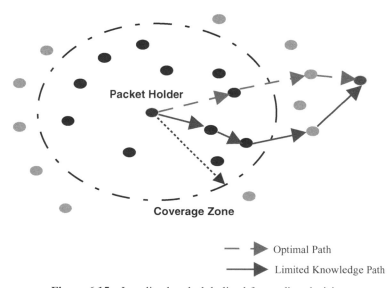

Figure 6.15 Localized and globalized forwarding decision.

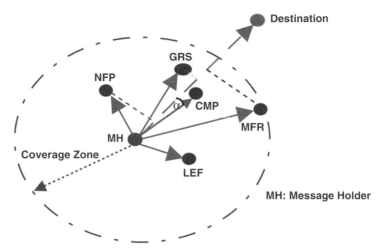

Figure 6.16 Geographical routing forwarding strategies.

forward the packet is node MFR. This greedy approach is myopic and does not necessarily minimize the remaining distance to the destination.

The nearest-forward-progress scheme selects the nearest node with forward progress. Based on this scheme, node NFP is selected by MH to forward the message to the destination. The nearest-closer node is an alternative to this approach, in which the node currently holding the message selects the nearest node among all its neighbors which are closer to the destination.

The compass routing scheme selects the node with the minimum angle between the straight line joining the current node and the destination and the straight line joining a neighbor and the destination. In Figure 6.16, node CMP will be selected as the next hop to forward the traffic to the destination.

The low-energy forward scheme selects the node that locally minimizes the energy required, expressed in terms of joules per meter, to progress forward toward the target. In the network configuration shown in Figure 6.16, node LEF is selected by MH to move the traffic forward toward the destination.

As stated previously, the scalability and data-centric attributes of geographical routing make it a feasible routing alternative in WSNs. Its applicability assumes, however, that the geographical locations of all neighboring nodes, or at least a subset thereof, are known to the message holder. Accurate information about the geographical location of nodes is typically available from a global positioning system (GPS) device [6.33–6.35]. It is possible that in certain settings, sensing nodes may be equipped with GPS devices. In most cases, however, the resource and energy limitation of sensor nodes prohibits the use of GPS devices. To address this shortcoming, strategies in which only GPS-augmented boundary nodes have access to exact location information have been suggested [6.36]. Nodes without GPS devices can use a variety of triangularization algorithms to determine their location and the location of their neighboring nodes.

Other strategies assume that sensor nodes do not need to have the capability of knowing their location coordinates. These strategies use virtual, as opposed to physical, coordinate systems [6.36–6.39]. Using a virtual polar coordinate system, for example, it is shown that a labeled graph can be embedded in the original network topology. In this system, each node is given a label that encodes its position in the original network topology in terms of a radius and an angle from a center location [6.38]. These virtual coordinates do not depend on physical coordinates and can therefore be used efficiently in geographical routing by using only node labels. It is worth mentioning that schemes based on virtual coordinate sensor nodes may need to know the distance, in terms of hop counts, to certain reference points. This, in turn, may require periodic updates to be exchanged among the sensor nodes and the reference points.

Despite its simplicity, the greedy approach to geographical routing may either fail to find a path, even when one exists, or produce inefficient routes. This typically occurs when, due to obstacles, for example, no neighboring node is closer to the destination than is the current packet holder. To illustrate this problem, consider Figure 6.17, where node S_1 needs to forward a packet to the destination D. Based on the greedy approach, S_1 must select the closest neighbor to destination as the next hop to forward the packet. However, nodes S_2 and S_3, are both farther away from the destination than is node S_1. The greedy approach is trapped in a local minimum (i.e., node S_1) and fails to make forward progress.

In WSN environments, where sensors are typically embedded in the environment or deployed in inaccessible areas, voids are likely to occur. To circumvent voids, the well-known graph traversal rule referred to as the *right-hand rule* has been

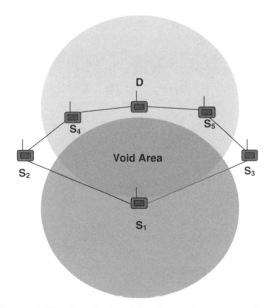

Figure 6.17 Greedy algorithm forward progress failure.

suggested [6.39–6.41]. The rule states that when a packet arrives at a given node N_i from node N_j, the next hop to be traversed by the packet is the node sequentially counterclockwise from node N_i with respect to the (N_i, N_j) edge. On graphs with edges that cross (i.e., nonplanar graphs), the right-hand rule may not traverse an enclosed face boundary. To remove crossing edges without partitioning the graph, the radio graph corresponding to the WSN is transformed into a planar subgraph in which all cross edges are eliminated. Upon the radio graph's transformation, perimeter traversal, in which packets are routed along the perimeter of the void, is used. This mode is also referred to as *face traversal*.

Combining greedy traversal with perimeter traversal, the routing algorithm can operate as follows. The routing algorithm begins in greedy mode, where the full graph is used. When the greedy approach fails, the node records its location in the packet and marks the packet to be in perimeter mode. The perimeter mode packet then follows a simple planar graph traversal. In this mode, a packet traverses successively closer faces of a planar subgraph of the full radio network connectivity graph. A *face* is defined as the largest possible region of the plane that is not cut by any edge of the graph. When the packet reaches a node closer to the destination, the mode reverts back to greedy.

The combined greedy and face traversal approach is illustrated in Figure 6.18. The figure shows the steps the packet follows to reach its destination. On each face, the traversal uses the right-hand rule to reach an edge that crosses the straight line connecting node S_1 to destination D. At that edge, the traversal moves to the adjacent face crossed by the straight line connecting S_1 to D. It is worth noting *that* the

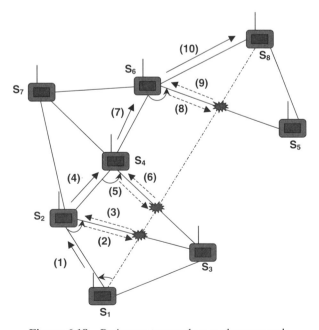

Figure 6.18 Perimeter traversal on a planner graph.

first two faces and the last face are interior faces, whereas the third face is an exterior face.

Geographical routing is an attractive approach for routing in WSNs because of its low overhead and localized interactions. The existence of asymmetric links, network partition, and cross-links increases the complexity of the approach considerably. Better planar graphs may be needed.

6.6 CONCLUSION

The attributes of WSNs and the characteristics of the environment within which sensor nodes are typically deployed make the routing problem very challenging. In this chapter we focused on issues central to routing in WSNs and describe different strategies used to develop routing protocols for these networks. In the first section of the chapter we discussed classes of sensor applications and highlighted the unique and distinctive features of the "nature" of their traffic models. In the second part of the chapter we provided a brief taxonomy of the basic routing strategies used to strike a balance between responsiveness and energy efficiency. In the third part of the chapter we presented a review of a number of protocols that address the problem of routing in today's WSNs. Multiple strategies have emerged as feasible solutions to the routing problem. As the application of WSNs to different fields become more apparent, advances in network hardware and battery technology will pave the way to practical cost-effective implementations of these routing protocols.

REFERENCES

[6.1] C. Y. Chong, S. P. Kumar, "Sensor Networks: Evolution, Opportunities, and Challenges," *Proceedings of the IEEE*, Aug. 2003.

[6.2] D. Estrin, D. Culler, K. Pister, "Connecting the Physical World with Pervasive Networks," *IEEE Pervasive Computing*, Vol. 1, No. 1, Jan.–Mar. 2002.

[6.3] G. J. Pottie, W. J. Kaiser, "Wireless Integrated Sensor Networks," *Communications of the ACM*, May 2000. An overview with more of a signal processing viewpoint.

[6.4] A. Mainwaring, R. Szewczyk, D. Culler, J. Anderson, "Wireless Sensor Networks for Habitat Monitoring," *Proceedings of the 1st ACM International Workshop on Wireless Sensor Networks and Applications* (WSNA'02), Atlanta, GA, Sept. 2002.

[6.5] L. Schwiebert, S. Gupta, J. Weinmann, "Research Challenges in Wireless Networks of Biomedical Sensors," *Proceedings of the 7th ACM International Conference on Mobile Computing and Networking* (MobiCom'01), Rome, Italy, July 2001, pp. 151–165.

[6.6] Q. Fang, F. Zhao, L. Guibas, "Lightweight Sensing and Communication Protocols for Target Enumeration and Aggregation," *Proceedings of the 4th ACM International Symposium on Mobile Ad Hoc Networking and Computing* (MobiHoc'03), Annapolis, MD, June 2003, pp. 165–176.

[6.7] I. F. Akyildiz, W. Su, Y. Sankarasubramaniam, E. Cayirci, "A Survey on Sensor Networks," *IEEE Communications*, Aug. 2002, pp. 102–114.

[6.8] Y. Yao, J. Gehrke, "The Cougar Approach to In-Network Query Processing in Sensor Networks," *SIGMOD Record*, Sept. 2002.

[6.9] M. Chu, H. Haussecker, F. Zhao, "Scalable Information-Driven Sensor Querying and Routing for Ad Hoc Heterogeneous Sensor Networks," *International Journal of High Performance Computing Applications*, Vol. 16, No. 3, Aug. 2002.

[6.10] J. N. Al-Karaki, A. E. Kamal, "On the Correlated Data Gathering Problem in Wireless Sensor Networks," *Proceedings of the 9th IEEE Symposium on Computers and Communications*, Alexandria, Egypt, July 2004.

[6.11] J. N. Al-Karaki, R. Ul-Mustafa, A. E. Kamal, "Data Aggregation in Wireless Sensor Networks: Exact and Approximate Algorithms," *Proceedings of the IEEE Workshop on High Performance Switching and Routing* (HPSR'04), Phoenix, AZ, Apr. 2004.

[6.12] K. Akkaya, M. Younis, "A Survey of Routing Protocols in Wireless Sensor Networks," *Ad Hoc Network Journal*, Vol. 3, No. 3, 2005, pp. 325–349.

[6.13] J. N. Al-Karaki, A. E. Kamal, "Routing Techniques in Wireless Sensor Networks: A Survey," *IEEE Wireless Communications*, Vol. 11, No. 6, Dec. 2004, pp. 6–28.

[6.14] C. Schurgers, M. B. Srivastava, "Energy Efficient Routing in Wireless Sensor Networks," *Proceedings of the IEEE Military Communications Conference* (MilCom'01): *Communications for Network-Centric Operations—Creating the Information Force*, McLean, VA, Oct. 2001.

[6.15] S. Hedetniemi, A. Liestman, "A Survey of Gossiping and Broadcasting in Communication Networks," *IEEE Networks*, Vol. 18, No. 4, 1988, pp. 319–349.

[6.16] D. Braginsky, D. Estrin, "Rumor Routing Algorithm for Sensor Networks," *Proceedings of the 1st Workshop on Sensor Networks and Applications* (WSNA'02), Atlanta, GA, Oct. 2002.

[6.17] J. Kulik, W. R. Heinzelman, H. Balakrishnan, "Negotiation-Based Protocols for Disseminating Information in Wireless Sensor Networks," *Wireless Networks*, Vol. 8, 2002, pp. 169–185.

[6.18] W. Heinzelman, A. Chandrakasan, H. Balakrishnan, "Energy-Efficient Communication Protocol for Wireless Microsensor Networks," *Proceedings of the 33rd Hawaii International Conference on System Sciences* (HICSS'00), Maui, HI, Jan. 2000.

[6.19] W. Heinzelman, J. Kulik, H. Balakrishnan, "Adaptive Protocols for Information Dissemination in Wireless Sensor Networks," *Proceedings of the 5th ACM/IEEE International Conference on Mobile Computing and Networking* (MobiCom'99), Seattle, WA, Aug. 1999, pp. 174–185.

[6.20] M. Handy, M. Haase, D. Timmermann, "Low Energy Adaptive Clustering Hierarchy with Deterministic ClusterHead Selection," IEEE MWCN, Stockholm, Sweden, Sep. 2002.

[6.21] S. Lindsey, C. Raghavendra, "PEGASIS: Power-Efficient Gathering in Sensor Information Systems," *IEEE Aerospace Conference Proceedings*, 2002, Vol. 3, No. 9–16 pp. 1125–1130.

[6.22] C. Intanagonwiwat, R. Govindan, D. Estrin, "Directed Diffusion: A Scalable and Robust Communication Paradigm for Sensor Networks," *Proceedings of the 6th ACM*

International Conference on Mobile Computing and Networking (MobiCom'00), Boston, MA, Aug. 2000, pp. 56–67.

[6.23] Y. Ko, N. Vaidya, "Location-Aided Routing (LAR) in Mobile Ad Hoc Networks," *Proceedings of the 4th ACM International Conference on Mobile Computing and Networking* (MobiCom'98), Dallas, Texas, Oct. 1998, pp. 66–75.

[6.24] Y. Xu, J. Heidemann, D. Estrin, "Geography-Informed Energy Conservation for Ad-Hoc Routing," *Proceedings of the 7th Annual ACM/IEEE International Conference on Mobile Computing and Networking* (MobiCom'01), Rome, Italy, July 2001, pp. 70–84.

[6.25] I. Stojmenovic, "Position-Based Routing in Ad Hoc Networks," *IEEE Communications*, Vol. 40, No. 7, July 2002, pp. 128–134.

[6.26] M. Mauve, J. Widmer, H. Hartenstein, "A Survey on Position-Based Routing in Mobile Ad Hoc Networks," *IEEE Network*, Vol. 15, 2001.

[6.27] T. Melodia, D. Pompili, I. F. Akyildiz, "Optimal Local Topology Knowledge for Energy Efficient Geographical Routing in Sensor Networks," *Proceedings of the 23rd Annual Joint Conference of the IEEE Computer and Communications Societies* (InfoCom'04), Mar. 7–11, 2004.

[6.28] H. Takagi, L. Kleinrock, "Optimal Transmission Ranges for Randomly Distributed Packet Radio Terminals," *IEEE Transactions on Communications*, Vol. 32, No. 3, 1984, pp. 246–257.

[6.29] T. C. Hou, V. O. K. Li, "Transmission Range Control in Multihop Packet Radio Networks," *IEEE Transactions on Communications*, Vol. 34, No. 1, Jan. 1986, pp. 38–44.

[6.30] I. Stojmenovic, X. Lin, "Power-Aware Localized Routing in Wireless Networks," *IEEE Transactions on Parallel Distributed Systems*, Vol. 12, No. 11, 2001, pp. 1122–1133.

[6.31] G. G. Finn, "Routing and Addressing Problems in Large Metropolitan-Scale Internetworks," Research Report ISU/RR- 87-180, Information Sciences Institute, Marina del Roy, CA, Mar. 1987.

[6.32] E. Kranakis, H. Singh, J. Urrutia, "Compass Routing on Geometric Networks," *Proceedings of the 11th Canadian Conference on Computational Geometry*, Vancouver, British Columbia, Canada, Aug. 1999.

[6.33] N. Bulusu, J. Heidemann, D. Estrin, "GPS-Less Low Cost Outdoor Localization for Very Small Devices," *IEEE Personal Communications*, Vol. 7, 2000.

[6.34] A. Savvides, C.-C. Han, M. Srivastava, "Dynamic Fine-Grained Localization in Ad-Hoc Networks of Sensors," *Proceedings of the 7th ACM Annual International Conference on Mobile Computing and Networking* (MobiCom'01), Rome, Italy, July 2001, pp. 166–179.

[6.35] S. Capkun, M. Hamdi, J. Hubaux, "GPS-Free Positioning in Mobile Ad-Hoc Networks," *Proceedings of the 34th Annual Hawaii International Conference on System Sciences*, (HICSS'01), 2001, pp. 3481–3490.

[6.36] A. Rao, C. Papadimitriou, S. Ratnasamy, S. Shenker, I. Stoica, "Geographic Routing Without Location Information," *Proceedings of the 9th ACM International Conference on Mobile Computing and Networking* (MobiCom'03), San Diego, CA, Sept. 2003.

[6.37] Newsome, D. Song, "GEM: Graph Embedding for Routing and Data-Centric Storage in Sensor Networks Without Geographic Information," *Proceedings of the 1st ACM International Conference on Embedded Networked Sensor Systems* (SenSys'03), Los Angeles, Nov. 2003.

[6.38] Fonseca, S. Ratnasamy, D. Culler, S. Shenker, I. Stoica, "Beacon Vector Routing: Scalable Point-to-Point in Wireless Sensornets," Technical Report IRBTR-04-12, Intel Research, Berkeley, CA, May 2004.

[6.39] P. Bose, P. Morin, I. Stojmenovic, J. Urrutia, "Routing with Guaranteed Delivery in Ad Hoc Wireless Networks," *Proceedings of the 3rd ACM International Workshop on Discrete Algorithms and Methods for Mobile Computing and Communications* (DIAL M99), Seattle, WA, Aug. 20, 1999, pp. 48–55; *Wireless Networks*, Vol. 7, No. 6, Nov. 2001, pp. 609–616.

[6.40] H. Huang, "Adaptive Algorithms to Mitigate Inefficiency in Greedy Geographical Routing," *IEEE Communications Letters*, Vol. 10, No. 3, Mar. 2006.

[6.41] B. Karp, H. T. Kung, "GPSR: Greedy Perimeter Stateless Routing for Wireless Networks," *Proceedings of the 6th ACM International Conference on Mobile Computing and Networking* (Mobicom'00), Boston, MA, Aug. 2000.

7

TRANSPORT CONTROL PROTOCOLS FOR WIRELESS SENSOR NETWORKS

7.1 TRADITIONAL TRANSPORT CONTROL PROTOCOLS

The architecture of computer and communication networks is often structured in layers: physical, data link, network (or internetworking), transport, and other higher layers, including session, presentation, and application. Each lower layer acts as a service provider to its immediate upper layer, which is a service user. Interactions between neighboring layers occur through service access points (SAPs). For example, the data link layer provides link services to the network layer, which is immediately above the link layer. The network layer provides addressing and routing services to the transport layer above it, which in turn provides message transportation service to the layers above it. In this model, the lower three layers exist almost exclusively in all nodes. But the transport and layers above it exist only at the endpoints or hosts, and perform as part of end-to-end protocol functions.

The transport layer provides end-to-end segment transportation, where messages are segmented into a chain of segments at the source and are reassembled back into the original message at the destination nodes. The transport layer does not concern itself with the underlying protocol structures for delivery and/or with the mechanisms used to deliver the segments to the destination nodes. Examples of transport protocols are the transport control protocol (TCP) [7.7], the user datagram protocol (UDP) [7.8], the sequenced packet exchange protocol (SPX), and NWLink (Microsoft's approach to implementing IPX/SPX). TCP and UDP are commonly deployed in the Internet.

Wireless Sensor Networks: Technology, Protocols, and Applications, by Kazem Sohraby, Daniel Minoli, and Taieb Znati
Copyright © 2007 John Wiley & Sons, Inc.

TCP can be classified as either connection-oriented and connectionless. The connection-oriented protocol operation consists of the following three phases:

1. *Connection establishment.* The sender issues a request message to establish a connection between itself and the destination. If the destination node is available and there is a path between source and the destination, a connection will be established. This connection is a logical link connecting the sender and the receiver.

2. *Data transmission.* After a connection has been established, data transmission commences between the sender and the receiver. During the information exchange, the rate at which either side is transmitting may be adjusted. This adjustment depends on the possible congestion (or lack thereof) in the network. Since data may be lost in the process of transmission, the transport protocol may support packet loss detection and loss recovery mechanisms.

3. *Disconnect.* After completion of data exchange between the source and the destination, the connection is torn down. In some cases, unexpected events such as the receiver becoming unavailable in the midst of data exchange may also lead to connection breakdown.

A connectionless protocol is very simple and its operation consists of only one phase. In essence, there is no need to request establishment of a connection between the sender and the receiver. When the source has information to send, it forwards it to the destination without the need for connection establishment.

Transport protocols can also be classified as elastic or nonelastic. TCP is an elastic protocol and UDP is a nonelastic protocol. *Elasticity* in a protocol means that the data transmission rate can be adjusted by the sender. *Nonelasticity,* on the other hand, implies that the transmission rate cannot be adjusted.

Due to their features, connection-oriented protocols often provide more services than do their connectionless counterparts. Therefore, when the underlying network layer lacks reliable and effective transmission services, and if the application has critical requirements for such delivery, connection-oriented protocols are preferred. Depending on the application, the transport protocol may or not support the following features:

1. *Orderly transmission.* Within a communications network in general, and in a wireless sensor network in particular, multiple paths may exist between a given source and destination. As a result, packets sent in a certain order by the source may not be received in the same order at the destination. For most applications, packets must be reordered at the destination to represent the same order as at the source. The transport protocol can provide the reordering. The common approach is for the protocol to include a field containing a *sequential number* of the segments transmitted. For each segment transmitted, the *sequence number* increases by one. As a result, the receiver can sort the received segments based on the *sequence number.*

2. *Flow and congestion control*. Hosts at each end of a particular connection can be diverse with different characteristics, such as capacity of communication and computation. If the sender transmits segments with a higher rate than the receiver is able to handle and process, the buffers at the receiver may overflow and congestion occurs. Congestion results in a loss of packets and reduction in overall system throughput. Therefore, some transport layers provide flow and congestion control service to coordinate the suitable transmission rate between senders and receivers. The key in the process of congestion control is *proper detection of congestion and notification of the sender about the congestion state*. Adjustment of the transmission rate is important after congestion detection. After the sender adjusts its rate and after congestion abates, the transmission rate should be increased to keep up with the link capacity. Factors such as link capacity when multiple heterogeneous links are in tandem, network topology, which may be static and/or variable, and the unpredictability in network traffic characteristics pose a serious challenge to the design of an effective flow and congestion control mechanism. For example, TCP provides window-based additive increase multiplicative decrease (AIMD) flow control, which deduces network congestion from the segment loss.

3. *Loss recovery*. Network congestion can lead to data loss due to limited resources at sensor nodes. However, some applications, such as the file transfer protocol (FTP), are loss-sensitive. Although in a wireless environment, the link layer can recover from lost data resulting from bit error, it cannot recover lost data as a result of congestion. Furthermore, the link layer may not provide loss recovery functions in all circumstances. Therefore, the transport protocol's support for loss recovery is a very helpful feature. An obvious approach to loss recovery is *retransmission after detection of loss*. But important concerns would be *how to detect loss* and *how to inform the sender about it*. The *sequence number* in the segment header can be used for this purpose. When there is a gap in the sequence number received, it is an indication that segments have been lost either from bit error or as a result of buffer overflow from congestion.

4. *Quality of service*. For real-time applications such as video on demand (VOD) and net meeting, which are real-time and delay-critical with a required bandwidth, the transport layer should provide high throughput within the constraints of allocated bandwidth. The transport protocol can incorporate QoS considerations into flow and congestion control.

The features described above are often used during the data transmission phase. But for connection-oriented protocols, these features and/or parameters related to these features may be negotiated and determined during connection establishment. For different applications, a subset of these features may sometimes be required.

7.1.1 TCP (RFC 793)

TCP is the commonly used connection-oriented transport control protocol for the Internet. Some applications, such as FTP and HTTP, reside on the TCP layer.

TCP uses network services provided by IP layer, with the objective of offering reliable, orderly, controllable, and elastic transmission. TCP operation consists of three phases:

1. *Connection establishment.* A logical connection for TCP is established during this phase. A logical connection is an association between the TCP sender and receiver, identified uniquely by the pair (*IP address, TCP port identifier*) of the TCP sender and receiver. There may be several connections between endpoints at the same time. These connections have the same IP address, but they will have different TCP port identifiers. TCP uses a three-way handshake to establish a connection. During the handshake, the TCP sender and receiver will negotiate parameters such as *initial sequence number*, window size, and others, and notify each other that data transmission can begin.

2. *Data transmission.* TCP provides reliable and orderly transmission of information between the sender and the receiver. TCP uses (accumulative) ACK to recover lost segments. The orderly transmission is realized through the sequence number in the segment header. Furthermore, TCP supports flow control and congestion control through adjustment of transmission rate by the sender. TCP uses a window-based mechanism to perform this task, where the sender maintains a variable *cwnd* (*congestion window*). The TCP sender can transmit a number of segments less than or equal to *cwnd*. *cwnd* is updated after receiving ACK from the receiver or after a timeout. Since ACK is used for both delivery notification and flow control, the two functions are somewhat coupled. There are three phases in the process of congestion control in TCP, and they will be explained later.

3. *Disconnect.* After completion of data transmission, the connection will be removed and the related resource released.

Flow and congestion control in TCP are performed through *cwnd*. There are three phases in the process:

1. *Slow start.* By default, all transmissions start with *slow start*. In this phase, the *cwnd* increases by one for each ACK that is received for a segment transmitted. *cwnd* therefore increases if ACK is not received due to segment loss.

2. *Congestion avoidance.* After *cwnd* reaches a maximum value (*threshold*), the system enters the *congestion avoidance* state. In this state, *cwnd* is incremented by only 1/*cwnd* after each ACK is received. For each segment transmitted, the sender maintains a timer. If the timer expires before an ACK corresponding to the segment is received, system enters the *slow start* phase again, and at the same time that *cwnd* is reset, the *threshold* is set to half of the current *cwnd*, and the segment timer is doubled. The timer is updated based on round-trip time (RTT), which is estimated through the ACK. If the sequence number acknowledged in two continuous ACKs are in sequence, TCP sender concludes that segments have been lost during transmission. In this case, the system state changes to *fast recovery and fast retransmission (FRFT)*, and *cwnd* will be halved at the same time.

3. *FRFT.* In the *fast recovery and fast retransmission* state, *cwnd* is updated using the same algorithm as that used in the *congestion avoidance* state. The reason for the *FRFT* state is that generally, sporadic segment loss does not necessarily mean that there is heavy congestion, and therefore there is no need to reset *cwnd*. However, a timeout usually indicates heavy congestion and/or link failure.

Therefore, TCP mechanisms allow flexible flow and congestion control. It should be noticed from the above that (1) If there is little or no congestion and few segment losses, *cwnd* will increase rather quickly and then will oscillate around a large value, which will result in high throughput. Conversely, *cwnd* will have a small value, and therefore low TCP throughput will result. (2) If RTT is small, ACKs are received rather quickly and *cwnd* will increase quickly as well. Therefore, the sender will achieve a high throughput. (3) It is obvious that large segment sizes also result in high throughput. These are also verified by the theoretical analysis of TCP [7.10].

7.1.2 UDP (RFC 768)

UDP is a connectionless transport protocol. It exchanges datagrams without a *sequence number,* and if information is lost in the process of exchange between the transmitter and the receiver, this protocol does not have the mechanisms to recover it. Since it does not offer a sequence number in the datagrams it therefore does not guarantee orderly transmission. It also does not offer capabilities for congestion or flow control. In circumstances where both TCP and UDP are present, since UDP does not perform congestion or flow control, it may turn out that it outperforms TCP. In recent years a *TCP-friendly rate control (TFRC)* [7.11] has been proposed for UDP to implement a certain level of control in this protocol. The basic idea behind TFRC is to provide almost identical throughput to both TCP and UDP when they are present on a connection.

7.1.3 Mobile IP

Mobile IP is proposed as a global mobility management technique in the network layer to provide terminal mobility in an all-IP network. The initial design of TCP/IP did not take mobility into consideration. The IP address is currently used both as a terminal identifier and to identify the terminal location in that network. Addresses are used in the routing process as well. However, mechanisms are required to separate the two. Mobile IP, which is designed to solve this problem, introduces two new entities and one new IP address. The new entities are (1) home agent (HA), the agent being located within the mobile terminal's home network (it is in charge of IP address management and packet forwarding on behalf of the mobile terminal), and (2) foreign agent (FA), the agent being located at the network visited by the mobile terminal. HA and FA have fixed IP addresses and can be addressed globally. The new IP address introduced for mobility is care of address (COA), the IP address obtained from FA after the mobile terminal enters a new network.

Mobile IP works as follows: When a terminal moves into a new network, it registers with the FA of the new network and subsequently receives a COA. At this time, either the terminal or the FA informs the terminal's HA of the COA. When a corresponding terminal sends packets to the mobile terminal, those packets are forwarded to the HA, which will, in turn, forward them to the mobile terminal's COA. Packets from the mobile terminal to the corresponding terminal are sent directly to the corresponding terminal. Therefore, there is an asymmetrical routing process between the corresponding terminal and the mobile terminal called the *triangular routing*, which leads to a longer path from the corresponding terminal to the mobile and therefore to low efficiency. In the process of mobility, since handoff results from movement and may cause packet loss and TCP timeout, the TCP sender is forced to reduce its rate, which may lead to low throughput even though physical link may offer sufficient bandwidth.

7.1.4 Feasibility of Using TCP or UDP for WSNs

Although TCP and UDP are popular transport protocols and deployed widely in the Internet, neither may be a good choice for WSNs. For the most part, there is no interaction between TCP or UDP and the lower-layer protocols. In wireless sensor networks, the lower layers can provide rich and helpful information to the transport layer and enhance the badly needed system performance.

Following are other problems that make either TCP or UDP unsuitable for implementation in WSNs:

- TCP is a connection-oriented protocol. However, in WSNs, the number of sensed data for event-based applications is usually very small. The three-way handshake process required for TCP is a large overhead for such a small volume of data.
- In TCP, segment loss can potentially trigger window-based flow and congestion control. This will reduce the transmission rate unnecessarily when, in fact, packet loss may have occurred as a result of link error and there may be no congestion. This behavior will lead to low throughput, especially under multiple wireless hops, which are prevalent in WSNs.
- TCP uses an end-to-end process for congestion control. Generally, this results in longer response to congestion, and in turn, will result in a large amount of segment loss. The segment loss, in turn, results in energy waste in the retransmission. Furthermore, a long response time to congestion results in low throughput and utilization of wireless channels.
- TCP uses end-to-end ACK and retransmission when necessary. This will result in much lower throughput and longer transmission time when RTT is long, as is the case in most WSNs.
- Sensor nodes may be within a different hop count and RTT from the sink. The TCP operates unfairly in such environments. The sensor nodes near the sink may receive more opportunities to transmit (which results in them depleting

their energy sooner). This may also result in a disconnect between more distant nodes and the sink.

As a connectionless transport control protocol, UDP is also not suitable for WSNs. Here are some reasons:

- Because of the lack of flow and congestion control mechanisms in UDP, datagram loss can result in congestion. From this point of view, UDP is also not energy efficient for WSNs.
- UDP contains no ACK mechanism; therefore, the lost datagrams can be recovered only by lower or upper layers, including the application layer.

7.2 TRANSPORT PROTOCOL DESIGN ISSUES

WSNs should be designed with an eye to energy conservation, congestion control, reliability in data dissemination, security, and management. These issues often involve one or several layers of the hierarchical protocol, and can be studied either separately in each layer or collaboratively in cross layers. For example, congestion control may involve only the transport layer, but energy conservation may be related to the physical, data link, network, and perhaps all other high layers. Generally, transport control protocols' design include two main functions: congestion control and loss recovery. For congestion control, one needs to detect the onset of congestion and to determine when and where it has occurred. Congestion can be detected, for example, by monitoring node buffer occupancy or link load (such as wireless channel). In the traditional Internet, methods to control congestion include selective packet dropping at a congestion point, such as is used in active queue management (AQM) schemes, rate adjustment at the source node, such as the technique of additive increase multiplicative decrease (AIMD) in TCP, and the use of routing techniques. For WSNs, one should consider carefully how to detect congestion and how to overcome it, because sensors have limited resources. These protocols must consider simplicity and scalability, to save energy, and ways to prolong the life of sensor batteries. For example, one may use an end-to-end mechanism such as that utilized in TCP or hop-by-hop backpressure such as that implemented in the asynchronous transfer mode (ATM) or frame relay networks. End-to-end approaches are very simple and robust, but they can result in additional traffic in the networks. However, hop-by-hop approaches usually detect congestion quickly, and as a result, introduce less additional network traffic. Due to energy constraint at the sensors, there is a clear trade-off between end-to-end and hop-by-hop mechanisms which should be considered carefully when designing congestion control algorithms for WSNs.

Packet loss in wireless sensor networks is usually due to the quality of the wireless channel, sensor failure, and/or congestion. WSNs must guarantee certain reliability at the packet or application level through loss recovery, in order to relay correct information. Certain critical applications need reliable transmission of each packet, and thus packet-level reliability is required. Other applications need only a proportionately reliable transmission of packets, and thus application reliability rather than packet-level reliability is important. The traditional methods used in packet-switched networks can be used to detect packet loss for wireless sensor networks as well. For example, each packet can piggyback a sequence number, and a receiver can detect packet loss though sequence numbers. After detecting packet loss, ACK and/or NACK can be used to recover missing packets based on an end-to-end or hop-by-hop control. With regard to energy, if there are few packets in transit and few retransmissions are required, efficiency is maintained. Effective congestion control can result in fewer in-transit packets. An effective loss recovery approach results in few retransmissions. In summary, the problem of transport control protocols for sensor networks boils down to the effective use of energy.

The design of transport protocols for WSNs should consider the following factors:

1. Perform congestion control and reliable delivery of data. Since most data are from the sensor nodes to the sink, congestion might occur around the sink. Although MAC protocol can recover packets loss as a result of bit error, it has no way handling packet loss as a result of buffer overflow. WSNs need a mechanism for packet loss recovery, such as ACK and selective ACK [7.9] used in TCP. Furthermore, reliable delivery in WSNs may have a different meaning than that in traditional networks, correct transmission of every packet is guaranteed. For certain sensor applications, WSNs only need to receive packets correctly from a fraction of sensors in that area, not from every sensor node in that area. This observation can result in an important input for the design of WSN transport protocols. Also, it may be more effective to use a hop-by-hop approach for congestion control and loss recovery since it may reduce packet loss and therefore conserve energy. The hop-by-hop mechanism can also lower the buffer requirement at the intermediate nodes.

2. Transport protocols for wireless sensor networks should simplify the initial connection establishment process or use a connectionless protocol to speed up the connection process, improve throughput, and lower transmission delay. Most applications in WSNs are reactive, which means that they monitor passively and wait for events to occur before sending data to the sink. These applications may have only a few packets to send as the result of an event.

3. Transport protocols for WSNs should avoid packet loss as much as possible since loss translates to energy waste. To avoid packet loss, the transport protocol should use an *active congestion control* (ACC) at the cost of slightly lower link utilization. ACC triggers congestion avoidance before congestion actually occurs. As an example of ACC, the sender (or intermediate nodes) may reduce its sending (or forwarding) rate when the buffer size of the downstream neighbors exceeds a certain threshold.

4. The transport control protocols should guarantee fairness for a variety of sensor nodes.

5. If possible, a transport protocol should be designed with cross-layer optimization in mind. For example, if a routing algorithm informs the transport protocol of route failure, the protocol will be able to deduce that packet loss is not from congestion but from route failure. In this case, the sender may maintain its current rate.

7.3   EXAMPLES OF EXISTING TRANSPORT CONTROL PROTOCOLS

Examples of several transport protocols designed for WSNs are shown in Table 7.1. Most examples can be grouped in one of the four groups: *upstream congestion control*, *downstream congestion control*, *upstream reliability guarantee*, and *downstream reliability guarantee*.

7.3.1   CODA (Congestion Detection and Avoidance)

CODA [7.1] is an *upstream congestion control* technique that consists of three elements: *congestion detection, open-loop hop-by-hop backpressure*, and *closed-loop end-to-end multisource regulation*. CODA attempts to detect congestion by monitoring current buffer occupancy and wireless channel load. If buffer occupancy or wireless channel load exceeds a threshold, it implies that congestion has occurred. The node that has detected congestion will then notify its upstream neighbor to reduce its rate, using an open-loop hop-by-hop backpressure. The upstream neighbor nodes trigger reduction of their output rate using methods such as AIMD. Finally, CODA regulates a multisource rate through a closed-loop end-to-end approach, as follows: (1) When a sensor node exceeds its theoretical rate, it sets a "regulation" bit in the "event" packet; (2) If the event packet received by the sink has a "regulation" bit set, the sink sends an ACK message to the sensor nodes and informs them to reduce their rate; and (3) if the congestion is cleared, the sink will send an immediate ACK control message to the sensor nodes, informing them that they can increase their rate. CODA's disadvantages are its unidirectional control, only from the sensors to the sink; there is no reliability consideration; and the response time of its closed-loop multisource control increases under heavy congestion since the ACK issued from the sink will probably be lost.

7.3.2   ESRT (Event-to-Sink Reliable Transport)

ESRT [7.2], which provides reliability and congestion control, belongs to the *upstream reliability guarantee* group. It periodically computes a reliability figure (r), representing the rate of packets received successfully in a given time interval. ESRT then deduces the required sensor reporting frequency (f) from the reliability figure (r) using an expression such as $f = G(r)$. Finally, ESRT informs all sensors of the values of (f) through an assumed channel with high power. ESRT uses an end-to-end approach to guarantee a desired reliability

TABLE 7.1 Several Transport Protocols for WSNs

Attributes	CODA	ESRT	RMST	PSFQ	GARUDA
Direction	Upstream	Upstream	Upstream	Downstream	Downstream
Congestion					
Support	Yes	Passive	No	No	No
Congestion detection	Buffer occupancy channel condition	Buffer occupancy	—	—	—
Open- or closed-loop congestion control	Both	No	—	—	—
Reliability					
Support	No	Yes	Yes	Yes	Yes
Packet or application reliability	—	Application	Packet	Packet	Packet
Loss detection	—	No	Yes	Yes	Yes
End-to-end (E2E) or hop-by-hop (H&H)	—	E2E	HbH	HbH	HbH
Cache	—	No	Option	Yes	Yes
In- or out-of-sequence NACK	—	N/A	In-sequence	Out-of-sequence	Out-of-sequence
ACK or NACK	—	ACK	NACK	NACK	NACK
Energy conservation	Good	Fair	—	—	Yes

figure through adjusting the sensors' reporting frequency. It provides overall reliability for the application. The additional benefit of ESRT is energy conservation through control of reporting frequency. Disadvantages of ESRT are that it advertises the same reporting frequency to all sensors (since different nodes may have contributed differently to congestion, applying different frequencies would be more appropriate) and considers mainly reliability and energy conservation as performance measures.

7.3.3   RMST (Reliable Multisegment Transport)

RMST [7.3] guarantees successful transmission of packets in the upstream direction. Intermediate nodes cache each packet to enable hop-by-hop recovery, or they operate in noncache mode, where only end hosts cache the transmitted packets for end-to-end recovery. RMST supports both cache and noncache modes. Furthermore, RMST uses selective NACK and timer-driven mechanisms for loss detection and notification. In the cache mode, lost packets are recovered hop by hop through the intermediate sensor nodes. If an intermediate node fails to locate the lost packet, or if the intermediate node works in noncache mode, it will forward the NACK upstream toward the source node. RMTS is designed to run above *directed diffusion* [7.12], which is a routing protocol, in order to provide guaranteed reliability for applications. Problems with RMST are lack of congestion control, energy efficiency, and application-level reliability.

7.3.4   PSFQ (Pump Slowly, Fetch Quickly)

PSFQ [7.4] distributes data from sink to sensors by pacing data at a relatively slow speed but allowing sensor nodes that experience data loss to recover any missing segments from immediate neighbors. This approach belongs to the group *downstream reliability guarantee*. The motivation is to achieve loose delay bounds while minimizing loss recovery by localizing data recovery among immediate neighbors. PSFQ consists of three operations: pump, fetch, and report. This is how PSFQ works: Sink broadcasts a packet to its neighbors every T time units until all the data fragments have been sent out. Once a sequence number gap is detected, the sensor node goes into fetch mode and issues a NACK in the reverse path to recover the missing fragment. The NACK is not relayed by the neighbor nodes unless the number of times that the NACK is sent exceeds a predefined threshold [7.4]. Finally, the sink can ask sensors to provide it with the data delivery status information through a simple and scalable hop-by-hop report mechanism. PSFQ has the following disadvantages: It cannot detect packet loss for single packet transmission; it uses a slow pump, which results in a large delay; and hop-by-hop recovery with cache necessitates larger buffer sizes.

7.3.5   GARUDA

GARUDA [7.5] is in the downstream reliability group. It is based on a two-tier node architecture; nodes with $3i$ hops from the sink are selected as *core sensor nodes* (i is an integer). The remaining nodes (noncore) are called *second-tier nodes*. Each non-core sensor node chooses a nearby core node as its core node. Noncore nodes use core nodes for lost packet recovery. GARUDA uses a NACK message for loss detection and notification. Loss recovery is performed in two categories: loss recovery among core sensor nodes [7.5], and loss recovery between noncore sensor nodes and their core node. Therefore, retransmission to recover lost packets looks like a hybrid scheme between pure hop by hop and end to end. GARUDA designs a repeated wait for first packet (WFP) pulse transmission to guarantee the success of single or first packet delivery. Furthermore, pulse transmission is used to compute the hop number and to select core sensor nodes in order to establish a two-tier node architecture. Disadvantages of GARUDA include lack of reliability in the upstream direction and lack of congestion control. Published results on GARUDA at the time of this writing did not include reports of any results on reliability or a performance comparison with other algorithms, such as PSFQ.

7.3.6   ATP (Ad Hoc Transport Protocol)

ATP [7.6] works based on a receiver-and network-assisted end-to-end feedback control algorithm. It uses selective ACKs (SACKs) for packet loss recovery. In ATP, intermediate network nodes compute the sum of exponentially distributed packet queuing and transmission delay, called D. The required end-to-end rate is set as the inverse of D. The values of D are computed over all packets that traverse a given sensor node, and if it exceeds the value that is piggybacked in each outgoing packet, it updates the field before forwarding the packet. The receiver calculates the required end-to-end rate (inverse of D) and feeds it back to the sender. Thus, the sender can intelligently adjust its sending rate according to the value received from the receiver. To guarantee reliability, ATP uses selective ACKs (SACKs) as an end-to-end mechanism for loss detection. ATP decouples congestion control from reliability and as a result, achieves better fairness and higher throughput than TCP. However, energy issues are not considered for this design, which raises the question of optimality of ATP for an end-to-end control scheme.

7.3.7   Problems with Transport Control Protocols

The major functions of transport protocols for wireless sensors networks that should be considered carefully in the design of these protocols are congestion control, reliability guarantee, and energy conservation. Most of the existing protocols reviewed here and reflected in the literature provide either congestion or reliability in either upstream or downstream (not both). Certain applications in wireless sensor networks require it in both directions: for example, re-tasking and critical time-sensitive monitoring and surveillance operations. Another problem with the existing transport protocols for wireless sensor networks is that they only control congestion either end-to-end or hop-by-hop. Although in CODA there are both end-to-end and hop-by-hop mechanisms for congestion control, it uses them simultaneously rather than adaptively. An adaptive congestion control that integrates end-to-end and hop-by-hop mechanisms may be more helpful for wireless sensor networks with diverse applications, and useful due to energy conservation and simplification of sensor node operation.

Transport protocols studied so far provide either packet- or application-level reliability (if reliability is provided at all). If a sensor network supports two applications, one that requires packet-level reliability and the other application-level reliability, the existing transport control protocols will face difficulty. Therefore, an adaptive recovery mechanism is required to support packet- and application-level reliability as well as for energy efficiency.

None of the existing transport protocols implement cross-layer optimization. As discussed earlier, lower layers, such as the network and MAC layers, can provide useful information to the transport layer.

7.4 PERFORMANCE OF TRANSPORT CONTROL PROTOCOLS

In this section a quantitative comparison of WSN congestion and loss performance is presented. The measure used for congestion comparison is energy consumption, which is calculated for end-to-end and hop-by-hop cases. The other measure is loss performance, which is based on cache and noncache approaches, as discussed earlier.

7.4.1 Congestion

As discussed earlier in the chapter, two general approaches to congestion control are end to end and hop by hop. In an end-to-end approach such as conventional TCP, it is the source node's responsibility to detect congestion in either the receiver-assisted (ACK-based loss detection) mode or the network-assisted mode (using explicit congestion notification). Therefore, rate adjustments occur only at the source node. In hop-by-hop congestion control, intermediate nodes detect congestion and notify the originating link node. Hop-by-hop control can potentially eliminate congestion faster than the end-to-end approach, and can reduce packet loss and energy consumption in sensor nodes.

A simple model is provided here to help understand the impact of congestion control on energy efficiency. The following assumptions are made:

- There are $h > 1$ hops between sources and sink nodes, and each hop introduces a delay d. The link capacity is C.
- Congestion occurs uniformly in the network. The frequency of congestion occurrence is f, which is dependent on network topology, traffic characteristics, and buffer size.
- When the total rate of source transmission exceeds $C(1 + a)$, congestion will be detected.
- e is the average energy consumed to send or receive a packet on each link.

In the end-to-end approach, the average time required to notify the source about the onset of congestion is $1.5hd$. During this interval (between the time that congestion occurs and the source is informed), all nodes can send up to $C(1 + a)(1.5hd)$ packets, except on the congested link, on which traffic may not exceed $C(1.5hd)$. Therefore, in this case, the number of packets lost due to congestion can be estimated as $n_e = aC(1.5hd)$.

In the hop-by-hop approach, the time required to trigger congestion control corresponds to only one-hop delay (d). Therefore, packet loss, before congestion is controlled, is approximately $n_b = aCd$.

Let $N_s(T)$ be the number of packets transmitted successfully through the congested link, and $N_d(T)$ be the number of packets dropped due to congestion during the time interval T. On average, each dropped packet has been through $0.5H$ hops. We define the energy efficiency of a congestion control mechanism as

$$E_c = \frac{N_s(T)He}{N_s(T)He + N_d(T)(0.5H)e} = \frac{N_s(T)}{N_s(T) + 0.5N_d(T)} \tag{7.1}$$

where E_c is the mean energy ratio required to send one packet successfully. In ideal situations, when there is no congestion, E_c would be 1.

Therefore, for end-to-end congestion control,

$$E_c = \frac{N_s(T)}{N_s(T) + 0.5N_d(T)} = \frac{TC}{TC + 0.5f\,Tn_e} = \frac{4}{4 + 3fahd} \tag{7.2}$$

and for the hop-by-hop control,

$$E_c = \frac{N_s(T)}{N_s(T) + 0.5N_d(T)} = \frac{TC}{TC + 0.5f\,Tn_h} = \frac{2}{2 + fad} \tag{7.3}$$

It can be seen from Eqs. (7.2) and (7.3) that the energy efficiency of an end-to-end mechanism is dependent on the path length (H), whereas hop-by-hop control is independent of path length and thus results in a higher efficiency ratio.

CODA defines the *energy tax* as the ratio of the total number of packets dropped in the sensor network to the total number of packets received at the sink for hop-by-hop congestion control. Therefore, the lower ratio is an indication of higher energy efficiency. Figure 7.1 shows CODA's energy efficiency.

7.4.2 Packet Loss Recovery

The question we deal with in this section is how to recover lost packets. Generally, two methods are available for this purpose: cache and noncache recovery. Noncache recovery is an end-to-end ARQ (automatic repeat request) similar to the traditional TCP. Cache-based recovery uses a hop-by-hop approach and relies on caching at the intermediate nodes, with retransmissions

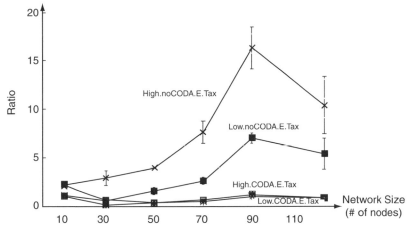

Figure 7.1 Energy tax in CODA as a function of network size for high- and low-data-rate traffic. The difference between the data points with and without CODA indicates the energy saving achieved by CODA. (Based on data from [7.1].)

between two neighboring nodes. In the noncache case, however, retransmissions may occur in h hops, and therefore more total energy is required. The *cache point* is defined as the node that copies transmitted packets locally for a certain time period; and the *loss point* is defined as the node at which packets are dropped due to congestion. Let's define the *retransmission path length* (l_p) as the number of hops from the caching node to the node where loss occurs. Therefore, in the noncache case, $l_p = h_1$, where h_1 is the number of hops from the loss point to the source node. In the cache case, l_p can be 1 if lost packets are found in neighboring nodes. Because sensor nodes have limited buffer space, packet copies can be stored for only a limited time period. Therefore, l_p in the cache case may be larger than 1 but still smaller than $h_1(1 < l_p < h_1)$. In cache-based recovery, different algorithms may have a different retransmission path length l_p and introduce different energy efficiency.

In cache-based recovery, each packet is stored at every intermediate node that it visits until its neighboring node receives the packet successfully, or when a timeout occurs (whichever is sooner). In this case it is likely that l_p is very close to 1. Another option is to distribute caching so that packet copies are scattered among intermediated nodes. Each packet is stored in only one or several intermediate nodes. Distributed caching might have a longer l_p than regular caching (but still smaller than in the noncache case) and requires less buffer space than regular caching.

RMST [7.3] investigated the performance of various loss recovery mechanisms that may provide reliability through the link, transport, and application layers. Figure 7.2 from [7.3] compares the performance of hop-by-hop and end-to-end loss recovery in the transport layer. The comparison is made in terms of the number of transmissions required to send 10 packets across a network in 10 hops. As shown

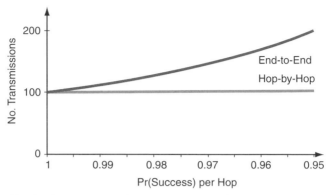

Figure 7.2   Hop-by-hop versus end-to-end: number of transmissions required to send 10 packets in 10 hops. (From [7.3].)

in this figure, when the success rate drops below 0.95, the number of end-to-end retransmissions doubles, which in turn leads to lower energy efficiency.

7.5   CONCLUSION

In this chapter we presented an overview of the transport control protocol for wireless sensor networks. The limitations of TCP and UDP protocols were discussed, and reasons for these two not to be suitable for wireless sensor networks were given. A review of several existing sensor transport control protocols was also provided, and several problems in the existing protocols were described. When designing transport control protocols for wireless sensor networks, one should consider carefully such issues as:

1. Protocol effectiveness and the efficiency of congestion control mechanisms. Effective mechanisms avoid packet loss as much as possible while providing high throughput.
2. Reliability in the transport layer, whether loss recovery is required at the transport layer, and which mechanism is effective and energy efficient. Preferably, any such mechanisms should have low buffering requirements.
3. Fairness among sensor nodes within different distances from the sink.
4. Utilization of some type of cross-layer optimization to improve performance.

REFERENCES

[7.1] C. Y. Wan, S. B. Eisenman, A. T. Campbell, "CODA: Congestion Detection and Avoidance in Sensor Networks," *Proceedings of the 1st ACM Conference on Embedded Networked Sensor Systems* (SenSys'03), Los Angeles, Nov. 2003.

[7.2] Y. Sankarasubramaniam, O. B. Akan, I. F. Akyidiz, "ESRT: Event-to-Sink Reliable Transport in Wireless Sensor Networks," *Proceedings of the 4th ACM International Symposium on Mobile Ad Hoc Networking and Computing* (MobiHoc'03), Annapolis, MD, June 2003.

[7.3] F. Stann, J. Heidemann, "RMST: Reliable Data Transport in Sensor Networks," *Proceedings of the 1st IEEE International Workshop on Sensor Network Protocols and Applications* (SNPA'03), Anchorage, AK, May 2003.

[7.4] C. Y. Wan, A. T. Campbell, "PSFQ: A Reliable Transport Protocol for Wireless Sensor Networks," *Proceedings of the ACM Workshop on Sensor Networks and Applications* (WSNA'02), Atlanta, GA, Sept. 28, 2002.

[7.5] S. J. Park, R. Vedantham, R. Sivakumar, I. F. Akyildiz, "A Scalable Approach for Reliable Downstream Data Delivery in Wireless Sensor Networks," *Proceedings of the 5th ACM International Symposium on Mobile Ad Hoc Networking and Computing* (MobiHoc'04), Roppongi, Japan, May 24–26, 2004.

[7.6] K. Sundaresan, V. Anantharaman, H.-Y. Hseeh, R. Sivakumar, "ATP: A Reliable Transport Protocol for Ad-Hoc Networks," *Proceedings of the 4th ACM International Symposium on Mobile Ad Hoc Networking and Computing* (MobiHoc'03), Annapolis, MD, June 2003.

[7.7] J. B. Postel, "Transmission Control Protocol," *RFC 793*, IETF, Sterling, VA, Sept. 1981.

[7.8] J. B. Postel, "User Datagram Protocol," *RFC 768*, IETF, Sterling, VA, Aug. 1980.

[7.9] M. Mathis, "TCP Selective Acknowledge Options," *RFC 2018*, IETF, Sterling, VA, Oct. 1996.

[7.10] J. Padhye, V. Firoiu, D. Towsley, J. Kurose, "Modeling TCP Throughput: A Simple Model and Its Empirical Validation," *Proceedings of the ACM SIGCOMM*, 1998.

[7.11] M. Handly et al., "TCP Friendly Rate Control (TFRC): Protocol Specification," *RFC 3448*, IETF, Sterling, VA, Jan. 2003.

[7.12] C. Intanagonwiwat, R. Govindan, D. Estrin, Directed diffusion: "A Scalable and Robust Communication Paradigm for Sensor Networks," *Proceedings of the 6th ACM/IEEE International Conference on Mobile Computing and Networking* (MobiCom'00), Boston, MA, Aug. 2000.

8

MIDDLEWARE FOR WIRELESS SENSOR NETWORKS

8.1 INTRODUCTION

There is a gap between network protocols, on the one hand, and applications in wireless sensor networks, on the other. We need to provide adaptation functions between applications and network protocols to satisfy the requirements of special features of wireless sensor networks and diversity of its applications. The adaptation functions should facilitate provision of quality of service to applications while using the limited resources of WSNs and extending their life span. Middleware [8.17,8.18] is an approach to satisfy the adaptation. In this chapter we examine the existing middleware for WSNs.

WSNs are constrained in resources such as bandwidth, computation and communication capabilities, and energy. WSN topology is variable due to node mobility, depletion of energy, switching between sleep and active states, radio range, and routing possibilities. A WSN may also need to support several applications simultaneously. Therefore, a WSN is a *wireless/mobile* and *resource-constrained* network with *diverse* applications. The problem in this resource-constrained environment is how to design middleware that is capable of adaptation between applications and network protocols.

Middleware is usually below the application level and on top of the operating systems and network protocols. It marshals the application requirements, hides details of lower levels, and facilitates application development and deployment and their management. WSNs have special requirements in this area since they

Wireless Sensor Networks: Technology, Protocols, and Applications, by Kazem Sohraby, Daniel Minoli, and Taieb Znati
Copyright © 2007 John Wiley & Sons, Inc.

are very different from traditional networks and/or distributed computing systems. After an introduction to the role and functions of middleware, we present a brief survey of the existing middleware for WSNs, a comparison, and the future direction of their development.

8.2 WSN MIDDLEWARE PRINCIPLES

Challenges in the design of middleware for WSNs are [8.5]: (1) topology control, to rearrange the sensor nodes into a connected network; (2) energy-aware data-centric computation; (3) application-specific integration, since integration of application information into the network protocol improves performance and conserves energy; (4) efficient utilization of computational and communications resources; and (5) support for real-time applications.

The basic middleware functions for WSNs are as follows [8.5]:

1. System services to diverse applications. To deploy current and future applications easily, middleware needs to provide a standardized system service.
2. An environment that coordinates and supports multiple applications; this is required to implement the diverse applications and to create new ones.
3. Mechanisms to achieve adaptive and efficient utilization of system resources; these mechanisms provide algorithms that dynamically manage limited and variable network resources of WSNs.
4. Efficient trade-offs between the multiple QoS dimensions; this can be used to adjust and optimize the required network resources.

References [8.5] and [8.6] propose design principles for WSN middleware as follows: (1) need for *localized algorithms* as distributed algorithms that achieve a global goal by communicating with nodes in some neighborhood; (2) need for *adaptive fidelity algorithms* to trade off between the quality of the results and resource utilization; (3) need for *data-centric mechanisms* for data processing and querying within the network and for decoupling data from the physical sensor; (4) need for Application knowledge, integrated into the services provided by the middleware, to improve resource and energy efficiencies; (5) need for *lightweight* middleware for both computation and communication; (6) and need to perform application QoS trade-offs since given the limited resources in WSNs, QoS for all applications cannot be satisfied simultaneously. In this way, middleware helps negotiate between applications and low-level network protocols in order to improve performance and save network resources. To perform this task, the middleware needs to know the features of both applications and network protocols. Specifically, it needs to analyze and abstract application-specific features as well as the network protocols. The remaining task is to construct an effective mapping between applications and network protocol based on the current network status and the required application QoS. This mapping may be implemented as middleware services that can be invoked by applications. Middleware services provide application knowledge and its current QoS, as well as the current network state, and in turn produce control to manage network resources. In certain cases, middleware informs applications to change their QoS requirements, but this requires that the applications be adaptive.

8.3 MIDDLEWARE ARCHITECTURE

The general middleware architecture is shown in Figure 8.1. The middleware gathers information from the application and network protocols and determines how to support the applications and at the same time adjust network protocol parameters. Sometimes the middleware interfaces with the operating system directly while bypassing the network protocol. The major difference between WSN and traditional middleware is that the former needs to dynamically adjust low-level network protocol parameters and configure sensor nodes for the purpose of performance improvement and energy conservation. The key is for the middleware to abstract the common properties of applications and to map application requirements into those actions that boil down to protocol parameter adjustment. For example, the middleware may consist of the following functional elements: resource management, event detection and management, and application programming interface (API). The resource management functional element monitors the network status and receives application requirements. It then produces the command to adjust the network resource. The event detection and management functional element is

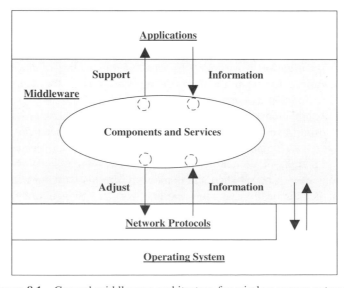

Figure 8.1 General middleware architecture for wireless sensor networks.

TABLE 8.1   APIs Defined in QueryAgent [8.21]

API	Description
PName unicast (QID, intst, dstn)	The application layer wants to get data from one sensor. The interface returns a suitable routing protocol name.
Boolean start_unicast (QID, PName, intst, dstn)	Ask the low-level layer to start the unicast process.
Data listen_unicast (QID, data)	The application layer listens to get the data from the sensor network.
Data finish_unicast (QID, data)	Explicitly finish the unicast process.
turnoff (QID)	Turn off the sensor.
turnon (QID)	Turn on the sensor.
move (QID, direction, Value)	Move the sensor to another location.
⋮	⋮

Source: [8.21].

used to detect and manage events such as sensing. The API can be invoked by applications to achieve better performance and network utilization. For example, QueryAgent [8.21] defines a general programming interface between application and lower layers. A subset of APIs used for unicast communication in QueryAgent is provided in Table 8.1 [8.21]. For example, for query of temperature from a sensor, QueryAgent works as follows: The application sends a *unicast*() message to the general interface and the interface returns a suitable routing protocol for the query; the API *start_unicast*() is then called and a message is sent to where the query is processed; the API *listen_unicast*() is then called in order to wait for the data from the sensor network; and finally, the application calls *finish_unicast*() and receives the query results. In applying these techniques, QueryAgent [8.21] consists of six modules, two of which are critical in improving system performance: Data Manager, which aggregates sensory data, and Intelligent Agent, which exploits the difference between consistency and timeliness of sensory data in order to optimize query processing. These also help reduce the number of reports and the path between a source sensor node and the sink. We discuss QueryAgent in more detail later in the chapter. Furthermore, as also shown in Table 8.1, QueryAgent provides two APIs, *turnon*(*QID*) and *turnoff*(*QID*), which turn the sensor nodes on and off to dynamically control energy consumption.

8.3.1   Data-Related Functions

Since WSN is a data-centric device, middleware would contain data management functions such as data dissemination, data compression, and data storage. For completeness, a brief overview of these functions is given.

Data Dissemination   In WSNs, the sensor nodes deployed produce data. The data sensed need to be transmitted to some special node or a sink for further analysis, management, and control. Therefore, a data dissemination protocol is required to provide effective data transmission from sensor nodes to the sink. Data dissemination protocols have a certain relation to the routing protocols. The routing protocols are general and are designed to find a path between the source and destination nodes. On the other hand, data dissemination protocols should guarantee successful transmission from nodes to the sink. Data dissemination protocols consist of at least two phases:

1. The initial phase of triggering data transmission, often initiated by the sink, by sending out a query to inform sensor nodes of its intent. The query contains information to guide data transmission from the node to the sink, the frequency of data reporting, the duration of interval in which data reporting should take place, and so on.
2. The data transmission phase; sensor nodes report data to the sink. Data dissemination protocols need to indicate whether the data are to be transmitted in broadcast or unicast mode. Routing protocols and other techniques, such as data replication and cache, may also be used for performance optimization.

Some protocols, such as directed diffusion (DD) [8.22], consider WSN with only one sink. Later protocols, such as two-tier data dissemination (TTDD) [8.23] and sinks accessing data from environments (SAFE) [8.24], consider multiple sinks. In DD, the query is flooded. The initial data are also broadcast to all neighbors to set up a *reinforced path*, but subsequent data are transmitted only on the *reinforced path*. TTDD proposes two-tier grid architecture for data dissemination. In TTDD, sensor nodes need to announce the process to build a grid structure. Then the query is flooded only in an area smaller than a grid cell in order to find a nearby dissemination node. The dissemination node is defined as the node closest to the crossing point of the grid. SAFE attempts to share and compress the data dissemination if there is the same part from a source node to the multiple sinks. In this way it avoids duplicate data and therefore conserves energy.

Data Compression   Communication components consume most of the energy in WSNs. Computation uses less. Therefore, it becomes attractive to deploy data compression techniques, which might increase computational energy somewhat, but decrease the number of packet transmissions. Several features of WSNs make it possible to implement effective data compression protocols: (1) Usually, the data collected in neighboring sensor nodes are correlated, especially when the deployment of sensor nodes is quite dense in the network; (2) due to the treelike logical topology of most WSNs, the correlation may become more apparent on the path from the sensor nodes to the sink; (3) the occurrence of an event may be assimilated with a continuous-time but random process, and sampling of the random processes helps extract information content from the process; (4) the application semantic may enable data aggregation or data fusion; and (5) the tolerance of applications for possible errors in data may make it possible to reduce data reading and reporting frequencies.

Compression techniques include the following:

1. Information theoretic–based techniques such as distributed source coding using syndromes (DISCUS) [8.25]. This is a distributed compression scheme for a dense microsensor network, is based on Slepian–Wolf coding [8.26], and does not require conversion. Since most WSNs consist of sensor nodes in a treelike topology where the root is the sink, information is compressed or encoded at each node incorporating the correlation with data from its parent node. The decompression or decoder process can be performed by the sink or jointly by the sensor nodes and the sink.

2. Data aggregation–based compression schemes such as tiny aggregation service for ad hoc sensor networks (TAG) [8.27]. TAG realizes several semantic-based aggregations such as MIN, MAX, and SUM, in an application-dependent manner. This approach would not be helpful for applications that have no such semantic expressions. A problem in this approach is the location of the aggregation point.

3. Sampling of a random process. If an application tolerates a certain level of error, sensor nodes can adaptively reduce sampling frequency.

Data Storage   Sensor nodes collect data related to the sensed events. Data need to be stored, usually for future use. Several questions that merit consideration for data storage are: What type of data need to be stored? Where should the data be stored? How and for how long should data be stored? The answers to these questions define the data storage requirements of WSNs. There are two types of data in WSNs: the raw data collected by the sensor nodes, and the results analyzed from the data collected initially, such as from an event and its location. Several data storage schemes have been proposed in the literature:

1. *External storage* (ES). The data sensed are transmitted to an external (centralized) host for storage. This approach is not energy-efficient, since while all the data are hauled to a center, not all the data are required for future query.

2. *Local storage* (LS). The data collected are stored locally in the sensor node itself. Although the LS scheme is more energy-efficient than ES, it is not efficient for query. For example, if frequent querying occurs for data in a distant sensor node, LS consumes more energy than if the data were centrally stored. An advantage of LS is that the data location will be known in the query process.

3. *Data-centric storage* (DCS) [8.28]. In DCS, the event data are stored based on their event type and at some special "home nodes" that may not be the original location of the data collected. Therefore, in DCS, a query can be routed to the corresponding home nodes according to the data type. This

approach can be energy efficient and it is easy to realize load balancing with such methods as a distributed hash table. But it is not possible in DCS to query the provenance of data.

4. *Provenance-aware data storage* (PADS) [8.29]. PADS emphasizes the necessity of being able to query the provenance of data for certain applications. The event data are stored locally in PADS, while the index or pointer of the data is stored in some "home hosts," as in DCS. Therefore, PADS realizes the advantages of both LS and DCS.

5. *Multiresolution storage* (MRS) [8.30]. In MRS, data are decomposed and classified into levels: for example, level 0 for raw data, level 1 for finer data, and level 2 for the coarsest data. Data of different levels will be stored for different time durations. The level 2 data will be stored for the longest term and raw data will be stored for the shortest. MRS is actually a differentiated storage scheme; it realizes better load balancing and incurs low communication overhead.

8.3.2 Architectures

Several middleware architectures for WSNs are proposed in [8.1]–[8.13]. Dynamic network configuration is considered in [8.1], [8.3], [8.5], and [8.10], database and query in [8.2] and [8.12], data fusion in [8.11], event detection in [8.4], monitoring in [8.7], and system platforms in [8.9] and [8.13]. The approaches in [8.1], [8.5], and [8.10] configure and adjust the network dynamically without violating application requirements, with the goal of conserving energy and/or maximizing the network life span. The assumption in these approaches is that either the application requirement is flexible or the network protocol or behavior of sensor nodes is adjustable. For example, it is assumed in [8.1] that an application can be involved in several types of data sensing and that the application performance can be described by the QoS of different variables of interests. Therefore, there are multiple choices (of sensor nodes), each of which can meet the application performance. The objective of MiLAN in [8.1] is to determine which of the choices is optimal in order to extend the network life span, to allow the application to last as long as possible, and furthermore, to configure the network dynamically. AMF in [8.3] attempts to trade off resource and application performance during information collection. The main idea in [8.3] is to reduce the frequency of communication at sensor nodes by lowering the sampling frequency without compromising the accuracy of results. For example, the sensor will send an update only when the actual measurement exceeds the previous value or the predicted value beyond a given error bound. In [8.5] it is assumed that each sensor node is equipped with discrete dynamic voltage [8.19] and modulation scaling [8.20], which can be used for efficient exploration of the energy–latency trade-offs for computation and wireless communication activities. This means that the energy dissipation of performing a specific computation or communication activity can be reduced at the cost of increased latency. These two techniques are used in [8.5] to adjust the behavior of sensor nodes in order to prolong network life while meeting the real-time constraint of the application. For example, when executing each task of an application, middleware in [8.5] will try to lower its current voltage (or modulation) to the next level if the real-time constraint of the application is not violated. However, Impala in [8.10] assumes that there are several protocol options and that each option has a different energy efficiency. Impala then attempts to use the protocol with the highest energy efficiency if the application requirement can be met. For example, Impala uses a history-based (unicast) routing protocol unless device failure and/ or degradation of application performance is detected. In that case, it will switch to a flooding routing protocol.

The middleware architectures proposed in [8.2] and [8.12] treat the sensor network as a distributed database where Structured Query Language-like (SQL) queries can be issued in order to collect data or control sensor nodes' activities. These are not designed specifically for wireless sensor networks. For example, Iris-Net [8.2] investigates a worldwide sensor web that can integrate a wide range of sensor data, while the Device Database System (DDS) [8.12] examines the device network. Both of these architectures try to achieve improved query performance, but not the network life span, which is an important metric in WSNs.

DSWare [8.4] is reliable and energy-efficient event detection middleware. In this architecture it is assumed that the event may include certain subevents. The event can be detected through joint detection of subevents with a certain level of confidence that is dependent on the application. DSWare uses features of events to improve the reliability of detection and the energy efficiency. For example, certain subevents may occur only in a "phase" and will last for a certain time (called an *absolute validity interval* in [8.4]). Then DSware uses these properties to lower the frequency of reporting while guaranteeing the reliability of detection.

DFuse, middleware proposed in [8.11], is for data fusion. It provides a data fusion API distributed algorithm for energy-aware role assignment. It also provides four cost functions for migrating fusion point. Based on the four functions, DFuse chooses a fusion point dynamically to minimize cost and provide energy efficiency.

In [8.7], middleware for monitoring wireless sensor networks is provided, and in [8.9] and [8.13], two systematical middleware platforms are proposed. Em* in [8.9] provides a series of tools to develop applications for WSNs. SensorWare in [8.13] is an agent-based middleware where an agent such as a small program or *mobile control script* can be injected into the network to collect local sensor data. The script can migrate or copy itself to other nodes and can communicate with remote copies. A complex distributed algorithm can be realized through such scripts.

8.4 EXISTING MIDDLEWARE

8.4.1 MiLAN (Middleware Linking Applications and Networks)

MiLAN [8.1] defines two classes of applications: data-driven applications (collect and analyze data) and state-based applications (in which application requirements may change with the data received). MiLAN states that middleware that enables applications to affect actively both the network and the sensors themselves is needed to support this new and growing class of applications. Each sensor node runs a version of MiLAN that receives information about applications in terms of their QoS requirements; the overall system with regard to the relative importance or desired interaction among applications; and the network with regard to available components and resources. MiLAN adjusts network characteristics to increase the application life span while meeting their QoS needs [8.1].

MiLAN receives the following information for its operation: (1) variables of interest to the application, (2) the QoS required for each variable, and (3) the level of QoS that data from each sensor or set of sensors can provide for each variable. In [8.1] it is assumed that for a given application, the QoS for each variable can be satisfied using data from one or more sensors. Then applications furnish information to MiLAN through a graph of sensor QoS which contains an *application feasible set* (f_a). MiLAN uses a service discovery protocol to obtain information about senor nodes, such as the type of data that can be provided by the sensor node, modes in which the node can operate with the transmission power levels, and the current residual energy level. Then it determines that set of sensors that can be supported by the network [called a *network feasible set* (f_n)]. Finally, MiLAN optimally chooses elements in the overlapped set of f_a and f_n such as to optimize network configuration and maximize application life. MiLAN takes a description of application requirements and checks the network conditions for dynamic network configuration to fulfill the performance requirements, with an emphasise on extending the runtime of the application rather than on the efficient utilization of sensor power. MiLAN runs different types of applications and suggests modifications in routing protocols for energy conservation according to the application. MiLAN is not well suited for optimization in applications that have only one variable of interest.

8.4.2   IrisNet (Internet-Scale Resource-Intensive Sensor Networks Services)

IrisNet [8.2] extends the traditional WSNs to a worldwide sensor web which can integrate a wide range of sensor data, from a high bit rate (such as Webcam-equipped PCs) to a low bit rate created by traditional WSNs. A worldwide sensor web can support many consumer-oriented services. IrisNet is a two-tier architecture comprising sensing agents (SAs) and organizing agents (OAs). SAs implement a generic data acquisition interface to access sensors [8.2]. OAs implement a distributed database to store service-specific data that SAs produce. Each OA participates in only one sensing service. IrisNet uses XML to represent sensor-produced data hierarchically. It also uses an adaptive data placement algorithm to reduce query response time and network traffic while it balances an OA's load. IrisNet designs the execution environment for the SA host, where an executable code (senselet) can be uploaded and executed in each SA for a service. A senselet tells SA to use the raw sensor data, and it also performs a specified set of processing steps and sends results to a nearby OA. In short, IrisNet is a general-purpose software infrastructure that supports the central tasks common to such services as collecting, filtering, and combining sensor data, and performing distributed queries within reasonable response times [8.2]. IrisNet is not designed specifically for resource-limited WSNs. For example, IrisNet has not considered localized algorithms or possible WSN application features.

8.4.3 AMF (Adaptive Middleware Framework)

The adaptive middleware framework (AMF) proposed in [8.3] exploits "resource and application QoS trade-offs" and "predictability of sensor readings" to reduce the energy consumed in the process of information collection. The assumption is that it is possible to collect approximate data at predetermined accuracy levels while satisfying an application's QoS. AMF has "sensor-side" and "server-side" components which bridge the application layer with the underlying sensor network infrastructure. It supports both precision- and prediction-based adaptation. Server-side components include application quality, data quality requirement translation, adaptive precision setting, sensor selection, sensor data management, and fault tolerance. Sensor-side components include sensor-state management and precision-driven adaptation. AMF has an energy-efficient message-updating mode, where the sensor sends an update to the server only when the measurement value exceeds the previous value or the value predicted beyond a given error level [8.3]. The server maintains a list of active sensors (active list) and a list of historic values for each sensor over a specified time period. To support prediction-based adaptations, a sensor and the server store a set of prediction models and choose the best one according to the network status. AMF attempts to trade off between resource and quality during information collection. In this context it reduces sampling frequency without compromising the accuracy of results [8.5].

8.4.4 DSWare (Data Service Middleware)

DSWare [8.4] resides between the application and network layers, integrates various real-time data services, and provides a database-like abstraction to applications. It includes several components: data storage, data caching, group management, event detection, data subscription, and scheduling. In DSWare [8.4], data are replicated in multiple physical nodes mapped to a single logical node using a hash-based mapping. Queries are directed to any of the nodes to avoid collision and to balance the load among the nodes. A data caching service in DSWare monitors the current use of copies and determines whether to increase or reduce the number of copies and whether to move some copies to another location by exchanging information in the neighborhood [8.4]. DSWare incorporates group management to provide localized cooperation among sensor nodes and to perform a global objective. It also performs real-time scheduling for queries in WSN. A data subscription service in DSWare minimizes communication among sensor nodes.

DSWare provides a novel event-detection mechanism that is reliable and energy efficient. As described earlier, a compound event is assumed to include subevents that may be correlated, and its occurrence can be measured by a *confidence function*. The result of the confidence function is called *confidence*. When the confidence is greater than a threshold minimal confidence, a compound event is assumed to have occurred. But when a compound event occurs, it is possible that not all subevents have been detected [8.4]. DSWare sends a report only when a compound event is determined to have occurred. DSWare also uses the properties *phase* and *absolute validity interval* to improve detection reliability and improve energy efficiency. Each subevent may occur only in a certain phase. DSWare uses SQL language to register and cancel events. This is event-driven middleware for real-time applications. Network protocol selection in DSWare is static and independent of applications. DSWare has the disadvantage of not being able to capture application requirements [8.7].

8.4.5 CLMF (Cluster-Based Lightweight Middleware Framework)

CLMF [8.5] is a virtual machine with two layers: a resource management layer and a cluster layer. The cluster layer is distributed among all sensor nodes and includes cluster forming and control protocols. The code for resource management resides at the cluster head. The cluster layer needs to distribute from the cluster head commands for resource management and cluster control purposes. The resource management layer commands the allocation and adaptation of resources such that the QoS requirements specified by the applications can be satisfied [8.5]. CLMF proposes a three-phase heuristic algorithm for resource allocation and adaptation for a simple environment. In this algorithm, a set of homogeneous sensor nodes are connected by a single-hop wireless network, using dynamic voltage and modulation scaling techniques [8.19,8.20]. CLMF considers no routing protocol since it is homed on an existing network stack. CLMF provides only a framework. Resource management mechanism(s) (if any) in CLMF need further investigation.

8.4.6 MSM (Middleware Service for Monitoring)

MSM [8.7] operates between the transport layer and applications. It divides a WSN into two regions: dominant and nondominant. The dominant part contains a gateway acting as a central access point and provides connectivity to the transit network. The gateway is an intelligent coordinator that keeps a log of all activities in a sensor network. The MSM core components include data distribution and monitoring services. Core components are used to communicate among devices in WSN. The data distribution service distributes information among sensor nodes, and the monitoring service uses the data distribution service to monitor sensor nodes. MSM uses an object request broker (ORB) as the interface to connect transport layer protocols. The current MSM is not adjustable to a variety of applications. Also, it does not consider communication and energy efficiency thoroughly.

8.4.7 Em*

Em* [8.9] is a software environment for developing and deploying wireless sensor network applications on Linux-class hardware platforms. It incorporates tools and services to create WSN applications. Em* tools can be used to support deployment, simulation, emulation, and visualization of live systems, and its services include link and neighborhood estimation, time synchronization, and routing. Em* supports flooding-based, geographical, and quad-tree-based routing protocols, and supports many devices and a variety of radio hardware. It does not provide information on how to adapt or manage network resources while utilizing application knowledge.

8.4.8 Impala

Impala [8.10] is lightweight middleware and an API for sensor application adaptation and update which can improve system reliability and energy efficiency. It is event-driven middleware that achieves effective application adaptation. It is intended to act as an operating system, resource manager, and event filter on top of which specific applications can be installed and run. This WSN middleware contains three middleware agents: an application adapter, an application updater, and an event filter. The application adapter adapts applications to various runtime conditions in order to improve performance, energy efficiency, and robustness. The application updater receives and propagates software updates through the wireless transceiver and installs them on the sensor node. The event filter captures and dispatches events to the application adapter and updater, and initiates chains of processing. Impala has five types of events: timer, packet, send done, data, and device failure. Applications, the application adapter, and the application updater are all programmed into a set of event handlers which are invoked by the event filter when events are received. Impala supports both parameter- and device-based adaptations. An example of the application adapter is: When a device failure is detected, history-based protocol is switched to flooding protocol.

8.4.9 DFuse

DFuse [8.11] is proposed for programming fusion applications, and it is middleware only for data fusion. Data fusion focuses on decision making based on data and information that is acquired, filtered, and correlated with other relevant information. That process would involve information conversion into an appropriate format, which may be acquired from one or multiple sources. Data fusion of multiple sources usually reduces uncertainty, improves the reliability of event detection, and enhances system tolerance and robustness. When performed systematically with an appropriate application in mind, it reduces volume, improves QoS, and reduces energy consumption. In WSNs, data fusion may occur in the sink or sensor nodes. If the fusion point is closer to the geographical area where the data have been generated, the data filtering/aggregation efficiency could be higher. If the data fusion point is too close to the area, the fusion operation will be limited to a few sources and therefore will be less immune to undetected errors. Therefore, the data fusion point may be variable and dependent on such factors as system parameters, network status, and performance requirements.

As middleware, DFuse comprises a data fusion API and a distributed algorithm for energy-aware role assignment. It supports distributed data fusion with automatic management of fusion point placement and migration to optimize a given cost function [8.11]. Application programmers only need to implement the fusion functions and provide a data flow graph. DFuse is suitable for applications that are hierarchical fusion functions intended for deployment in a heterogeneous ad hoc sensor network environment. It offers four sample cost functions for moving the fusion point: (1) minimize transmission cost without node power considerations, (2) minimize power variance, (3) minimize the ratio of transmission cost to power, and (4) minimize transmission cost with node power considerations. DFuse provides a heuristic role assignment algorithm that works as follows: First, run and deploy a naive role assignment to the network nodes from the root node to the source node, then allow every node to decide locally if it wants to transfer the role to any of its neighbors [8.11].

8.4.10   DDS (Device Database System)

DDS [8.12] enables distributed query processing over a device network. It defines three types of queries: historical, snapshot, and long-running. DDS is more suitable for queries than are the traditional warehousing approaches. Each device is a miniserver capable of supporting a set of functions and able to process portions of the queries. This device capability results in improved aspects of query performance such as throughput, response time, resource use, and time delay. Due to its resource requirements, DDS is not effective in a resource-constraint WSN, yet it can be useful for WSNs without resource limitations, such as when the batteries can be recharged. DDS considers only the problem of queries.

8.4.11   SensorWare

SensorWare [8.13] provides a language and runtime environment to support WSN programming. The language model is used to implement distributed algorithms while hiding unnecessary details from the application programmer and to enable sharing node resources among several applications [8.13]. A distributed algorithm is a set of programs executed in a set of nodes. SensorWare calls these programs *mobile control scripts*. The scripts are defined at the node level and can be recognized by SensorWare at each node. SensorWare has event-driven behavior. It resides on the top of operating system and uses functions and services of the operating system.

SensorWare provides a compact runtime environment and script (180 kB). It targets a specific type of distributed algorithm for a collaborative signal processing task. It does not provide adaptation between applications and node resources or among applications. SensorWare has a fixed addressing scheme.

8.5 CONCLUSION

In this chapter the purpose of middleware in WSNs and several existing architectures were discussed. There are more architectures than those covered here, and some are very important. For example, a temporal adaptive next-generation query optimizer and processor (TANGO) is proposed by Slivinskas et al. [8.14] as middleware used on top of a conventional DBMS to optimize query performance. He et al. [8.15] proposed a programmable routing framework to provide a universal routing service through tunable parameters and programmable components. Wolenetz et al. [8.8] studied energy use and performance of DFuse-like [8.11] middleware and examined some guidelines on the design of middleware for WSNs.

Envisioning that the future sensor node will be resource-rich, new applications involving high data rates, complex processing, and strict QoS requirements will appear. Middleware for supporting such applications would comprise more functions, such as fusion, migration of fusion points, the ability to change the device behavior/operation mode dynamically, and effective adaptation between applications and sensor nodes. The reflective middleware described by Kon et al. [8.16] can be used to perform such tasks, since its components can be (re-)configured by the applications. Reflective middleware is flexible and can be adapted to changes in the environment and devices, and is therefore more suitable for WSNs.

REFERENCES

[8.1] W. B. Heinzelman, A. L. Murphy, H. S. Carvalho, M. A. Perillo, "Middleware to Support Sensor Network Applications," *IEEE Network*, Jan.–Feb. 2004, pp. 6–14.

[8.2] P. B. Gibbons, B. Karp, Y. Ke, S. Nath, S. Seshan, "IrisNet: An Architecture for a Worldwide Sensor Web," *IEEE Pervasive Computing*, Oct.–Dec. 2003, pp. 22–33.

[8.3] X. Yu, K. Niyogi, S. Mehrotra, N. Venkatasubramanian, "Adaptive Middleware for Distributed Sensor Environments," *IEEE Distributed Systems Online*, May 2003.

[8.4] S. Li, S. H. Son, J. Stankovic, "Event Detection Services Using Data Services Middleware in Distributed Sensor Networks," *Proceedings of the Workshop on Information Processing in Sensor Networks* (IPSN'03), Palo Alto, CA, Apr. 2003.

[8.5] Y. Yu, B. Krishnamachari, V. K. Prasanna, "Issues in Designing Middleware for Wireless Sensor Networks," *IEEE Network*, Jan.–Feb. 2004, pp. 15–21.

[8.6] K. Romer, O. Kasten, F. Mattern, "Middleware Challenges for Wireless Sensor Networks," *Mobile Computing and Communications Review*, Vol. 6, No. 4, October 2002, pp. 59–61.

[8.7] S. I. Ahamed, A. Vyas, M. Zulkernine, "Towards Developing Sensor Networks Monitoring as a Middleware Service," *Proceedings of the International Conference on Parallel Processing Workshops* (ICPPW'04), Montreal, Quebec, Canada, Aug. 2004, pp. 465–471.

[8.8] M. Wolenetz, R. Kumar, J. Shin, U. Ramachandran, "Middleware Guidelines for Future Sensor Networks," *Proceedings of the Workshop on Broadband Advanced Sensor Networks*, San Jose, CA, Oct. 2004.

[8.9] L. Girod, J. Elson, A. Cerpa, T. Stathopoulos, N. Ramanathan, D. Estrin, "Em*: A Software Environment for Developing and Deploying Wireless Sensor Networks," *Proceedings of the USENIX 2004 Annual Technical Conference*, Boston, MA, June–July 2004, pp. 283–296.

[8.10] T. Liu, M. Martonosi, "Impala: A Middleware System for Managing Autonomic Parallel Sensor Systems," *Proceedings of the 9th ACM SIGPLAN Symposium on Principles and Practice of Parallel Programming* (*PPoPP'03*), San Diego, CA, June 2003, pp. 107–118.

[8.11] R. Kumar, M. Wolenetz, B. Agarwalla, J. K. Shin, "DFuse: A Framework for Distributed Data Fusion," *Proceedings of the 1st ACM Conference on Embedded Networked Sensor Systems* (SenSys'03), Los Angeles, Nov. 2003, pp. 114–125.

[8.12] P. Bonnet, J. Gehrke, P. Seshadri, "Querying the Physical World," *IEEE Personal Communications*, Oct. 2000, pp. 10–15.

[8.13] A. Boulis, C. C. Han, M. B. Srivastava, "Design and Implementation of a Framework for Efficient and Programmable Sensor Networks," *Proceedings of the ACM International Conference on Mobile Systems, Applications, and Services* (MobiSys'03), San Francisco, CA, May 2003, pp. 187–200.

[8.14] G. Slivinskas, C. S. Jensen, R. T. Snodgrass, "Adaptable Query Optimization and Evaluation in Temporal Middleware," *Proceeding of the ACM SIGMOD*, Santa Barbara, CA, May 2001, pp. 127–138.

[8.15] Y. He, C. S. Raghavendra, S. Berson, B. Braden, "A Programmable Routing Framework for Autonomic Sensor Networks," *Proceedings of the Autonomic Computing Workshop at the 5th Annual International Workshop on Active Middleware Services* (AMS'03), Seattle, WA, June 2003, pp. 60–68.

[8.16] F. Kon, F. Costa, G. Blair, R. H. Campbell, "The Case for Reflective Middleware," *Communications of the ACM*, Vol. 45, No. 6, June 2002, pp. 33–38.

[8.17] P. A. Bernstein, "Middleware: A Model for Distributed System Services," *Communications of the ACM*, Vol. 39, No. 2, Feb. 1996, pp. 86–98.

[8.18] D. G. Schmidt, "Middleware for Real-Time and Embedded Systems," *Communications of the ACM*, Vol. 45, No. 6, June 2002, pp. 43–48.

[8.19] M. Weiser et al., "Scheduling for Reduced CPU Energy," *Proceedings of the USENIX Symposium on Operating Systems Design and Implementation*, Nov. 1994, pp. 13–23.

[8.20] B. Prabhakar, E. Uysal-Biyikoglu, A. E. Gamal, "Energy-Efficient Transmission over a Wireless Link via Lazy Packet Scheduling," *Proceedings of the 20th Annual Joint Conference of the IEEE Computer and Communications Societies* (InfoCom'01), 2001.

[8.21] W. Shi, S. Sellmuthu, K. Sha, L. Schwiebert, "QueryAgent: A General Query Processing Tool for Sensor Networks," *Proceedings of the International Conference on Parallel Processing Workshops* (ICPPW'04), Montreal, Quebec, Canada, Aug. 2004, pp. 488–495.

[8.22] C. Intanagonwiwat, R. Govindan, D. Estrin, "Directed Diffusion: A Scalable and Robust Communication Paradigm for Sensor Networks," *Proceedings of the 6th ACM/IEEE International Conference on Mobile Computing and Networking*, (MobiCom'00), Boston, MA, Aug. 2000, pp. 56–67.

[8.23] F. Ye, H. Luo, J. Cheng, S. Lu, L. Zhang, "A Two-Tier Data Dissemination Model for Large-Scale Wireless Sensor Networks," *Proceedings of the 8th ACM International Conference on Mobile Computing and Networking* (MobiCom'02), Atlanta, GA, Sept. 2002.

[8.24] S. Kim, S. H. Son, J. A. Stankovic, Y. Choi, "Data Dissemination over Wireless Sensor Networks," *IEEE Communication Letters*, Vol. 8, No. 9, Sept. 2004, pp. 561–563.

[8.25] S. S. Pradhan, J. Kusuma, K. Ramchandran, "Distributed Compression in a Dense Microsensor Network," *IEEE Signal Processing*, Mar. 2002, pp. 51–60.

[8.26] D. Slepian, J. K. Wolf, "Noiseless Coding of Correlated Information Sources," *IEEE Transactions on Information Theory*, Vol. 19, July 1973, pp. 471–480.

[8.27] S. Madden, M. J. Franklin, J. Hellerstein, W. Hong, "TAG: A Tiny Aggregation Service for Ad-Hoc Sensor Networks," *ACM SIGOPS Operating Systems Review*, Vol. 36, Issue SI (Winter 2002), pp. 131–146.

[8.28] S. Shenker, S. Ratnasamy, B. Karp, R. Govindan, D. Estrin, "Data-Centric Storage In Sensornets," *Proceedings of the 1st ACM SIGCOMM Workshop on Hot Topics in Networks*, Oct. 2002.

[8.29] J. Ledlie, C. Ng, D. A. Holland, "Provenance-Aware Sensor Data Storage," *Online Manuscript*, Dec. 2004.

[8.30] D. Ganesan, B. Greenstein, D. Perelyyubskiy, D. Estrin, J. Heidemann, "An Evaluation of Multi-resolution Storage for Sensor Networks," *Proceedings of the 1st ACM Conference on Embedded Networked Sensor Systems* (SenSys'03), Los Angeles, Nov. 2003.

9

NETWORK MANAGEMENT FOR WIRELESS SENSOR NETWORKS

9.1 INTRODUCTION

In this chapter we first briefly introduce traditional network management models, then identify issues and requirements of a network management system for WSNs and present an existing network management architecture. Issues of naming and localization as they relate to WSN management are also discussed.

9.2 NETWORK MANAGEMENT REQUIREMENTS

A computer communication network generally consists of three components: physical devices, including links (wireless or wired link), network nodes (hub, bridge, switch, or router), and terminals and servers; protocol; and information that is being carried, including applications. Protocols are used to transport information efficiently, preferably in a correct, secure, reliable, and understandable manner. They consist of a set of software residing at physical devices. The collaboration of physical devices and network protocols forms the underpinning support for the applications. However, the physical devices and protocols are not sufficient to support effective operation of a communications network; network management (NM) tools and techniques are also required to help provision network services and ensure cooperation of entities in the network. In general,

Wireless Sensor Networks: Technology, Protocols, and Applications, by Kazem Sohraby, Daniel Minoli, and Taieb Znati
Copyright © 2007 John Wiley & Sons, Inc.

the reasons for management functions are manifold and may be summarized as follows:

1. There are many heterogeneous devices and software entities that comprise the network, and some may fail. It is the NM responsibility to determine when, where, and why the fault had occurred and how to restore these entities.
2. Optimization of system performance as a distributed system require NM to collaborate in the process. For example, in some networks, congestion control through admission control, by changing routes, or through device upgrade occurs by NM functions.
3. For most networks, NM functions can be used to gather and analyze the behavior of user interaction during network interface, which is very important in planning the long-term evolution of network capacity and its performance.

Generally speaking, network management consist of a set of functions to monitor network status, detect network faults and abnormalities, manage, control, and help configure network components, maintain normal operation, and improve network efficiency and application performance. To perform these tasks, NM needs to collect real-time information in network devices, analyze the information, and apply control based on the information. Information is often organized as a management information base (MIB) in each network device. Usually, there is an agent in each device to collect the information and report to a network management center that has a view of the entire network information. Therefore, network management can be considered as an application.

9.3   TRADITIONAL NETWORK MANAGEMENT MODELS

9.3.1   Simple Network Management Protocol

The simple network management protocol (SNMP) for managing networks is in broad use today. It includes three components: a network management system (NMS), managed elements, and agents. NMS is a set of applications that monitor and/or control managed elements. It can request management information (or attributes) from the agent and present the results to NM users in the form of figures or tables. It can also set attributes within the agent. The managed elements are the network devices that are managed. SNMP agents run on each managed element. The managed elements collect and store management information in the MIB and provide access through SNMP to the MIBs. Examples of managed elements include routers, switches, servers, and hosts. SNMP agents are management software modules that reside on managed elements. Agents collect and store the state of the managed elements and translate this information into a form compatible with SNMP MIB. Exchanges of network management information are through messages called protocol data units (PDUs). These are sent to nodes and contain variables that have both attributes and values. The SNMP defines five types of messages or PDUs: Two deal with the reading terminal, another two handle terminal configuration, and the fifth is *Trap*, used to monitor events in the managed elements. Each PDU contains both attributes and values. NM information can be exchanged through the PDUs in order to monitor the managed elements.

An advantage of SNMP is its simplicity and wide deployment. However, it consumes considerable bandwidth since it often gets only one piece of management information at a time: *GetRequest* (*GetNextRequest*) and *GetResponse*. Although in SNMP version 3 it can obtain more information by a pair of PDUs such as (*GetBulkRequest* and *GetResponse*), due to the usually large number of managed elements, large bandwidth consumption still exists. The other disadvantage of SNMP is that it only manages network elements; it does not support network-level management.

9.3.2   Telecom Operation Map

The telecom operation map (TOM), proposed by TeleManagement Forum [9.15], is based on the service management and network management process models. TOM presents a model for telecommunications management for network and service management and a view of "operations." The idea behind TOM is to introduce processes comprising operations and their automation. There are three vertical layers for service management: network and systems management, service development and operations, and customer care process. Horizontally, the service management is divided in service fulfillment, service assurance, and service billing. TOM only provides a framework for service management.

Neither SNMP nor TOM is designed particularly for wireless sensor networks. However, one can utilize the simplicity of SNMP and the layered framework of TOM to design effective and efficient network management architecture for wireless sensor networks as well.

9.4   NETWORK MANAGEMENT DESIGN ISSUES

WSN is a special type of wireless network, possibly with ad hoc structure and probably with limited resources. Due to these WSN constraints, networking protocols, the application model, middleware, and sensor node operating systems should be designed very carefully. Network management for WSNs is required to use those limited resources effectively and efficiently. Network management is much more important for WSNs than for traditional networks for the following reasons:

1. In order to deploy an adaptive and resource-efficient algorithm in WSNs, the current resource level needs to be gathered through network management. For

example, the power availability should be known before switching a sensor node from active (or sleep) mode to sleep (or active) mode. Most traditional networks do not have these requirements.

2. Most WSN applications need to know the coverage area so that they ensure that the entire space is being monitored. Topology management can be used in case an uncovered area is detected. Generally, there are three approaches to increasing the coverage area: (1) increase the node's radio power, (2) increase the density of deployment of senor nodes, and (3) move the sensor nodes around to achieve equal distribution.

3. Nodes in WSNs are usually arranged in an ad hoc manner. The parameters of this ad hoc network are obtained by the network management system.

4. Collaboration and cooperation between sensor nodes are required to optimize system performance. Network management is an effective tool to provide the platform required for this purpose.

So far, very little attention has been paid to the management of WSNs. An issue is whether in the meanwhile, any of the existing network management solutions (e.g., SNMP [9.14], TOM [9.15]) can be used for WSNs. SNMP is often used to manage network elements such as switches and routers. It uses *GetResponse* and *GetResponse* PDUs to collect information from network elements. In SNMP, a local management agent should run in each managed element. The local agent is a static and passive agent that receives commands from a manager and returns the corresponding response. It can also issue *Trap* messages to the manager when the managed element encounters a preconfigured event. Agents in different network elements are independent, and there is not collaboration among them. TOM is a new operation and management model that provides a layered architecture for management and administration. Each layer has a different management function and set of managed objects. TOM can be used to manage most tasks, from the underlying physical network element to the entire network, as well as the services provided. However, SNMP is just a simple protocol that only manages network elements. Given that WSNs are data centric, resource constrained, and ad hoc, SNMP and TOM, which were designed for traditional networks, may not provide the right tool.

Several issues must be addressed carefully before designing network management tools for WSNs. To begin with, the management functions required for WSNs should first be identified. SNMP provides five management functions: fault management, configuration management, accounting management, performance management, and security management; and in TOM, the management functions are layered in network element management, network management, and service management. In each layer, different management functions are embodied. WSNs need some of these management functions. Therefore, WSNs need layered management architecture with different management functions in each layer. For example, WSNs do not need all the capabilities of the five basic management functions, such as billing capability or the accounting management function. WSNs might require new management capabilities: for example, network coverage information and a sensor node power distribution map. WSNs also need energy-efficient key management algorithms [9.13] for security. WSNs also need new management functions for data management [9.4,9.10] since the type and purpose of data collected in WSNs is quite different from those of the traditional networks.

The issue of management architecture for WSNs should also be considered carefully. A network management platform consists of three major components: manager, agent, and MIB. The manager is used to manage and control the entire network and works as an interface to other systems. The agent is located in managed elements. MIB is an object-oriented structured tree that informs the manager and agent about the organization of management information. A standardized MIB guarantees that the management products from different vendors interconnect. The manager receives management information and commands the managed elements using a SNMP-like method or mobile-agent-based entities [9.9]. Sometimes a network management system would include several distributed managers, each of which manages part of the entire network. The method of accessing management information and the placement of the manager or agent usually determines the management architecture. The agent-based method can save bandwidth since it can report only final management information. Although WSNs have a centralized data collecting point (sink), they are more like distributed networks. As a result, agent-based hybrid management architectures might be more suitable for WSNs.

In WSNs, management information can be used to improve system performance. For example, if the network management system detects a dysfunctional sensor node, it can command another sensor node to take over. So the issue of integration of network management with the functions of network protocols and algorithms becomes critical.

Network management functions should therefore consider all the special features of WSNs. Some of these considerations follow:

- Management solutions should be energy efficient, using as little wireless bandwidth as possible since communication is highly energy demanding.
- Management solutions should be scalable. This is especially important since it future WSNs may consist of tens to thousands of nodes.
- Management solutions should be simple and practical since WSNs are resource-constrained distributed systems.
- MIB for WSNs should contain a general information model for sensor nodes, features of WSNs, and WSN applications.
- Management solutions for WSNs should provide a general interface to the applications since applications can perform better when able to access management information.
- Management solutions should be implementable as middleware.

9.5   EXAMPLE OF MANAGEMENT ARCHITECTURE: MANNA

Several references, notably [9.1–9.13], have extensive discussions and some results on WSN network management. Specifically, [9.1] has an initial discussion of the topics, management architecture is discussed in [9.2] and [9.3], monitoring management in [9.5], resource management is discussed in [9.6] and [9.11], secure management in [9.7] and [9.13], topology management in [9.8], and data stream management in [9.10]. An optimization problem for monitoring management formulated in [9.5] provides the monitoring regions given that the battery and energy consumption rate for each sensor are known beforehand. In topology management scheme called sparse topology and energy management (STEM) proposed in [9.8], the nodes only need to be awake when there is data to forward. Golab and Ozsu [9.10] present an overview of data stream management.

MANNA is a management architecture for WSNs proposed by Ruiz et al. [9.2,9.3]. The architecture considers three management dimensions: function areas, management levels, and WSN functionalities (see Figure 9.1). The management function areas contain five types of traditional management functions similar to SNMP: fault, configuration, performance, security, and accounting management. But configuration management has a notably more important role in MANNA, where all other functions depend on it. The management levels in MANNA are similar to those in TOM: network element, network element management, network management, service management, and business management. A number of other functions are proposed by MANNA: configuration, maintenance, sensing, processing, and communication. With the aim of promoting productivity and integrating the functions of configuration, operation, administration, and maintenance of all

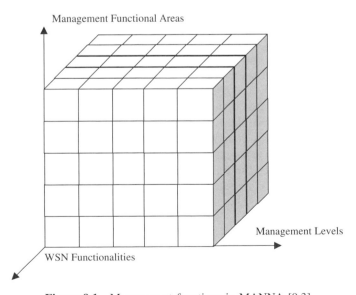

Figure 9.1   Management functions in MANNA [9.3].

elements and services in a WSN, MANNA architecture includes three architectural elements: functional, physical, and informational architectures. The functional architecture provides functions executed in the management entities (manager, agent, and MIB) and the location scheme for managers and agents. The physical architecture is where functional architecture is implemented. MANNA uses a light-weight protocol as a communication interface between management entities. The information architecture element provides an object-oriented model for mapping manageable resources and supporting object classes. MANNA defines the following managed object classes: (1) network (information on network behavior and features such as data delivery model, network structure, and mobility), (2) managed elements (such as sensor nodes), (3) equipment (the physical components of sensor nodes), (4) system (information on operating system), (5) environment (the environment the WSN is running), (6) phenomenon, and (7) connection.

MANNA lists several common management functions for WSNs: environment monitoring functions, a coverage area supervision function, a topology map discovery function, an energy-level discovery function, an energy map generation function, and several others. It also provides a dynamic MIB model for WSNs: a sensing coverage area map, a communication coverage area map, a WSN behavior model, a node dependence model, network topology, residual energy, and so on. In MANNA, the management functions have the lowest granularity and can be combined into management services.

9.6   OTHER ISSUES RELATED TO NETWORK MANAGEMENT

There are several other issues related to sensor network management, the most important being naming, localization, maintenance, and fault tolerance. Naming is the scheme used to identify a sensor node. An efficient naming scheme can lower computation overhead and make routing protocol energy efficient. Localization schemes determine the location of sensor nodes since such information is important for some sensor applications. The maintenance issue may involve actions such as replacing batteries, keeping connectivity [9.18], and configuring sensor nodes. The maintenance activity is used to maintain normal operation of the entire network for as long as possible. A maintenance model is discussed in [9.19]. Several factors can cause faults in network operation, including hardware and software error. Therefore, different schemes must be implemented to provide fault tolerance. Hardware backup schemes can be used to overcome hardware problems. Software techniques can be used to provide fault detection and fault tolerance for hardware. For example, multipath routing or provisioning of redundant connections can guarantee network connectivity when a node is not operational. A fault tolerance technique has been proposed in [9.2] for wireless sensor networks using multimodal sensor fusion. Using multimodal sensor fusion and a suitable resource allocation algorithm, fault tolerance can be provided at the cost of hardware backup. For WSNs, the networking protocol and algorithms should be capable of providing fault tolerance. Naming and localization are discussed below in more detail.

9.6.1 Naming

A node in a networked system is identified through naming. This identifier is then used for communication between nodes. Generally, there are two traditional approaches to naming: low level and high level. *Low-level naming* such as node addresses is typically application independent but topology and location dependent. On the other hand, *high-level naming* is usually application dependent and location independent. High-level naming is built on the top of low-level naming. Communication between applications uses high-level naming only, whereas physical communication relies on low-level naming. Therefore, a binding mechanism is required to realize mapping between high- and low-level naming. For example, the domain name system (DNS) in the Internet uses two types of naming: a domain name and an IP address. The domain name is used by applications such as Internet browers, and the IP address is used by routing protocol to guarantee packet forwarding. Domain name and IP addresses are often directed to the same host. The DNS servers map between domain names and IP addresses. When a Web site is accessed using a domain name, the application program requests a corresponding IP address from the DNS so as to set up low-level communications.

Although the traditional hierarchical naming approaches can be used for wireless sensor networks, those approaches are not efficient compared with application-oriented low-level naming [9.16], which has the following advantages:

- It avoids the overhead resulting from mapping between high- and low-level naming. This feature is attractive for a sensor since it has limited resources.
- Location-dependent addressing is not required. Since the topology of WSNs is highly variable due to node mobility, node life span, and wireless channel quality, a location-dependent address would cause additional problems.
- It enables application-specific processing in the network, such as data compression and data fusion, which in turn reduces data transmission.

Sensor nodes are usually classified by the type of data they gather. For sensor nodes that gather only one type of data or can have differing personalities and gather multiple types of data, one name as their identifier would be sufficient. The objective of low-level application naming is to realize energy efficiency and fault tolerance in a variety of environments.

9.6.2 Localization

Sensor nodes are distributed all over the place for sensing and data collection. It is usually helpful if the locations of sensor nodes are also known. Advantages of this knowledge are that (1) some applications, such as those for tracking of objects, are highly location dependent; (2) location-based routing, which may also result in energy conservation is enabled; (3) knowledge of location usually enhances security; (4) locations are helpful for sensor network management and monitoring;

(5) locations stimulate the creation of new applications; (6) sensor nodes that move can be controlled through knowledge of their location; and (7) for applications with low-level naming and/or data-centric WSNs, knowledge of location information is absolutely necessary.

Although global positioning systems (GPSs) can provide precise location information, deployment of a GPS receiver in every sensor node is expensive and unaffordable for most WSN applications. Non-GPS localization schemes are more practical for WSNs. The existing non-GPS approaches are either hardware or topology dependent [9.17]. Hardware-dependent algorithms need sensor hardware to provide information such as signal strength. Topology-dependent algorithms for localization do not need hardware support but do require support from special "seed nodes," with exact knowledge of their location.

On the other hand, localization algorithms can be classified into centralized and distributed schemes. In the centralized scheme, sensor nodes send control messages to a central node whose location is known. The central node then computes the location of every sensor node and informs the nodes of their locations. In the distributed scheme, each sensor node determines its own location independently. The distributed localization can be further grouped into range-based and range-free schemes. In the range-based approach, some range information, such as time of arrival, angle of arrival, or time difference of arrival is required. The range-free algorithms works as follows: Several seed nodes are distributed in WSNs. Seed nodes know their own locations, and they periodically broadcast a control message with their location information. Sensor nodes that receive these control messages can then estimate their own locations.

9.7 CONCLUSION

In this chapter we discussed network management for wireless sensor networks, including traditional network management models such as SNMP and TOM. Then issues and requirements of network management system for WSNs were identified. Finally, an overview of MANNA as an example of NM for WSN was provided. Management functions provide a major challenge to the design of WSN NM. This includes an effective and practical management architecture, an effective MIB, and an approach to utilize network management to increase productivity. The final objectives of management are to prolong the life span of WSNs and to guarantee the performance of their applications.

REFERENCES

[9.1] F. Wang, Q. Tian, Q. Gao, Q. Pan, "A Study of Sensor Management Based on Sensor Networks," *Proceedings of the 2003 IEEE International Conference on Robotics, Intelligent Systems and Signal Processing*, Changsha, China, Oct. 2003, pp. 1058–1062.

[9.2] L. B. Ruiz, F. A. Silva, T. R. M. Braga, J. M. S. Nogueira, A. A. F. Loureiro, "On Impact of Management in Wireless Sensor Networks," *Proceedings of the IEEE/IFIP Network Operations and Management Symposium* (NOMS'04), Vol. 1, April 2004, pp. 19–23.

[9.3] L. B. Ruiz, J. M. Nogueira, A. A. F. Loureiro, "MANNA: A Management Architecture for Wireless Sensor Networks," *IEEE Communications*, Feb. 2003, pp. 116–125.

[9.4] J. C. Navas, M. Wynblatt, "The Network Is the Database: Data Management for Highly Distributed Systems," *Proceedings of the ACM SIGMOD*, Santa Barbara, CA, May 2001, pp. 544–551.

[9.5] P. Berman, G. Calinescu, C. Shah, Z. Zelikovsky, "Power Efficient Monitoring Management in Sensor Networks," *Proceedings of the IEEE Wireless Communications and Networking Conference* (WCNC'04), Mar. 2004, pp. 2329–2334.

[9.6] J. Zhang, K. Premaratne, P. H. Bauer, "Local Resource Management of Distributed Sensor Networks Via Static Output Feedback Control," *Proceedings of the IEEE International Symposium on Circuits and Systems* (ISCAS), Vol. 3, 2002, pp. 25–28.

[9.7] J. Zachary, "A Decentralized Approach to Secure Management of Nodes in Distributed Sensor Networks," *Proc. of IEEE MILCOM 2003*, Vol. 1, Oct. 2003, pp. 579–584.

[9.8] C. Schurgers, V. Tsiatsis, S. Ganeriwal, M. Srivastava, "Topology Management for Sensor Networks: Exploiting Latency and Density," *Proceedings of the 2nd ACM International Symposium on Mobile Ad Hoc Networking and Computing* (MobiHoc'02), EPFL, Lausanne, Switzerland, June 2002, pp. 135–145.

[9.9] A. Bivens, R. Gupta, I. Mclean, B. Szymanski, J. White, "Scalability and Performance of an Agent-Based Network Management Middleware," *International Journal of Network Management*, 2004, pp. 131–146.

[9.10] L. Golab, M. T. Ozsu, "Issues in Data Stream Management," *Proceedings of the ACM SIGMOD Record*, Vol. 32, No. 2, June 2003, pp. 5–14.

[9.11] J. Liu, P. Cheung, L. Guibas, F. Zhao, "A Dual-Space Approach to Tracking and Sensor Management in Wireless Sensor Networks," *Proceedings of the 1st Workshop on Sensor Networks and Applications* (WSNA'02), Atlanta, GA, Sept. 2002, pp. 131–139.

[9.12] B. Tierney, B. Crowley, D. Gunter, M. Holding, J. Lee, M. Thompson, "A Monitoring Sensor Management System for Grid Environments," *Proceeding. of the 9th International Symposium on High-Performance Distributed Computing,* 2000, pp. 97–104.

[9.13] L. Eschenauer, V. D. Gligor, "A Key-Management Scheme for Distributed Sensor Networks," *Proceedings of the 9th ACM Conference on Computer and Communications Security* (CCS'02), Washington, DC, Nov. 2002, pp. 41–47.

[9.14] J. Case et al., "Simple Network Management Protocol," *RFC 1157*, IETF, Sterling, VA, May 1990.

[9.15] TeleManagement Forum Specification, Telecomm Operations Map (GB910), Approved Version 2.1, March 2000.

[9.16] J. Heidemann et al., "Building Efficient Wireless Sensor Networks with Low-Level Naming," *Proceedings of the 18th ACM Symposium on Operating Systems Principle* (SOSP'01), Banff, Alberta, Canada, pp. 146–159.

[9.17] L. Hu, D. Evans, "Localization for Mobile Sensor Networks," *Proceedings of the 10th ACM International Conference on Mobile Computing and Networking* (MobiCom'04), Philadelphia, PA, Sept. 2004.

[9.18] D. Tian, N. D. Georganas, "Connectivity Maintenance and Coverage Preservation in Wireless Sensor Networks," *Canadian Conference on Electrical and Computer Engineering*, Vol. 2, May 2004, pp. 1097–1100.

[9.19] A. Barroso, U. Roedig, C. J. Sreenan, "Maintenance Awareness in Wireless Sensor Networks," Short Paper, *Proceedings. of the 1st European Workshop on Wireless Sensor Networks* (EWSN), Jan. 2004.

[9.20] F. Koushanfar, M. Potkonjak, A. S. Vincentelli, "Fault Tolerance Techniques for Wireless Ad Hoc Sensor Networks," *Proceedings. of IEEE Sensors 2002*, Vol. 2, 2002, pp. 1491–1496.

10

OPERATING SYSTEMS FOR WIRELESS SENSOR NETWORKS

10.1 INTRODUCTION

WSNs can be used to monitor and/or control physical environment in a space where it is difficult or impossible to do so manually. A WSN is generally composed of a centralized station (sink) and tens, hundreds, or perhaps thousands of tiny sensor nodes such as Mote [10.1] and Mica2 [10.1]. With the integration of information sensing, computation, and wireless communication, these devices can sense the physical phenomenon, (pre-)process the raw information, and share the processed information with their neighboring nodes. The sensor nodes can form a WSN either ad hoc or with, for example, a cluster-based architecture. The sink node can query information and sometimes control the behavior of the sensor nodes. The information is often unidirectional flow from the sensor nodes to the sink. Since a WSN has a centralized sink and unidirectional information flow, it acts as a centralized system. But the sensor nodes are distributed and behave collaboratively. At the same time, a WSN is not only a database system but also a resource-constrained network with most networking functions, so they are often used to monitor events and collect data. Therefore, the environment is event driven and data centric. Therefore, WSNs are a special type of distributed network system that is similar to database, real-time, and embedded systems.

Wireless Sensor Networks: Technology, Protocols, and Applications, by Kazem Sohraby, Daniel Minoli, and Taieb Znati
Copyright © 2007 John Wiley & Sons, Inc.

The basic function of WSNs is to collect information and to support certain applications specific to the task of WSN deployment. Commercially available sensor nodes are categorized into four groups [10.13]:

1. Specialized sensing platforms such as the Spec [10.13] node designed at the University of California–Berkeley. This sensor node has a single chip with low-power low-cost, operation.
2. Generic sensing platforms such as Berkeley motes [10.1]. This node can perform generic sensing tasks.
3. High-bandwidth sensing platforms such as iMote [10.13]. This node can handle sensed data flow with high bandwidth.
4. Gateway platforms such as Stargate [10.13]. This node can be used as a sink and can connect low-level senor nodes directly to the Internet.

The differences in the sensor types above are in the function of the sensor, frequency of the microprocessor, memory size, and transceiver bandwidth. Although these nodes have different characteristics, their basic hardware components are the same: a physical sensor, a microprocessor or microcontroller, a memory, a radio transceiver, and a battery. Therefore, these hardware components should be organized in a way that makes them work correctly and effectively without a conflict in support of the specific applications for which they are designed. Each sensor node needs an operating system (OS) that can control the hardware, provide hardware abstraction to application software, and fill in the gap between applications and the underlying hardware.

The traditional OS is system software that operates between application software and hardware and is often designed for workstations and PCs with plenty of resources. This is usually not the case with sensor nodes in WSNs. There are also embedded operating systems such as VxWorks [10.21] and WinCE [10.22], none of which is specially designed for data-centric WSNs with constrained resources. Sensors usually have a slow processor and small memory, different from most current systems. In this chapter, parameters that should be kept in mind in the process of OS design for WSN nodes are considered.

10.2 OPERATING SYSTEM DESIGN ISSUES

Traditional operating systems [10.17,10.20] are system software, including programs that manage computing resources, control peripheral devices, and provide software abstraction to the application software. Traditional OS functions are therefore to manage processes, memory, CPU time, file system, and devices. This is often implemented in a modular and layered fashion, including a lower layer of kernels and a higher layer of system libraries. Traditional OSs are not suitable for wireless sensor networks because WSNs have constrained resources and diverse data-centric applications, in addition to a variable topology. WSNs need a new type of operating system, considering their special characteristics. There are several issues to consider when designing operating systems for wireless sensor networks. The first issue is process management and scheduling. The traditional OS provides process protection by allocating a separate memory space (stack) for each process. Each process maintains data and information in its own space. But this approach usually causes multiple data copying and context switching between processes. This is obviously not energy efficient for WSNs. For some real-time applications in WSNs, a real-time scheduler such as *earliest deadline first* (EDF) or its variants may be a good choice, but the number of processes should be confined since that would determine the time complexity of the EDF scheduler.

The second issue is memory management. Memory is often allocated exclusively for each process/task in traditional operating systems, which is helpful for protection and security of the tasks. Since sensor nodes have small memory, another approach, sharing, can reduce memory requirements.

The third issue is the kernel model. The event-driven and finite state machine (FSM) models have been used to design microkernels for WSNs. The event-driven model may serve WSNs well because they look like event-driven systems. An event may comprise receiving a packet, transmitting a packet, detection of an event of interest, alarms about energy depletion of a sensor node, and so on. The FSM-based model is convenient to realize concurrency, reactivity, and synchronization.

The fourth issue is the application program interface (API). Sensor nodes need to provide modular and general APIs for their applications. The APIs should enable applications access the underlying hardware.

The fifth issue is code upgrade and reprogramming. Since the behavior of sensor nodes and their algorithms may need to be adjusted either for their functionality or for energy conservation, the operating system should be able to reprogram and upgrade.

Finally, because sensor nodes generally have no external disk, the operating system for WSNs cannot have a file system. These issues should be considered carefully in the design of WSN OSs and to meet their constrained resources, network behavior, and data-centric application requirements.

Sensor operating systems (SOS) should embody the following functions, bearing in mind the limited resource of sensor nodes:

1. Should be compact and small in size since the sensor nodes have very small memory. The sensor nodes often have memories of only tens or hundreds of kilobytes.
2. Should provide real-time support, since there are real-time applications, especially when actuators are involved. The information received may become outdated rather quickly. Therefore, information should be collected and reported as quickly as possible.
3. Should provide efficient resource management mechanisms in order to allocate microprocessor time and limited memory. The CPU time and limited memory must be scheduled and allocated for processes carefully to guarantee fairness (or priority if required).

4. Should support reliable and efficient code distribution since the functionality performed by the sensor nodes may need to be changed after deployment. The code distribution must keep WSNs running normally and use as little wireless bandwidth as possible.

5. Should support power management, which helps to extend the system lifetime and improve its performance. For example, the operating system may schedule the process to sleep when the system is idle, and to wake up with the advent of an incoming event or an interrupt from the hardware.

6. Should provide a generic programming interface up to sensor middleware or application software. This may allow access and control of hardware directly, to optimize system performance.

10.3   EXAMPLES OF OPERATING SYSTEMS

10.3.1   TinyOS

The design of TinyOS [10.1,10.3] allows application software to access hardware directly when required. TinyOS is a tiny microthreaded OS that attempts to address two issues: how to guarantee concurrent data flows among hardware devices, and how to provide modularized components with little processing and storage overhead. These issues are important since TinyOS is required to manage hardware capabilities and resources effectively while supporting concurrent operation in an efficient manner. TinyOS uses an event-based model to support high levels of concurrent application in a very small amount of memory. Compared with a stack-based threaded approach, which would require that stack space be reserved for each execution context, and because the switching rate of execution context is slower than in an event-based approach, TinyOS achieves higher throughput. It can rapidly create tasks associated with an event, with no blocking or polling. When CPU is idle, the process is maintained in a sleep state to conserve energy.

TinyOS includes a tiny scheduler and a set of components. The scheduler schedules operation of those components. Each component consists of four parts: command handlers, event handlers, an encapsulated fixed-size frame, and a group of tasks [10.3]. Commands and tasks are executed in the context of the frame and operate on its state. Each component will declare its commands and events to enable modularity and easy interaction with other components. The current task scheduler in TinyOS is a simple FIFO mechanism whose scheduling data structure is very small, but it is power efficient since it allows a processor to sleep when the task queue is empty and while the peripheral devices are still running. The frame is fixed in size and is assigned statically. It specifies the memory requirements of a component at compile time and removes the overhead from dynamic assignment [10.1]. Commands are nonblocking requests made to the low-level components. Therefore, commands do not have to wait a long time to be executed. A command provides feedback by returning status indicating whether it was successful (e.g., in the case of buffer overrun or of timeout). A command often stores request parameters into its frame and conditionally assigns a task for later execution. The occurrence of a hardware event will invoke event handlers. An event handler can store information in its frame, assign tasks, and issue high-level events or call low-level commands. Both commands and events can be used to perform a small and usually fixed amount of work as well as to preempt tasks. Tasks are a major part of components. Like events, tasks can call low-level commands, issue high-level events, and assign other tasks. Through groups of tasks, TinyOS can realize arbitrary computation in an event-based model. The design of components makes it easy to connect various components in the form of function calls.

This WNS operating system defines three type of components: hardware abstractions, synthetic hardware, and high-level software components. Hardware abstraction components are the lowest-level components. They are actually the mapping of physical hardware such as I/O devices, a radio transceiver, and sensors. Each component is mapped to a certain hardware abstraction. Synthetic hardware components are used to map the behavior of advanced hardware and often sit on the hardware abstraction components. TinyOS designs a hardware abstract component called the radio-frequency module (RFM) for the radio transceiver, and a synthetic hardware component called *radio byte*, which handles data into or out of the underlying RFM.

An evaluation of TinyOS shows that it achieves the following performance gains or advantages:

- It requires very little code and a small amount of data.
- Events are propagated quickly and the rate of posting a task and switching the corresponding context is very high.
- It enjoys efficient modularity.

10.3.2 Mate

Mate [10.2] is designed to work on the top of TinyOS as one of its components. It is a byte-code interpreter that aims to make TinyOS accessible to nonexpert programmers and to enable quick and efficient programming of an entire sensor network. Mate also provides an execution environment, which is helpful for the UC–Berkeley mote (see Chapter 7 for an overview of the mote) since in this system there is no hardware protection mechanism. In Mate, a program code is made up of capsules. Each capsule has 24 instructions, and the length of each instruction is 1 byte. The capsules contain type and version information, which makes code injection easy. Mate capsules can deploy themselves into the network. Mate implements a beaconless (BLESS) ad hoc routing protocol as well as the ability to implement new routing protocols. A sensor node that receives a newer version of a capsule installs it. Through hop-by-hop code injection, Mate can program the entire network. Capsules are classified into four categories: message send, message receive, timer, and subroutine. An event can trigger Mate to run. It can be used not only as a virtual machine platform for application development, but also as a tool to manage and control the entire sensor network.

10.3.3. MagnetOS

MagnetOS [10.4] is a distributed adaptive operating system designed specifically for application adaptation and energy conservation. Other operation systems do not provide a network-wide adaptation mechanism or policies for application to effectively utilize the underlying node resources. The burden of creating adaptation mechanisms (if any) is on the application itself. This approach is usually not energy efficient. The goals of MagnetOS are (1) to adapt to the underlying resource and its changes in a stable manner, (2) to be efficient with respect to energy conservation, (3) to provide general abstraction for the applications, and (4) to be scalable for large networks.

MagnetOS is a single system image (SSI) or a single unified Java virtual machine that includes static and dynamic components. The static components rewrite the application in byte-code level and add necessary instructions on the semantics of the original applications. The dynamic components are used for application monitoring, object creation, invocation, and migration. SSI abstraction provides more freedom in object placement and simplifies application development. MagnetOS provides an interface to programmers for explicit object placement and override of the automatic object placement decisions. This OS also provides two online power-aware algorithms (NetPull and NetCenter) for use in moving application components within the entire network so as to reduce energy consumption and extend network lifetime. Netpull works hop by hop at the physical layer, and NetCenter runs multihop at the network level. The difference between traditional ad hoc routing and NetPull (NetCenter) is that the communication endpoints in ad hoc routing are fixed, whereas NetPull tries to move the communication endpoints in order to conserve energy [10.4].

10.3.4 MANTIS

MANTIS [10.5] is a multithread embedded operating system, which with its general single-board hardware enables flexible and fast deployment of applications. With the key goal of ease for programmers, MANTIS uses classical layered multithreaded structure and standard programming language. The layered structure contains multithreading, preemptive scheduling with time slicing, I/O synchronization via mutual exclusion, a network protocol stack, and device drivers. The current MANTIS kernel realizes these functions in less than 500 bytes of RAM. MANTIS uses standard C to implement the kernel and API.

In the current implementation of MANTIS, the RAM size allocated to each new thread is fixed. The thread table stored in a global data structure has a capacity for of items, each of which is 10 bytes and is used to store thread-related information. The thread scheduler in MANTIS is priority based and round robin within each priority level. The scheduler is triggered only by timer interrupts from hardware to perform context switching. In MANTIS, other interrupts are handled by device drivers.

The network protocol stack in MANTIS has four layers: application, network, MAC, and physical. MANTIS implements these as one or more user-level threads, which would allow a trade-off between flexibility and performance. The network stack is realized with a standard API between layers. MANTIS implements flooding as a routing protocol and a simple stop-and-wait protocol for flow and congestion control [10.4]. The total code size of the kernel, scheduler, and network stack is smaller than 500 bytes and 14 kB flash. MANTIS supports certain advanced features, such as a multimodal prototyping environment for testing sensor networking applications, dynamic binary update-based reprogramming [10.5], and a remote shell and command server enabling the user to log in and inspect the sensor node's memory and status.

10.3.5 OSPM

OSPM (or dynamic power management, DPM), proposed in [10.6], is directed at power management techniques. The general dynamic power management is based on a greedy algorithm that will switch the system to a sleep state as soon as it is idle. It considers the following factors [10.6]:

- Transitioning to a sleep state has the overhead of storing the processor state and shutting off the power supply.
- Waking up takes a finite amount of time.
- The deeper the sleep state, the less the power consumption will be lower and the wake-up time will take longer [10.6].

Then, based on a given event arrival model, transition time, and power consumption rate, it reduces the energy savings. If the energy savings is positive, it will trigger a state transition; otherwise, the current state is maintained. This adaptive shutdown algorithm is a trade-off between energy savings and the cost of delay and possibly missed events.

10.3.6 EYES OS

As indicated earlier, the operating system for WSNs should be very small in terms of memory requirement and coding, should enjoy power awareness, and should be capable of distribution and reconfiguration. EYES OS [10.7,10.8] uses an event-driven model and task mechanism to realize these objectives. It works in a simple sequence as follows: perform a computation, return a value, and enter the sleep mode. The task can be scheduled using a FIFO-, priority-, or deadline-based approach (such as EDF), and is triggered by events in a nonblocking manner. EYES OS defines an application programming interface (API) locally and for the network components. The local information component provides functions such as access to sensor node data, availability of resources and their status, and setting of parameters or variables in sensor nodes. The network component provides functions to transmit and receive data and to retrieve network information. In summary, EYES OS realizes two groups of functions: those that can be executed at boot time to upload software module, and those that can provide node localization information.

EYES OS also provides an efficient code distribution mechanism with the following objectives: (1) to update the code on the sensor node, including the operating system; (2) to be resilient in case of packet loss during update; (3) to use as few communications and local resources as possible, and (4) to halt the application for a short period when updating. The procedure to distribute code is performed in four steps [10.8]: initialization, code image building, verification, and loading. There are three options for updating the running code: halved memory, a two-phase approach, and built-in EEPROM, which is used by EYES OS.

10.3.7 SenOS

SenOS [10.9] is a finite state machine (FSM)–based operating system. It has three components:

1. A kernel that contains a state sequencer and an event queue. The state sequencer waits for an input from the event queue (a FIFO queue).
2. A state transition table that keeps the information on state transition and the corresponding callback functions. Each state transition table defines an application. Using multiple state transition tables and switching among them, SenOS supports multiple applications in a concurrent manner [10.9].
3. A callback library of call functions. An incoming event will be queued in the event queue. The first event in the event queue is scheduled, which triggers a state transition and correspondingly, invokes the associated functions.

The kernel and callback library are statically built and stored in the flash ROM of a sensor node, whereas the state transition table can be reloaded or modified at runtime since it is application dependent. Since SenOS is FSM-based, it can easily realize concurrency and reconfiguration. It can also be extended to network management.

10.3.8 EMERALDS

EMERALDS [10.11] is an extensible microkernel written in C++ for embedded, real-time distributed systems with embedded applications running on slow processors (15 to 25 MHz) and with limited memory (32 to 128 kB). It supports multithreaded processes and full memory protection, which are scheduled using combined earliest deadline first (EDF) and a rate-monotonic (RM) scheduler. The device drivers are implemented at the user level, whereas interrupt handling takes place at the kernel level. EMERALDS uses semaphores and condition variables for synchronization with priority inheritance at the same time and provides full semaphore semantics to reduce the amount of context switching. Interprocessor communication (IPC) is realized based on message passing, mailboxes, and shared memory, optimized especially for intranode, intertask communication. EMERALDS does not use a mailbox; it uses global variables to exchange information between tasks, to avoid message sending. EMERALDS does not consider networking issues [10.11].

10.3.9   PicOS

One property of OS microcontrollers with limited RAM is to try to allocate as little memory as possible to a process or thread. PicOS [10.14] is written in C for a microcontroller with limited on-chip RAM (e.g., 4 kB). In PicOS, all tasks share the same global stack and act as coroutines with multiple entry points and implicit control transfer, which is different from classical multitasking approaches. In PicOS, each task is like a FSM where the state transition is triggered by events. The FSM approach is effective for reactive applications whose primary role is to respond to events rather than to process data or crunch numbers [10.14]. The CPU cycle is multiplexed among multiple tasks, but the tasks can be preempted only at the FSM state boundary. It has few resource requirements and supports multitasking, a flat structure for processes—but perhaps not good for real-time applications.

10.4   CONCLUSION

In this chapter we discussed operating systems for wireless sensor networks and presented design guidelines and objectives for a WSN operating system. A survey of some existing operating systems is also provided. The major issues for the design of operation systems for WSNs are size (memory requirement), energy-efficient IPCs and task scheduling, effective code distribution and upgrades, and finally, generic application programming interfaces.

REFERENCES

[10.1] J. Hill, R. Szewczyk, A. Woo, S. Hollar, D. Culler, K. Pister, "System Architecture Directions for Networked Sensors," *ACM SIGOPS Operating Systems Review*, Vol. 34, No. 5, December 2000, pp. 93–104.

[10.2] P. Levis, D. Culler, "Mate: A Tiny Machine for Sensor Networks," *Proceedings of* (ASPLOS'02), San Jose, CA, Oct. 2002, pp. 85–95.

[10.3] S. Coleri, M. Ergen, T. J. Koo, "Lifetime Analysis of a Sensor Network with Hybrid Automata Modeling," *Proceedings of the 1st Workshop on Sensor Networks and Applications* (WSNA'02), Atlanta, GA, Sept. 2002, pp. 98–104.

[10.4] R. Barr et al., "On the Need for System-Level Support for Ad Hoc and Sensor Networks," *ACM Operating Systems Review*, Vol. 36, No. 2, Apr. 2002, pp. 1–5.

[10.5] H. Abrach et al., "MANTIS: System Support for Multimodal Networks of In-Situ Sensors," *Proceedings of the 2nd Workshop on Sensor Networks and Applications* (WSNA'03), San Diego, CA, Sept. 2003, pp. 50–59.

[10.6] A. Sinha, A. P. Chandrakasan, "Operating System and Algorithmic Techniques for Energy Scalable Wireless Sensor Networks," *Proceedings of the 2nd International Conference on Mobile Data Management*, Hong Kong, Jan. 2001, pp. 199–209.

[10.7] S. Dulman, P. Havinga, "Operating System Fundamentals for the EYES Distributed Sensor Network," *Proceedings of Progress'02*, 2002.

[10.8] N. Reijers, K. Langendoen, "Efficient Code Distribution in Wireless Sensor Networks," *Proceedings of the 2nd Workshop on Sensor Networks and Applications* (WSNA'03), San Diego, CA, Sept. 2003, pp. 60–67.

[10.9] S. Hong, T. H. Kim, "Designing a State-Driven Operating System for Dynamically Reconfigurable Sensor Networks," *Proceedings of the 2003 System on Chip (SoC) Design Conference*, Nov. 2003, pp. 40–42.

[10.10] J. Heidemann et al., "Building Efficient Wireless Sensor Networks with Low-Level Naming," *Proceedings of ACM Symposium on Operating Systems Principles* (SOSP'01), Banff, Alberta, Canada, 2001, pp. 146–159.

[10.11] K. M. Zuberi, P. Pillai, K. G. Shin, "EMERALDS: A Small-Memory Real-Time Microkernel," *Proceedings of ACM Symposium on Operating Systems Principles* (SOSP'99), Kiawah Island, SC, 1999, pp. 277–291.

[10.12] J. Hill, M. Satyanarayanan, "Energy-Aware Adaptation for Mobile Applications," *Proceedings of ACM Symposium on Operating Systems Principles* (SOSP'99), Kiawah Island, SC, 1999, pp. 48–63.

[10.13] J. Hill, M. Horton, R. Kling, L. Krishnamurthy, "The Platforms Enabling Wireless Sensor Networks," *Communications of the ACM*, Vol. 47, No. 6, June. 2004, pp. 41–46.

[10.14] E. Akhmetshina, P. Gburzynski, F. Vizeacoumar, "PicOS: A Tiny Operation System for Extremely Small Embedded Platforms," *Proceedings of the Conference on Embedded Systems and Applications* (ESA'02), Las Vegas, NV, June 2002, pp. 116–122.

11

PERFORMANCE AND TRAFFIC MANAGEMENT

11.1 INTRODUCTION

In Chapter 7, we discussed the performance of the transport protocol and the impact on energy of hop-by-hop versus end-to-end control. Aside from the transport protocol, the performance of WNSs is affected strongly by other parameters in two groups: basic models and network models. Basic models form the elementary blocks based on which the network models can be analyzed and the overall system performance studied. In this chapter we review existing work [11.17] on the performance modeling of WSNs and provide a simple model to compute system lifetime that explains factors that influence the longevity of WSNs. We introduce several special characteristics and describe the impact of networking protocols on the performance of WSNs. Then performance metrics used in the evaluation of WSNs are presented, and some of the existing models to analyze them are discussed.

11.2 BACKGROUND

WSNs usually consist of hundreds or thousands of sensor nodes scattered in a geographical area and one or multiple sink(s) collecting information and transmitting it through wireless channels (Figure 11.1). The special design and character of sensors and their applications make WSNs different from traditional networks.

Wireless Sensor Networks: Technology, Protocols, and Applications, by Kazem Sohraby, Daniel Minoli, and Taieb Znati

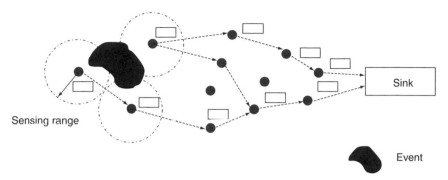

Figure 11.1   A WSN.

These characteristics pose great challenges for architecture and protocol design, performance modeling, and implementation. For example, sensor nodes need small simplified operating systems and energy-efficient communication protocols. Performance modeling and evaluation should consider new metrics for WSNs, such as *system lifetime* and *energy efficiency*, and the introduction of new traffic attributes.

Sensor nodes have resource constraints: limited energy, limited communication and computational capabilities, and limited memory. A sensor node may belong in one of four groups [11.2]: (1) a *specialized sensing platform* such as Spec, designed at the University of California–Berkeley, which is small in size and memory, and has a narrow communication bandwidth and short radio distance; (2) a *generic sensing platform* such as the Bekeley mote, which is designed using off-the-shelf components and has a bandwidth of 100 kbps or so and more memory than Spec; (3) a *high-bandwidth sensing* device such as iMote, developed by Intel Research, which has a much broader bandwidth than the earlier ones (Bluetooth-based radio) as well as a larger memory; and (4) a *gateway-like sensor node* such as Stargate, which is a gateway to directly connect mote (or iMote)-based devices. These sensor nodes have different levels of resources within them, but they all contain at least the following physical units: a radio unit with a transceiver, a processing unit with a microcontroller and a memory, a sensing unit with a sensor (or multiple sensors), and a power supply unit, usually with a battery. Sensor nodes may have an additional unit to support mobility or be equipped with a GPS-based unit. With the development of micro electromechanical systems (MEMSs) and the new battery technologies, sensor nodes might be able to carry more resources. However, the resources of sensor nodes are still constrained compared to the practically unconstrained physical and networking interfaces and other resources in traditional network nodes. These constraints have a direct impact on system and protocol design.

WSNs usually have a multihop physical topology, even when single-hop topology is possible. For the multihop case, the topology can be well structured or ad hoc. The first type organizes all sensor nodes in a hierarchical structure such as a two-tier architecture [11.3], where the sensor nodes at the first layer perform only sensing, the sensor nodes in the second layer perform sensing and data relaying, and so on. This topology can result in more efficient routing, but the topology formation is an energy-consuming task and also increases the complexity of sensor nodes. In ad hoc mode, sensor nodes self-organize into a flat and unstructured topology, and all nodes perform the task of sensing and relaying. A sensor node will become dysfunctional if its energy is depleted and recharging is not possible. In a multihop environment, multiple paths exist between a source and the sink, which lead to route redundancy and therefore flexible routing. When the number of dysfunctional nodes accumulates to a certain level, the topology will be disjointed and service failure may occur. The movement of sensor nodes in some cases (such as when installed in tanks on the battlefield) still makes this topology variable. In summary, the topology of sensor networks can be well structured or ad hoc. The topology is usually variable and has multiple paths from the source nodes to the sink. These attributes influence the design of routing protocols in WSNs.

Most traffic in WSNs flows starlike from sensor nodes to the sink. If there are multiple sinks, multiple traffic flows will be generated between sensor nodes and the sink. The sensor nodes gather data and report to the sink according to the preconfigured rules. This many-to-one traffic flow is called *convergecast* [11.24], which means many-to-one traffic flow from sensor nodes to the sink. Therefore, sensor nodes closer to the sink have the heavier burden for relaying, and due to higher energy consumption they might become dysfunctional sooner. A helpful way to get around this problem is to deploy more densely around the sink or to perform in-network processing (e.g., data aggregation) to reduce traffic flow. The traffic flow and specific functional requirements of the sensor deployment can be used to optimize networking protocols.

The basic service provided by WNSs is to detect certain events and report them. The data related to the events are usually small, usually just a few bytes and in many cases just a few bits. Therefore, it may be possible to transmit more than one event in a single data unit if the application reporting frequency allows it.

Other factors that affect WSN design are listed in (Table 11.1). These factors have a direct impact on the system performance of WSNs.

TABLE 11.1 Design Factors for Wireless Sensor Networks [11.1]

Factor	Options
Node deployment	Random, manual, one-time, iterative
Mobility	Immobile, partly, all; occasional, continuous; active, passive
Network topology	Single-hop, star, networked stars, tree, graph
Coverage	Sparse, dense, redundant
Connectivity	Connected, intermittent, sporadic
Network size	Hundred, thousand, more
Communications	Laser, infrared, radio-frequency (narrowband, spread spectrum, UWB)

Source: [11.1].

The remainder of this chapter is organized as follows. In Section 11.3 several design issues are described as they affect system performance. In Section 11.4 we present the metrics of system performance and in Section 11.5, a simple model to compute system lifetime. Section 11.6 concludes the chapter. In this section we highlight briefly networking protocols for wireless sensor networks, including MAC, routing, and transport protocols, from a performance point of view. These protocols heavily influence the overall performance of WSNs.

11.3 WSN DESIGN ISSUES

11.3.1 MAC Protocols

MAC protocols affect the efficiency and reliability of hop-by-hop data transmission. Existing MAC protocols such as the IEEE 802 series standard may not be completely suitable for WSNs because of energy efficiency. General MAC protocols can result in a waste of energy in the following ways [11.12]:

- Since a wireless channel is shared in a distributed manner, packet collision cannot be avoided. The collided packets require retransmission and result in energy waste.
- Most distributed wireless MAC protocols require control messages for data transmission (e.g., request-to-send/clear-to-send in the IEEE 802.11 distributed coordination function). Control messages consume energy.
- Overhearing and idle listening can also result in energy waste. Overhearing means that a node receives packets destined for other nodes. Idle listening refers to a situation where nodes there need to listen on the channel to get its status.

MAC protocols for wireless sensor networks emphasize energy efficiency through design of effective and practical approaches to deal with the foregoing problems. For example, S-MAC [11.13] designs an adaptive algorithm to let sensor nodes sleep at a certain time. The approach of Tay et al. [11.14] devises a nonuniform contention slot assignment algorithm to speed up collision resolution and reduce latency while in the idle state. Typical parameters used to measure performance of MAC protocols include collision probability, control overhead, delay, and throughput.

11.3.2 Routing Protocols

As we have seen in earlier chapters, routing protocols in WSNs are for setting up one or more path(s) from sensor nodes to the sink. Since sensor nodes have limited resources, routing protocols should have a small overhead, which may result from control message interchange and caching. Therefore, the traditional address-centric routing protocols for Internet (e.g., the routing information

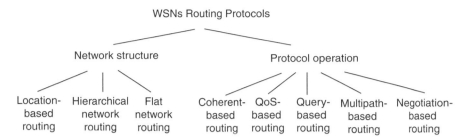

Figure 11.2 Classification of routing protocols for wireless sensor networks. (From [11.15].)

protocol, open shortest path first, border gateway protocol) do not meet the requirements of WSNs. Data-centric routing is more suitable for WSNs because it can be deployed easily, and due to data aggregation, it saves energy. Traffic models and system characteristics can be utilized to design efficient routing protocols. To conserve energy, most routing protocols for WSNs employ certain technique to minimize energy consumption (e.g., data aggregation and in-network processing, clustering, node role assignment). Al-karaki and Kamal [11.15] classify routing protocols in several categories shown in Figure 11.2. For example, directed diffusion [11.16] is a data-centric routing scheme with three phases in its operation:

1. A sink broadcasts its interest across the network in query messages with a special query semantic at a low rate.
2. All the nodes cache the interest. When a node senses that an event matches the interest, it sends the data relevant to the event to all the interested nodes. Sink will also get the initial data and "reinforce" one of source nodes by resending the interest at a higher rate.
3. After the reinforcement propagation, the source nodes send data directly on the reinforced path. The performance of a routing protocol can be expressed through such measures as computational overhead, communications overhead, path reliability, path length, convergence rate, and stability.

11.3.3 Transport Protocols

The following factors should be considered carefully in the design of transport protocols: a congestion control mechanism and especially, a reliability guarantee. As discussed in Chapter 7, since most data streams are convergent toward the sink, congestion is likely to occur at nodes around the sink. Although a MAC protocol can recover packet loss as a result of bit error, it has no way to handle packet loss as a result of buffer overflow. Therefore, transport protocols should have mechanisms for loss recovery; to guarantee reliability, mechanisms such as ACK and selective ACK [11.4] used in the TCP would be helpful. At the same time, reliability in WSNs has a different meaning than that of traditional networks where correct transmission of every packet is guaranteed. As discussed in Chapter 7, for some applications, WSNs need to receive packets correctly only from a certain area, not necessarily from every sensor in that area. For certain applications, only a certain ratio of successful transmissions from a sensor node is sufficient. These observations can be utilized to design more efficient transport protocols. As observed in Chapter 7, it is more efficient to have a hop-by-hop mechanism for congestion control and loss recovery since packet loss can be reduced and energy may be conserved. The hop-by-hop mechanism can also lower the buffer requirement at the intermediate nodes. Transport control protocols for WSNs should also avoid packet loss as much as possible since packet loss translates to waste of energy. Furthermore, it should guarantee fairness so that individual nodes can achieve their fair throughput.

11.4 PERFORMANCE MODELING OF WSNs

Two important performance metrics, *system lifetime* and *energy efficiency*, are discussed in this section. Both of these metrics relate to energy consumption. In WSNs, new models are required to capture special characteristics of these networks which are different from the traditional networks. In this chapter we review an approach to study the overall system performance based on [11.17].

11.4.1 Performance Metrics

As discussed earlier, wireless sensor networks are different from the traditional communication networks, and therefore different performance measures may also be required to evaluate them. Among them are the following [11.17]:

1. *System lifetime.* This term can be defined in several ways: (a) the duration of time until some node depletes all its energy; or (b) the duration of time until the QoS of applications cannot be guaranteed; or (c) the duration of time until the network has been disjoined.

2. *Energy efficiency.* Energy efficiency means the number of packets that can be transmitted successfully using a unit of energy. Packet collision at the MAC layer, routing overhead, packet loss, and packet retransmission reduce energy efficiency.

3. *Reliability.* In WSNs, the event reliability is used as a measure to show how reliable the sensed event can be reported to the sink. For applications that can tolerate packet loss, reliability can be defined as the ratio of successfully received packets over the total number of packets transmitted.

4. *Coverage.* Full coverage by a sensor network means the entire space that can be monitored by the sensor nodes. If a sensor node becomes dysfunctional

due to energy depletion, there is a certain amount of that space that can no longer be monitored. The coverage is defined as the ratio of the monitored space to the entire space.

5. *Connectivity.* For multihop WSNs, it is possible that the network becomes disjointed because some nodes become dysfunctional. The connectivity metric can be used to evaluate how well the network is connected and/or how many nodes have been isolated.

6. *QoS metrics.* Some applications in WSNs have real-time properties. These applications may have QoS requirements such as delay, loss ratio, and bandwidth.

11.4.2 Basic Models

Traffic Model   The applications and corresponding traffic characteristics in WSNs are different from those of traditional networks. For example, whereas the widely used applications for Internet include e-mail, Web-based services, the file transfer protocol, and peer-to-peer services, wireless sensor networks have totally different ones. As a result, traffic and data delivery models are also different. Currently, four traffic models are used in WSNs: event-based delivery, continuous delivery, query-based delivery, and hybrid delivery. Traffic model greatly influences protocol design and affects performance. The four models and the related performance aspects are discussed below.

Event-Based Delivery   In this case, sensor nodes monitor the occurrence of events passively and continuously. When an event occurs, the sensor node begins to report the event, and possibly an associated value, to the sink. When delivering event data to the sink, a routing protocol is often triggered in order to find a path to the sink. This routing method is called *routing on-demand*. If an event appears frequently, at a node or a group of nodes, the routing function is executed frequently, which results in more energy consumption. An alternative approach is to set up in advance a frequently used path. Therefore, the routing efficiency for this delivery model is heavily dependent on the frequency of occurrence of the events. An adaptive routing protocol may be required to set up a path dynamically in advance if events occur frequently; otherwise, the path is set up on-demand.

Continuous Delivery   The data collected by the sensors need to be reported regularly, perhaps continuously, or periodically. For example, in [11.11] a WSN is be used to observe the breeding behavior of a small bird on Great Duck Island. In this situation, sensor nodes deployed inside the burrows and on the surface measure humidity, pressure, temperature, and ambient light level. Once a minute, sensors report sample values to the sink.

Query-Based Delivery   Sometimes, the sink may be interested in a specific piece of information that has already been collected in sensor nodes. The sink will issue query messages to sensor nodes to get the up-to-date value for the information. Query messages may also carry a command from the sink to the sensors about the information, reporting frequency and other parameters of interest to the sink. In this delivery model, the sink broadcasts the query message, a path is constructed automatically when the query arrives at the sensor nodes, and the sensor nodes report their findings according to the request in the query message.

Hybrid Delivery   In some WSNs, the types of sensors and the data they sense may be very diverse. For example, data may be reported continuously by some nodes, and the sink may need to query information from other sensor nodes.

Energy Models   The radio communication function of sensor nodes is the most energy-intensive function in the node. Compared with that, the actual sensing operation consumes the least energy (see Figure 11.3 from [11.7]). There are two approaches to reducing consumption for sensor communications. The first approach is to design a communication scheme that conserves energy inherently: for example, turning off the transceiver for a period of time. The second approach is to reduce the volume of communications through in-network processing. These would entail functions such as data aggregation and data compression since computation tasks usually require less energy than do communication tasks.

Model for Sensing   Usually, the least amount of energy is consumed for sensing. Let the sensing range be r_s. It can be assumed that the power consumed to perform sensing over a circle with radius r_s is proportional to r_s^2 or r_s^4 [11.10].

Model for Communication   An energy model for communications is as follows [11.4,11.5]. The energy for transmitting l-bit data over a distance d is $E_{tx}(l,d)$

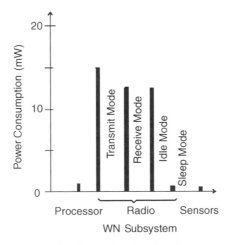

Figure 11.3   Energy consumption for each subsystem in sensor nodes. (From [11.7].)

and the energy for receiving l-bit data over a distance d is

$$E_{tx}(l, d) = lE_c + led^s \quad \text{where } e = \begin{cases} e_1 & s = 2, \quad d < d_{cr} \\ e_2 & s = 4, \quad d > d_{cr} \end{cases} \quad (11.1)$$

$$E_{rx}(l, d) = lE_c \quad (11.2)$$

where E_c is the base energy required to run the transmitter or receiver circuitry. A typical value for E_c is 50 nJ/bit for a 1-Mbps transceiver; d_{cr} is the crossover distance, and its typical value is 86.2 m; e_1 (or e_2) is the unit energy required for the transmitter amplifier when $d < d_{cr}$ (or $d > d_{cr}$). Typical values for e_1 and e_2 are 10 pJ/bit$\cdot$m^2 and 0.0013 pJ/bit$\cdot$m^4, respectively. Therefore, the total energy for transmitting l-bit data from source node i to destination node j within a distance of d is

$$E_{i,j}(l, d) = E_{tx}(l, d) + E_{rx}(l, d) = l(a_1 + a_2 d^s)$$
$$(\text{let } a_1 = 2E_c, a_2 = e_1 \text{ or } e_2, s = 2 \text{ or } 4) \quad (11.3)$$

The optimal distance between relay nodes (d_m) is computed as follows [11.6]:

$$d_m = \sqrt[s]{\frac{a_1}{a_2(s-1)}} \quad (11.4)$$

Then the optimal hop count (H) can be calculated as $H = d/d_m$.

Model for Computation   A sensor node usually has a microcontroller or micro-CPU performing computations. As is pointed out in [11.8,11.9], it is worth noticing that low power is different from energy efficiency. Low power means that the CPU consumes low energy per clock cycle; energy efficiency represents energy consumed per instruction. For example, for ATMega 128L at 4 MHz consuming 16.5 mW, the efficiency is 242 MIPS/W or 4 nJ/instruction, whereas in ARM Thumb at 4 MHz consuming 75 mW, the efficiency is 480 MIPS/W or 2.1 nJ/instruction. Other microcontrollers and their properties are listed in Table 11.2.

TABLE 11.2   Properties of Various Microcontrollers [11.8], [11.9]

Name	Voltage (v)	Frequency	Efficiency (nJ/instruction)
Cygnal C8051 F300	3.3	25 MHz	0.5
	3.3	32 kHz	0.2
IBM 405LP	1.8	380 MHz	1.3
	1.0	152 MHz	0.35
TMS320VC5510	1.5	200 MHz	0.8
Xscale PXA250	1.3	400 MHz	1.3
	0.85	130 MHz	1.9

Source: [11.8].

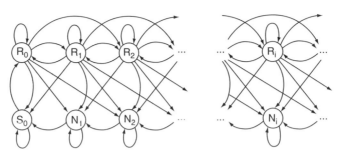

Figure 11.4 DTMC sensor node model. (From [11.17].)

Node Model To conserve energy, a common approach is to let nodes sleep when they have no need to transmit or receive. This behavior is modeled in [11.17] as follows. The sensor nodes have two states: *active* (*A*) and *sleep* (*S*). The length of the active and sleep period are geometrically distributed random variables with a mean value of p and q time slots, respectively. The active phase is divided further into R and N states. In the R state, sensor nodes can transmit or receive data and/or generate data according to a Poisson process with an average rate of g. In the N state, nodes can transmit data only if there are packets backlogged in their buffer [11.17].

Reference [11.17] presents a two-state discrete-time Markov chain (DTMC) model for the next-hop nodes, where the next-hop nodes represent the neighboring nodes relative to the node in mind. The two states defined for the next-hop node are *wait* (*W*) and *forwarding* (*F*). State *W* means that all the next-hop nodes are in either the *S* or *N* state and unable to receive data from the node in mind. *F* represents that there is at least one next-hop node in state *R* and it can receive the data from the node in mind. The transition probabilities from *W* to *F and* vice versa are assumed to be *f* and *w*.

Based on the description above, a Markov chain model for the sensor node model is created by [11.17] and shown in Figure 11.4, where the subscript represents the number of packets in the buffer. Using this model, the stationary distribution of a node's state (p) can be calculated given the successful transmission probability of data (b) and the probability that data are received in a time slot (a). Other metrics can be calculated based on p: for example, the average number of data generated in a time slot, the sensor throughput or average number of data forwarded by the sensor in a time slot, and the average buffer occupancy in sensor node.

11.4.3 Network Models

MAC Model Channel access is controlled and allocated by MAC protocols. In a decentralized environment, packet collision may occur on the channel and should be handled by the MAC. One of the important performance parameters that one needs to capture from a MAC protocol is the probability that data are transmitted successfully in a time slot (b). Gupta and Kumar [11.18] defines successful transmission conditions as follows. Let r be the common maximum radio range.

The one-hop transmission from transmitter i to the receiver j is successful if (1) the distance between i and j is less than or equal to r: $d_{i,j} \geq r$; (2) for every other node, k, which is also receiving at the same time, $d_{i,k} > r$; and (3) for every other node, l, which is transmitting at the same time, $d_{l,j} > r$.

Based on the definition above and the term $l_{n,m}$, defined as the average transmission rate between node n and its generic receiver m, Chiasserini and Garetto [11.17] have constructed an *interference model* to calculate a successful transmission probability (b) using $l_{n,m}$ as an input, under the assumption that the WSN employs a CSMA/CA mechanism with handshaking as in an IEEE 802.11 DCF. Also, if an IEEE 802.11 DCF or its variants is used as the MAC protocol, one may apply many existing performance analysis results directly (e.g., [11.19,11.20]) to obtain performance metrics such as throughput, delay, and collision probability, given the number of competing nodes.

Routing Model   Based on the energy model introduced earlier in the chapter, energy consumed for a generic route P $[E(P)]$ can be computed as follows [11.17]:

$$E(P) = \sum_{i \to P} E_{i,n_p(i)}(l_i, d_i) \tag{11.5}$$

where $n_p(i)$ is the next hop of node i on path P. $E_{i,n_p(i)}(l, d)$ is the energy from node i to node $n_p(i)$. Assuming that the data size is l_i bits and the distance between them is d_i, the total energy consumed can be written using Eq. (11.3).

The advantages of data-centric routing over address-centric routing in supporting data aggregation were found from analysis [11.21]. These results show that:

1. If the diameter of the set of source nodes (X) is shorter than the minimal length of the shortest path from any source node to the sink (D_{min}), the total number of transmissions under data-centric routing is smaller than with address-centric routing. Therefore, data-centric routing is more energy efficient.

2. The larger the distance between X and D_{min}, the more energy is conserved by the data-centric routing.

System Model   Analysis of the overall performance of the sensor network is presented in [11.17], where a closed-loop model has been constructed to consider the sensor node model, MAC protocol, and routing policy all at the same time. This model consists of three submodels, as shown in Figure 11.5. The *sensor node model* was shown earlier in the chapter and in Figure 11.4. The *interference model* was also described earlier. The *network model* is used to model routing policy and to determine the average transmission rate between nodes, which is an input to the *inference model*. In [11.17], a simple routing policy is assumed as follows: When transmitting data to the next hop, a sensor node chooses the neighboring node that will result in the lowest energy consumption. The solution of the system model in Figure 11.5 has been obtained through fixed-point approximation [11.17], and

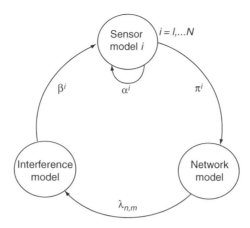

Figure 11.5 Closed-loop model for the system [11.17].

system performance metrics such as average energy consumption and average delay have also been calculated in the following manner:

1. Construct a DTMC sensor node model, represented by the leftmost box for each sensor node i to get the stationary distribution p_i and the probability that data are received in a time slot (a_i) in node i.

2. Solve the network model using queuing network analysis to calculate the average data transmission rate between any pair of sensor nodes n and m in the network $(l_{n,m})$ as well as the average throughput for each sensor node.

3. Given $l_{n,m}$ as input to the interference model, compute the value of the probability that data are transmitted successfully in a time slot in node $i(b_i)$.

4. b_i is used as input to the sensor node model, iterating through steps 1 to 3. The worst relative error for two successive estimates of the sensor throughput is used as the stopping criterion. It is stated in [11.17] that 10 iterations result in an error below 0.0001.

11.5   CASE STUDY: SIMPLE COMPUTATION OF THE SYSTEM LIFE SPAN

In this section we present a simple model to compute the system life span. The following assumptions are made:

1. All sensor nodes (N) in the network organize a two-tiered topology. The sensor nodes in the lower layer are called *leaf nodes*. The sensor nodes in the high layer are called *leader nodes*. At the high layer, there are N_1 leader nodes forming a k-tree topology with $h + 1$ levels (or h hops) from the sink, where each leader node in level i connects k child nodes in level $i + 1$ to its parent node at layer $i - 1$ (see Figure 11.6). Each leader node in the higher layer that

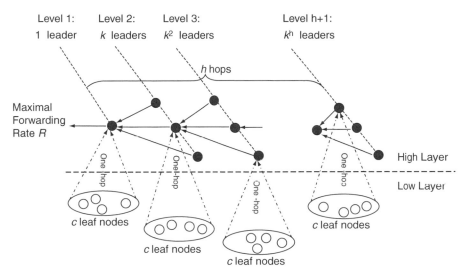

Figure 11.6 Two-tiered topology of a WSN.

is generating local data flow plays a role in relaying data from its leaf nodes. Each leader node has c leaf nodes. The leaf node in the lower layer (N_2) generates data flow and forwards it directly to the leader node within one hop. If a leader node lies at level i, it is assumed that its leaf nodes also lie at the level i.

2. All sensor nodes are distributed equally and densely in a space for monitoring events. Each sensor node is assumed to have a large enough buffer size so that data loss due to buffer overflow can be neglected. Each sensor node generates event reporting with an independent and identically distributed Poisson process. The reporting frequency is f, and the total information to be forwarded in each reporting instant is B bits. Let O (bits) represent the number of overhead in each packet, and L the packet length. Therefore, the number of packets in each reporting instant is $n_p = B/(L - O)$. The corresponding data transmission rate in each sensor node i is $r_i = n_p f = [B/(L - O)]f$.

3. Each higher-layer node (leader node) receives data from other leader nodes (k), its leaf nodes (c), as well as locally generated data. The leader node aggregates or compresses only the input data from itself and leaf nodes. The raw data rate from sensor node $i(r_i)$ is compressed to $0 \le r'_i \le r_i$. Let's define the aggregation efficiency as $a = (r_i - r'_i)/r_i$.

4. There is a congestion and flow control mechanism between sink and sensor nodes to guarantee that in steady-state congestion will not occur; therefore, the total converged data rate (r_c) of all sensor nodes should be smaller than the maximum forwarding rate (R) of the leader node at level 1 (see Figure 11.6): $r_c > R$. Congestion and flow control may lead to two types of fairness:

per-node fairness and max-min fairness. *Per-node fairness* ensures that all sensor nodes have the same data transmission rate; however, *max-min fairness* provides proportional fairness in that the nodes closer to the sink can be allocated a higher data rate. For example, TCP produces max-min fairness where the connection with smaller RTT gets a higher transmission rate.

5. Each sensor node has a maximum energy of E. Let us define the system lifetime as the time duration from the beginning to when the leader node at level 1 consumes all its energy.

6. All nodes are stationary, and no adaptive power control would be assumed.

11.5.1 Analysis

Total Node Number   Based on the assumed two-tiered topology, the total number of sensor nodes is

$$N = N_1 + N_2 = (1 + c)N_1 = (1 + c) \sum_{i=1}^{h+1} k^{i-1} \qquad (11.6)$$

Average Number of Retransmissions   Let b_e and p_e be bit-error probability and packet error probability, respectively. p_e is as follows, assuming a packet length of L bits and ignoring the error-correction mechanism:

$$p_e = 1 - (1 - b_e)^L \qquad (11.7)$$

Packet error may cause retransmission at the MAC layer. Retransmission improves system reliability, but increases delay. Let's assume that the maximum number of retransmissions is K. The average number of retransmissions can be calculated as follows:

$$n_r = \sum_{i=1}^{K-1} i p_e^i (1 - p_e) + K p_e^K \qquad (11.8)$$

If there are no limits on the maximum retransmissions, that is, if $K = \infty$, n_r becomes

$$n_r = \sum_{i=1}^{\infty} i p_e^i (1 - p_e) = \frac{p_e}{1 - p_e} \qquad (11.9)$$

Average Cost of Packet Forwarding in One Hop   In each hop, data consume the following amount of energy: transceiver (e_i), MAC-layer collision and idle and overhearing (e_m), and computation (e_c). e_i is dependent on physical distance d and packet size L; e_m is determined by the number of neighboring nodes $(c + 1)$ within one hop and traffic intensity (assuming CSMA-like MAC protocol); e_c may be a constant. Therefore, the total energy consumed by forwarding a packet within one hop is about

$$e = e_t + e_m + e_c \tag{11.10}$$

Converged Data Rate   For per-node fairness, each sensor node has the same data rate, $r_i = n_p f$; therefore, r_c is given as follows:

$$r_c = r(1 - a)N = n_p f(1 - a)N \tag{11.11}$$

For max-min fairness, the sensor nodes at the same level j has the same date rate r^j, and the data rate for nodes at different levels is proportional to their level number:

$$\frac{j+1}{j} \approx \frac{r^j}{r^{j+1}} = \frac{n_p f_j}{n_p f_{j+1}} = \frac{f_j}{f_{j+1}} \, (1 \leq j \leq h+1) \tag{11.12}$$

where f_j is the reporting frequency of the sensor nodes at level j. Therefore, the converged rate r_c can be stated as

$$r_c = \sum_{j=1}^{h+1} r^j (1-a)[k^{j-1}(c+1)] = \sum_{j=1}^{h+1} \frac{r^1}{j}(1-a)[k^{j-1}(c+1)]$$

$$= n_p f^1 (1-a)(c+1) \sum_{j-1}^{h+1} \frac{k^{j-1}}{j} \tag{11.13}$$

System Lifetime   Let's assume that all sensor nodes become active at time $t_0 = 0$ and let's assume that at time t_1 the energy of the highest node at level 1 is depleted first. System lifetime T_l can be approximately $T_l = t_1 - t_0$. Here, the effects of node mobility and/or power control have been ignored. The quantity T_l can be used as the lower bound of the system lifetime.

For per-node fairness,

$$(r_c t_1)(1+n_r)e = E \Rightarrow t_1 = \frac{E}{r_c e(1+n_r)} = \frac{E}{[n_p f(1-a)N](1+n_r)(e_t+e_m+e_c)}$$

$$\Rightarrow T_l = \frac{E}{[n_p f(1-a)N](1+n_r)(e_t+e_m+e_c)} \tag{11.14}$$

For max-min fairness,

$$(r_c t_1)(1+n_r)e = E \Rightarrow t_1 = \frac{E}{r_c(1+n_r)e}$$

$$\Rightarrow t_1 = \frac{E}{(e_t+e_m+e_c)(1+n_r)[n_p f^1(1-a)(c+1)\sum_{j=1}^{h+1} k^{j-1}/j]}$$

$$\Rightarrow T_l = \frac{E}{(e_t+e_m+e_c)(1+n_r)[n_p f^1(1-a)(c+1)\sum_{j=1}^{h+1} k^{j-1}/j]} \tag{11.15}$$

It can be seen from Eqs. (11.14) and (11.15) that system life span is dependent on several factors, including the energy model, reporting frequency f, aggregation efficiency a, packet length L, and packet error probability p_e. When designing and deploying a wireless sensor network, it is possible to choose the appropriate values that will extend system lifetime.

11.5.2 Discussion

Packet Length   If packet length L increases, the number of packets used in each reporting instant (n_p) will decrease, which results in a longer system life span (T_l). However, increasing L may lead to higher packet error probability and therefore a higher number of retransmissions (n_r), which in turn will increase T_l. Depending on the bit-error probability and packet overhead, packet length L can be optimized so that an optimal system lifetime can be achieved.

From the expression of the system life span (T_l) in Section 11.5.1, the energy index E_i is defined as

$$E_i = \frac{L - O}{(1 + n_r)(e_t + e_m + e_c)} \tag{11.16}$$

For example, suppose that $e_c = 100\,\text{nJ}$, $e_m = 200\,\text{nJ}$, and $e_t = 100L\,\text{nJ}$; then E_i becomes

$$E_i = \frac{(L - O)(1 - b_e)^L}{100L + 300} \tag{11.17}$$

Figure 11.7 presents numerical results of E_i as a function of packet length. It can be observed that in fact there is an optimal value of packet length (L) which maximizes E_i. The optimal value of L increases with a lower bit-error rate (BER) and/or with an increase in the packet overhead (O).

Reporting Frequency   A smaller reporting frequency f results in a longer system life since lesser energy will be consumed. But at the same time, the smaller f will result in a lack of time correlation of events, and the aggregation efficiency may decrease. This can, in turn, result in a shorter system lifetime. If the occurrence of events can be assumed as bandlimited signals, the Nyquist theorem can be applied to decide about the minimum reporting frequency f. If applications can tolerate a certain distortion, the reporting frequency can be reduced.

Aggregation Efficiency   Increasing aggregation efficiency a can also result in extending the system lifetime. The spatiotemporal correlation shows that continuous data flow from several neighboring nodes compared with a single node contains redundant information. This observation can be explored for data aggregation. Characteristics of applications can be explored further to reduce the number of data reported. The number of nodes (c) within a distance of one hop will influence

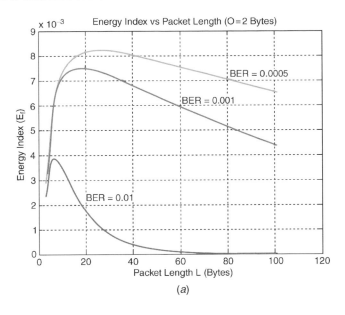

(a)

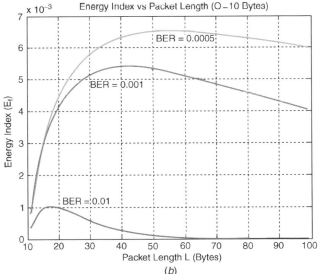

(b)

Figure 11.7 Energy index versus packet length: (a) packet overhead (O) is 2 bytes; (b) O is 10 bytes.

aggregation efficiency. For example, if c is reduced through choosing more leader nodes, the spatial correlation will increase and the aggregation efficiency should improve; however, this will increase the forward hop count and end-to-end packet delay and therefore the energy consumed for each packet. Moreover, the smaller c can reduce the energy consumption (e_m) in the MAC layer by lowering the collision probability.

11.6   CONCLUSION

In this chapter we discussed performance modeling for wireless sensor networks. First, we described briefly the characteristics of wireless sensor networks for performance modeling. Several basic design issues were surveyed briefly in Section 11.2. Performance metrics and basic models for wireless sensor networks were then investigated, and a systematic approach for performance modeling of wireless sensor networks in [11.17] was reviewed. Finally, a simple model to compute system lifetime was presented. It should be pointed out that there are several other performance modeling issues related to capacity, coverage, and connectivity in [11.22] and [11.23], but not discussed in this chapter.

REFERENCES

[11.1] K. Romer, F. Mattern, "The Design Space of Wireless Sensor Networks," *IEEE Wireless Communications*, Dec. 2004.

[11.2] J. Hill, M. Horton, R. Kling, L. Krishnamurthy, "The Platforms Enabling Wireless Sensor Networks," *Communications of the ACM*, Vol. 47, No. 6, June 2004, pp. 41–46.

[11.3] F. Ye, H. Luo, J. Cheng, S. Lu, L. Zhang, "A Two-Tier Data Dissemination Model for Large-Scale Wireless Sensor Networks," *Proceedings of the 8th ACM International Conference on Mobile Computing and Networking* (MobiCom'02), Atlanta, GA, Sept. 2002.

[11.4] W. Heinzelman, A. Chandrakasan, H. Balakrishnan, "An Application-Specific Protocol Architecture for Wireless Microsensor Networks," *IEEE Transactions on Wireless Communications*, Vol. 1, No. 4, Oct. 2002, pp. 660–670.

[11.5] A. A. Ahmed, H. Shi, Y. Shang, "A Survey on Network Protocols for Wireless Sensor Networks," *Proceedings of Information Technology Research and Education (ITRE)*, 2003.

[11.6] M. Bhardwaj, T. Garnett, A. P. Chandrakasan, "Upper Bounds on the Lifetime of Sensor Networks," *Proceedings of the IEEE International Conference on Communications*, June 2001.

[11.7] C. Schurgers, V. Tsiatsis, S. Ganerival, M. Srivastava, "Optimizing Sensor Networks in Energy–Latency–Density Design Space," *IEEE Transactions on Mobile Computing*, January 2002, pp. 70–80.

[11.8] M. Srivastava, "Sensor Node Platforms and Energy Issues," *Proceedings of the 8th ACM International Conference on Mobile Computing and Networking* (MobiCom'02), Atlanta, GA, Sept. 2002, Tutorial 2.

[11.9] M. A. M. Vieira et al., "Survey on Wireless Sensor Network Devices," *Proceedings of IEEE Conference on Emerging Technologies and Factory Automation* (ETFA'03), Sept. 16–19, 2003, pp. 537–544.

[11.10] J. Wu, S. Yang, "Coverage Issue in Sensor Networks with Adjustable Ranges," *Proceedings of the International Conference on Parallel Processing Workshops* (ICPPW'04), Montreal, Quebec, Canada, Aug. 2004.

[11.11] A. Mainwaring, J. Polastre, R. Szewczyk, D. Culler, J. Anderson, "Wireless Sensor Networks for Habitat Monitoring," *Proceedings of the 1st Workshop on Sensor Networks and Applications* (WSNA'02), Atlanta, GA, Sept. 2002.

[11.12] I. Demirkol, C. Ersoy, F. Alagoz, "MAC Protocols for Wireless Sensor Networks: A Survey," accepted by *IEEE Communications*, 2005.

[11.13] W. Ye, J. Heidemann, D. Estrin, "Medium Access Control with Coordinated Adaptive Sleeping for Wireless Sensor Networks," *IEEE/ACM Transactions on Networking*, Vol. 12, No. 3, June 2004, pp. 493–506.

[11.14] Y. C. Tay, L. Jamieson, H. Balakrishnan, "Collision-Minimizing CSMA and Its Applications to Wireless Sensor Networks," *IEEE Journal on Selected Area in Communications*, Vol. 22, No. 6, Aug. 2004, pp. 1048–1057.

[11.15] J. N. Al-karaki, A. E. Kamal, "Routing Techniques in Wireless Sensor Networks: A Survey" *IEEE Wireless Communications*, Vol. 11, No. 6, Dec. 2004, pp. 6–28.

[11.16] C. Intanagonwiwat, R. Govindan, D. Estrin, "Directed Diffusion: A Scalable and Robust Communication Paradigm for Sensor Networks," *Proceedings of the 6th ACM/IEEE International Conference on Mobile Computing and Networking* (MobiCom'00), Aug. 2000.

[11.17] C. F. Chiasserini, M. Garetto, "Modeling the Performance of Wireless Sensor Networks," *Proceedings of the 23rd Annual Joint Conference of the IEEE Computer and Communications Societies* (InfoCom'04), Mar. 2004.

[11.18] P. Gupta and P. R. Kumar, "The Capacity of Wireless Sensor Networks," *IEEE Transactions on Information Theory*, Vol. 46, Mar. 2000.

[11.19] G. Bianchi, "Performance Analysis of the IEEE 802.11 Distributed Coordination Function," *IEEE Journal on Selected Areas in Communications*, Vol. 18, No. 3, Mar. 2000, pp. 535–547.

[11.20] Z. Hadzi-Velkov, B. Spasenovski, "Saturation Throughput-Delay Analysis of IEEE 802.11 DCF in Fading Channel," *Proceedings of the IEEE International Conference on Communications* (ICC'03), 2003, pp. 121–126.

[11.21] B. Krishnamachari, D. Estrin, S. Wicker, "Modelling Data-Centric Routing in Wireless Sensor Networks," *Proceedings of the 21st Joint Conference of the IEEE Computer and Communications Societies* (InfoCom'02), June 2002.

[11.22] C. Florens and R. McElicec, "Packet Distribution Algorithm for Sensor Networks," *Proceedings of the 22nd Joint Conference of the IEEE Computer and Communications Societies* (InfoCom'03), Mar. 2003.

[11.23] S. Shakkottai, R. Srikant, N. B. Shroff, "Unreliable Sensor Grids: Coverage, Connectivity and Diameters," *Proceedings of the 22nd Joint Conference of the IEEE Computer and Communications Societies* (InfoCom'03), Mar. 2003.

[11.24] H. Zhang, A. Arora, Y. Choi, M. G. Gouda, "Reliable Bursty Convergecast in Wireless Sensor Networks," Proceedings of the 6th ACM International Symposium on Mobile Ad Hoc Networking and Computing (MobiHoc'05), Urbana-Champaign, IL, May 2005.

INDEX